Working

Red Dot Design Yearbook 2013/2014
Edited by Peter Zec

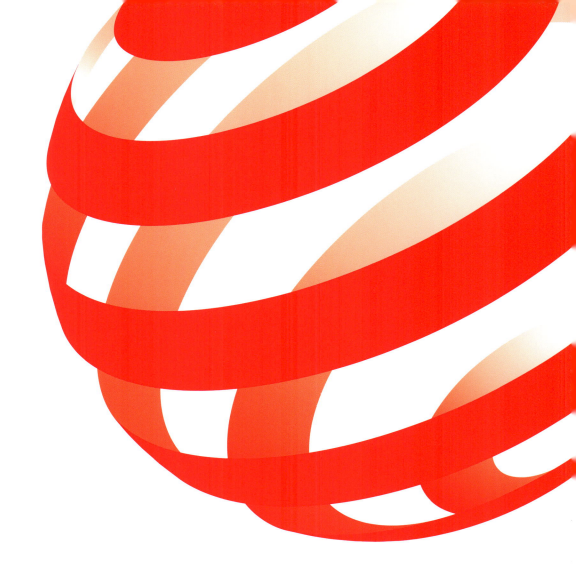

reddot design award
product design 2013

Working

Red Dot Design Yearbook 2013/2014
Edited by Peter Zec

Prof. Dr. Peter Zec
Preface of the editor
Vorwort des Herausgebers

Dear Readers,

Have you ever thought about how much time you spend at your work place or in your home office every day? These presumably many hours are all the more reason to make this part of your day as pleasant as possible. The volume "Working" offers you an enormous variety of innovations and inspirations for a comfortable, sophisticated and secure working environment.

This book presents a selection of the best current products created by designers from around the world. Each work has received a distinction from the internationally renowned Red Dot Design Award competition – widely recognised as a hallmark of excellence. The book features spectacular developments and intelligent technologies, including novel ways to significantly improve the functionality of already existing products. In the medical sector, we observed a trend towards user-friendly devices that greatly facilitate work processes. Aesthetically, these professional devices hardly differ anymore from personal lifestyle products. In the tools and machines category, new user concepts allow for simpler handling and improved precision.

"Working" has a clear, intuitive structure and features high-quality illustrations. Readers will easily locate the areas and themes that interest them most. All product categories also put a spotlight on the products awarded with the coveted "Red Dot: Best of the Best" award. Demonstrating excellent design and certain to set new standards, these works captivated and won over the jury in every respect.

The "Working" volume of the Red Dot Design Yearbook is a veritable source of inspiration for optimising the custom setup of your work station, and for taking part in exciting new technological developments. In association with the other two volumes ("Living" and "Doing"), "Working" offers an exceptional overview of the best and latest design of our time.

I wish you an engaging and inspiring read.

Sincerely, Peter Zec

Liebe Leserin, lieber Leser,

haben Sie schon einmal darüber nachgedacht, wie viel Zeit des Tages Sie an Ihrem Arbeitsplatz oder in Ihrem Homeoffice verbringen? Diese mutmaßlich vielen Stunden sind Grund genug, diesen Bereich so angenehm wie möglich zu gestalten. Der vor Ihnen liegende Band „Working" bietet Ihnen eine enorme Vielfalt an Innovationen und Anregungen für ein komfortables, stilvolles und auch sicheres Arbeiten.

Dieses Buch präsentiert eine Auswahl der aktuell besten Produkte, gestaltet von Designern aus aller Welt. Jedes einzelne von ihnen wurde im Rahmen des international renommierten Red Dot Design Award von einer Jury prämiert und ist dabei für sich einzigartig. So finden sich in diesem Band spektakuläre Entwicklungen und intelligente Technologien ebenso wie neue Ansätze, die etwa die Funktionalität eines bereits bekannten Produktes derart verbessern, dass es sich im Arbeitsalltag hervorragend handhaben lässt. Im Medizinbereich geht der Trend etwa hin zu freundlich gestalteten Geräten, die die Arbeitsabläufe erheblich verbessern – auch die Hilfsmittel unterscheiden sich heute kaum mehr von modernen Lifestyleprodukten. Bei den Werkzeugen und Maschinen ermöglichen neue Bedienkonzepte ein einfaches Handling für noch mehr Präzision.

„Working" ist übersichtlich gegliedert und reich bebildert – es bietet dem Leser eine gute Orientierung. Rasch findet er die Bereiche und Themen, die ihn besonders interessieren. Hervorzuheben sind in jeder Kategorie die Produkte mit der begehrten Auszeichnung „Red Dot: Best of the Best". Diese verfügen über eine herausragende Designqualität, sie setzen neue Maßstäbe und konnten die Jury in jeder Hinsicht begeistern.

In dem Band „Working" des Red Dot Design Yearbook finden Sie zahlreiche Anregungen für eine individuell optimierte Ausstattung Ihres Arbeitsbereichs und haben teil an spannenden technologischen Entwicklungen. In Verbindung mit den beiden Bänden „Living" und „Doing" bietet „Working" einen einzigartigen Überblick über das beste und neueste Design unserer Zeit.

Ich wünsche Ihnen eine spannungsreiche und inspirierende Lektüre.

Ihr Peter Zec

Red Dot: Design Team of the Year

People who change the world through design

Menschen, die mit Design die Welt verändern

In recognition of its feat, the Red Dot: Design Team of the Year receives the "Radius" trophy. This sculpture was designed and crafted by the Weinstadt-Schnaidt based designer, Simon Peter Eiber. Like all the years before, the Radius changes hands again in 2013. This year Michael Mauer and Style Porsche will pass on the coveted trophy to Yao Yingjia and the Lenovo Design & User Experience Team.

Als Anerkennung erhält das Red Dot: Design Team of the Year den Wanderpokal „Radius". Die Skulptur wurde entworfen und angefertigt von dem Designer Simon Peter Eiber aus Weinstadt-Schnaidt. Auch im Jahr 2013 wechselt der Radius seinen Besitzer: Michael Mauer und Style Porsche geben die begehrte Trophäe an Yao Yingjia und das Lenovo Design & User Experience Team weiter.

For the 26th time in the long history of the competition, one design team will be honoured for its exceptional achievements in the field of design. This year Lenovo Design & User Experience Team led by Yao Yingjia are the recipients of the "Red Dot: Design Team of the Year" award. This is a unique honour and one which is very highly regarded, both within the design scene and beyond, due to the fact that the award is given in recognition of the overall design output of a design team that has generated exceptional products and groundbreaking design over a number of years. Since 1988 the Radius has been presented to a design team as part of the Red Dot Design Award. It has become a tradition that all the prize winners have their names engraved on the trophy before they pass it on to the next year's winner.

Zum 26. Mal wird in der langen Geschichte des Wettbewerbs ein Designteam ausgezeichnet, das durch besondere Leistungen auf sich aufmerksam gemacht hat. In diesem Jahr geht die Ehrenauszeichnung „Red Dot: Design Team of the Year" an das Lenovo Design & User Experience Team unter der Leitung von Yao Yingjia. Diese Würdigung ist einzigartig auf der Welt und genießt über die Designszene hinaus hohes Ansehen, da die gestalterische Gesamtleistung von Designteams honoriert wird, die über mehrere Jahre hinweg mit außergewöhnlichen Produkten und wegbereitenden Gestaltungsleistungen in Erscheinung getreten sind. Seit 1988 wird der Radius im Rahmen des Red Dot Design Award an ein Designteam vergeben. Es ist gute Tradition, dass die Preisträger ihre Namen eingravieren lassen, bevor sie die Trophäe wie einen Wanderpokal weiterreichen.

2013   Lenovo Design & User Experience Team
2012   Michael Mauer & Style Porsche
2011   The Grohe Design Team led by Paul Flowers
2010   Stephan Niehaus & Hilti Design Team
2009   Susan Perkins & Tupperware World Wide Design Team
2008   Michael Laude & Bose Design Team
2007   Chris Bangle & Design Team BMW Group
2006   LG Corporate Design Center
2005   Adidas Design Team
2004   Pininfarina Design Team
2003   Nokia Design Team
2002   Apple Industrial Design Team
2001   Festo Design Team
2000   Sony Design Team
1999   Audi Design Team
1998   Philips Design Team
1997   Michele De Lucchi Design Team
1996   Bill Moggridge & Ideo Design Team
1995   Herbert Schultes & Siemens Design Team
1994   Bruno Sacco & Mercedes-Benz Design Team
1993   Hartmut Esslinger & Frogdesign
1992   Alexander Neumeister & Neumeister Design
1991   Reiner Moll & Partner & Moll Design
1990   Slany Design Team
1989   Braun Design Team
1988   Leybold AG Design Team

»Design today is about being able to predict and present the look of tomorrow.«
»Im Design geht es heute darum, sich vorstellen zu können, wie die Produkte von morgen aussehen.«

Yao Yingjia, Vice President
Lenovo Design & User Experience

Red Dot: Design Team of the Year 2013
Lenovo Design & User Experience Team led by Yao Yingjia
A new legend on the international design map

For the first time in the history of the competition, the honorary title "Red Dot: Design Team of the Year" goes to China. The Lenovo Design & User Experience Team, under the leadership of Yao Yingjia, has placed the company and the Lenovo brand firmly in the spotlight. In the last two years alone, the international jury of the Red Dot Award: Product Design has honoured the Chinese manufacturer of PCs, notebooks, tablets, smart TVs and smartphones with the Red Dot a total of 15 times, two of these were the Red Dot: Best of the Best.

Erstmals in der Geschichte des Wettbewerbs geht die Ehrenauszeichnung „Red Dot: Design Team of the Year" nach China. Mit dem Lenovo Design & User Experience Team unter der Leitung von Yao Yingjia rücken auch das Unternehmen und die Marke Lenovo in den Blickpunkt. Allein in den letzten beiden Jahren zeichnete die internationale Jury des Red Dot Award: Product Design den chinesischen Hersteller von PCs, Notebooks, Tablets, Smart TVs und Smartphones fünfzehn Mal mit dem Red Dot aus, darunter zwei Mal mit dem Red Dot: Best of the Best.

Good design keeps the balance between innovation and evolution. Lenovo design captivates with its simple and reserved elegance.
Gutes Design hält die Balance zwischen Innovation und Evolution. Lenovo Design besticht durch eine schlichte und zurückhaltende Eleganz.

The story of the Chinese company, Lenovo, reads just like the American dream. Reminiscent of the legendary garage-starts of Hewlett-Packard and Apple, Lenovo is following closely behind these legends, not only in terms of the company's history, but also regarding the claim to leadership in computer design.

A legend out of the garage

In a world of bits and bytes, of binary codes and logical commands, the computer industry inspires a certain romanticism when strong characters such as William Hewlett, David Packard or Steve Jobs are seen against the simple backdrop of the humble garage, setting the scene as the origin and ultimate symbol of the American entrepreneurial spirit. So it isn't much of a surprise that the Silicon Valley landmark is no architectural masterpiece or artistic sculpture, rather it is a simple garage in a Palo Alto suburb, not far from Stanford and the time-honoured Stanford University. The garage, "the birthplace of Silicon Valley" as it is referred to on the plaque outside, is the very place where Hewlett and Packard tinkered on their first product together at the end of the 1930s. Hewlett-Packard was the founding technology company in Silicon Valley and this garage is the enduring symbol of inventiveness and passion. At the same time, the birth of Silicon Valley is inseparably linked with the name of Frederick Terman, a professor and later dean of Stanford University, who encouraged his students to start their own businesses; William Hewlett and David Packard among them. In 1951, Terman breathed life into the Stanford Industrial Park, starting countless state-funded research projects, which then served as a seedbed for countless high-tech companies to follow. Whilst reading such captivating stories such as these, in which a successful company is built up without any outside involvement, it must be noted that this success would never have been possible without the support of the university and the public research funds.

Die Geschichte des chinesischen Unternehmens Lenovo liest sich wie ein amerikanischer Traum. Sie erinnert an die Garagenlegende von Hewlett-Packard und Apple. Lenovo ist diesen Legenden auf der Spur, nicht nur mit Blick auf die eigene Firmengeschichte, auch im Hinblick auf den Führungsanspruch im Computerdesign.

Die Legende jenseits der Garage

In einer Welt von Bits und Bytes, von binären Codes und logischen Befehlen haftet der Computerbranche bis heute etwas Romantisches an, wenn der Charakter starker Persönlichkeiten wie William Hewlett und David Packard oder Steve Jobs beschrieben und das Bild der Garage als Ursprung und Symbol des amerikanischen Unternehmergeistes beschworen wird. So verwundert es nicht, dass das Wahrzeichen des Silicon Valley kein architektonisches Meisterwerk und keine künstlerische Skulptur ist, sondern eine schlichte Garage in einem Wohngebiet von Palo Alto, unweit von Stanford und der altehrwürdigen Stanford-Universität: die Garage als „Geburtsort des Silicon Valley", wie es auf der Gedenktafel vor dem Anwesen heißt, der Ort, an dem Hewlett und Packard Ende der 1930er Jahre ihr erstes Produkt zusammenschrauben. Hewlett-Packard ist das erste Technologieunternehmen in Silicon Valley und die Garage bis heute das Symbol für Erfindergeist und Leidenschaft. Gleichwohl bleibt die Geburtsstunde des Silicon Valley untrennbar mit dem Namen Frederick Terman verbunden, der als Professor und späterer Dekan an der Stanford-Universität seine Studenten ermutigt, eigene Unternehmen zu gründen; auch William Hewlett und David Packard. 1951 ruft Terman den Stanford Industrial Park ins Leben und zieht zahlreiche staatlich finanzierte Forschungsprojekte an, die den Boden für viele Technologiefirmen bereiten. Auch wenn man nur allzu gerne die packenden Geschichten liest, in denen ohne fremdes Zutun, nur mit den eigenen Händen ein erfolgreiches Unternehmen aufgebaut wird, wäre dieser Erfolg ohne die Universität und ohne öffentliche Forschungsgelder kaum möglich gewesen.

Every product tells a story. The IdeaPad U430s is simply designed and exhibits the design language of a portable and practical journal.
Jedes Produkt erzählt eine Geschichte. Das IdeaPad U430s ist schlicht gestaltet und trägt die Formensprache eines portablen und praktischen Journals.

Lenovo founder Liu Chuanzhi made his start in a garage workshop just like William Hewlett and David Packard – with a slight difference: He developed a technology, together with ten colleagues at the Chinese Academy of Sciences in Beijing, which enabled an American computer to be used using Chinese characters. Because it is impossible to create a keyboard that includes all of the Chinese characters, the characters that can be seen on the screen itself must be represented using a phonetic transcription; hence the idea of a "Chinese Character Card" was born. In 1984, with the help of the Chinese Academy of Sciences in Beijing and start-up capital of RMB 200,000 (which at the time was worth around US$ 25,000), Liu Chuanzhi founded New Technology Development Inc., the forerunner of the Legend Group. In 1988, Legend's "Chinese Character Card" was awarded the highest national honour for scientific and technical innovation in China, the National Science and Technology Progress Award.

In 1989, the Beijing Legend Computer Group was born and began with the production of its own computer, which, just one year later appeared on the Chinese market under the name "Legend". The company evolved from a computer dealership into a computer manufacturer and, in 1994, was listed on the Hong Kong stock exchange. Two years later Legend introduced its first laptop under its own brand and became Number 1 on the Chinese market. 1998 also marked the production of the millionth Legend computer and Liu Chuanzhi had become, by this stage, an icon of the Chinese computer industry.

As history shows, legends such as these do not function merely as interesting sideshows, but instead play a decisive role in how an organisation perceives itself and how it is seen from the outside. In spite of their ready reference to facts and data, such companies are actually bursting with ideas, inventions and stories just as the American organisational psychologist Karl E. Weick proclaimed: "They pay a lot of attention to it, working out legends, developing myths, telling stories about their past and recreating them in episodes, which they have sorted out from their experiences with special attention."[1]

Wie William Hewlett und David Packard startet auch Lenovo-Gründer Liu Chuanzhi in einer Garagenwerkstatt – mit einem feinen Unterschied: Gemeinsam mit zehn Kollegen entwickelt er an der chinesischen Akademie der Wissenschaften in Beijing eine Technologie, die es ermöglicht, amerikanische Computer mit chinesischen Schriftzeichen zu betreiben. Da eine Tastatur unmöglich mit allen chinesischen Schriftzeichen belegt werden kann, müssen die auf dem Bildschirm sichtbaren chinesischen Zeichen über eine phonetische Umschrift dargestellt werden – die Idee der „Chinese Character Card" ist geboren. 1984 gründet Liu Chuanzhi mit Unterstützung der chinesischen Akademie der Wissenschaften in Beijing und einem Startkapital von 200.000 RMB, was zur damaligen Zeit etwa 25.000 US-Dollar entspricht, die New Technology Development Inc., den Vorläufer der Legend Group. 1988 wird die „Chinese Character Card" von Legend mit der höchsten nationalen Auszeichnung für wissenschaftlichen und technischen Fortschritt in China bedacht, dem National Science and Technology Progress Award.

1989 wird die Beijing Legend Computer Group ins Leben gerufen und beginnt mit der Produktion eigener Computer, die nur ein Jahr später unter dem Namen „Legend" auf den chinesischen Markt kommen. Das Unternehmen wandelt sich vom Computerhändler zum Computerhersteller und geht 1994 in Hongkong an die Börse. Zwei Jahre später stellt Legend den ersten Laptop unter eigenem Namen vor und ist erstmals die Nummer eins auf dem chinesischen Markt. 1998 wird der millionste Legend-Computer produziert. Zu diesem Zeitpunkt ist Liu Chuanzhi bereits eine Ikone der chinesischen Computerindustrie.

Wie die Geschichte zeigt, haben diese Legenden durchaus nicht den Charakter störender Nebengeräusche, sie spielen vielmehr eine entscheidende Rolle in der Selbst- und Fremdwahrnehmung von Organisationen. So sind Unternehmen trotz ihrer Inanspruchnahme von Daten und Fakten in Wirklichkeit voller Ideen, Erfindungen und Geschichten, wie der amerikanische Organisations-Psychologe Karl E. Weick weiß: „Sie verwenden einen großen Teil ihrer Zeit darauf, Legenden auszuarbeiten, Mythen zu entwickeln, Geschichten über ihre Vergangenheit zu erzählen und in Episoden auszuschmücken, die sie aus ihrem Erleben zur besonderen Beachtung ausgesondert haben."[1]

The Lenovo Yoga Concept won the Red Dot: Luminary in 2005.
Das Lenovo Yoga Concept wurde 2005 mit dem Red Dot: Luminary ausgezeichnet.

The S2110 is a hybrid tablet laptop providing high mobility, for it is very slim and lightweight.
Das S2110 ist ein Tablet-Laptop-Hybrid und bietet eine hohe Mobilität, da es sehr dünn und leicht ist.

The birth of a new brand

In preparation for its expansion strategy, Legend presented its new brand and company name in April 2003: Lenovo, an artificial word consisting of the initial syllable "Le" from "Legend" and the Latin word "novo" meaning "new". As other companies in many different countries had already registered the "Legend" name as their brand, the new "Lenovo" brand was meant to prepare the ground for and symbolise the move of the Chinese computer and electronics group into the international market.

The overnight emergence of a global player

The strategy was deemed to be successful with the surprise announcement by Lenovo on 8 December 2004 that it wanted to take over IBM's PC division for US$ 1.25 billion. Lenovo was the first Chinese company to make an acquisition of this magnitude in the USA. It was a milestone in the history of the company and sent out a signal to the global computer industry. Yang Yuanqing, CEO of Lenovo, knew that Lenovo would have to face two distinct challenges: computers and people. On the one hand there was the issue of integrating the "ThinkPad" brand with IBM computer technology, on the other was the question of whether established IBM customers and employees would accept Lenovo. Yang Yuanqing was certain that Lenovo had the capability of becoming the leader in the computer industry.

With the consent of the American authorities, the acquisition of IBM's PC division was officially completed towards the end of April 2005. With the aim of accessing global markets, Lenovo became a global player overnight and advanced to third largest computer manufacturer in the world behind Hewlett-Packard and Dell. Lenovo not only positioned itself as a global brand, but simultaneously used the "Think" brand to underline its own product and design quality standards. The acquisition of the IBM computer division and the closely linked Research and Development Centre gave Lenovo the rights to the "Think" brand and to five years usage of the IBM brand. In future, Lenovo under the leadership of Yang Yuanqing wanted above all to present itself as a global brand. In 2008, Lenovo appeared as sponsor of the Summer Olympics in Beijing and won the design contest for the Olympic torch.

Die Geburt einer neuen Marke

In Vorbereitung der Expansionsstrategie stellt Legend im April 2003 seinen neuen Marken- und Firmennamen vor: Lenovo, ein Kunstwort aus der englischen Silbe „Le" von „legend" und dem lateinischen Wort „novo" für „neu". Da der Name „Legend" bereits in vielen Ländern von anderen Unternehmen als Marke geschützt ist, soll mit der neuen Marke „Lenovo" auch der internationale Aufbruch des chinesischen Computer- und Elektronikkonzerns vorbereitet und symbolisiert werden.

Über Nacht zum Global Player

Die Strategie geht auf, als Lenovo am 8. Dezember 2004 überraschend bekannt gibt, die IBM PC Division für 1,25 Milliarden US-Dollar übernehmen zu wollen; ein Meilenstein in der Geschichte des Unternehmens und ein Signal für die weltweite Computerindustrie. Lenovos CEO, Yang Yuanqing, ist klar, dass sich das Unternehmen zwei unterschiedlichen Herausforderungen wird stellen müssen: dem Computer und den Menschen. Einerseits geht es um die Integration der Marke „ThinkPad" und der IBM-Computertechnologie, andererseits stellt sich die Frage, ob die früheren IBM-Kunden und IBM-Mitarbeiter Lenovo akzeptieren werden. Yang Yuanqing ist sich sicher, dass Lenovo zum besten Unternehmen in der Computerindustrie aufsteigen kann.

Mit Zustimmung der amerikanischen Behörden ist die Übernahme der IBM PC Division Ende April 2005 offiziell abgeschlossen. Verbunden mit dem Ziel, globale Märkte zu erschließen, wird Lenovo über Nacht zum Global Player und nach Hewlett-Packard und Dell zum drittgrößten Computerhersteller der Welt. Lenovo positioniert sich nicht nur als globale Marke, sondern unterstreicht mit der Marke „Think" den eigenen Anspruch an Produkt- und Designqualität. Mit der Übernahme der IBM-Computersparte und des damit eng verbundenen Forschungs- und Entwicklungszentrums von IBM erwirbt sich Lenovo auch die Rechte an der Marke „Think" und an der fünfjährigen Nutzung der Marke IBM. Zukünftig möchte sich Lenovo unter der Leitung von Yang Yuanqing vor allem als globale Marke präsentieren. Im Jahr 2008 tritt Lenovo als Sponsor der Olympischen Sommerspiele in Beijing in Erscheinung und gewinnt den Wettbewerb für die Gestaltung der olympischen Fackel. Der Entwurf erinnert an ein aufgerolltes Blatt Papier und schlägt eine Brücke zur chinesischen Geschichte und zu einer der

The design was reminiscent of a rolled up sheet of paper thereby creating a link to Chinese history and to one of mankind's most important inventions of all time. Lenovo also sponsored Formula 1 events, thus building up recognition of its young brand.

A question of attitude

"The expansion of Chinese company brands into the rest of the world is based on a systematic, sustained strategy by the Chinese central government... Its name is 'Zhou Chu Qu – Go global!'"[2], China expert Hans Joachim Fuchs believes. In his opinion, Chinese entrepreneurs and politicians don't see expansion as a one-way street, but rather consider foreign investment and the globalisation of Chinese businesses as two sides of the same coin. "The aim is relatively quickly to develop world class Chinese companies that can hold their own in a globally competitive market"[3], Fuchs writes in his book "Die China AG".

»I do believe attitude is the key to achieving anything in life, so I try to face everything in my life with a positive attitude.«

Yao Yingjia, vice president and chief designer at the Design & User Experience Team at Lenovo interprets the supposed "Strategy of going global" as a form of "openness", as an attitude towards the rest of the world. "I do believe attitude is the key to achieving anything in life," he admits in an interview with us. Lenovo was the first Chinese company to carry out an acquisition on this scale in the USA. This development was impressive because of the approach taken by Lenovo, one that other companies had not expected and that even industry insiders had not been able to imagine. In retrospect, it all makes sense and economic success appears to be a given. However, in order to truly appreciate Lenovo's achievement, an understanding of the link between American and Chinese business culture is needed. The Lenovo company and Lenovo brand embody the link between two worlds: the East and the West, Chinese and North-American company culture. What counts is not the distinction between and contrasting of Eastern and Western culture, but rather the ability to combine and make the most of different ways of thinking and behaving in a single company in order to allow the best of both worlds to come to fruition. In Lenovo, we have a company that possesses enterprising and creative expertise, which comes to the fore particularly in monitoring current market conditions, evaluating their potential and taking advantage of opportunities that present themselves.

Whereas our Western concept of management and design is often governed by the logic resulting from the means to an end, the unique effect of Lenovo's design and management strategy appears to arise from the monitoring of market situations and conditions, as well as the opportunities that arise from them. Its worldwide presence gives Lenovo detailed insight into the current state and different conditions of regions and markets, particularly emerging economies. This helps the company to develop innovative and relevant solutions for diverse markets.

wichtigsten Erfindungen der Menschheit überhaupt. Auch in der Formel 1 präsentiert sich Lenovo als Sponsor und steigert damit die globale Bekanntheit der noch jungen Marke.

Eine Frage der Einstellung

„Hinter dem Aufbruch der chinesischen Markenunternehmen ins Ausland steht eine konsequente und nachhaltige Strategie der chinesischen Zentralregierung ... Ihr Name ist ‚Zou Chu Qu – Schwärmt aus!'"[2], glaubt der China-Experte Hans Joachim Fuchs. Seiner Ansicht nach verstehen die chinesischen Unternehmer und Politiker die Globalisierung nicht als Einbahnstraße. Stattdessen gehen ausländische Investitionen und die Globalisierung chinesischer Unternehmen Hand in Hand. „Das Ziel ist, relativ schnell chinesische Weltklassefirmen zu entwickeln, die sich im internationalen Wettbewerb durchsetzen können"[3], schreibt Fuchs in seinem Buch „Die China AG".

»Ich glaube, dass die Einstellung der Schlüssel zum Erfolg ist, und versuche deshalb, alles in meinem Leben mit einer positiven Einstellung anzugehen.«

Yao Yingjia, Vizepräsident und Designchef des Design & User Experience Teams bei Lenovo, versteht die vermeintliche „Strategie des Ausschwärmens" eher als eine Form der „Offenheit", als Einstellung gegenüber der Welt. „Ich glaube, dass die Einstellung der Schlüssel zum Erfolg ist", verrät er uns im Interview. Lenovo ist das erste chinesische Unternehmen, das eine Übernahme dieser Größenordnung in den USA tätigt. Beeindruckend an dieser Entwicklung ist der Weg, den Lenovo geht; ein Weg, den andere Unternehmen nicht einmal gesehen haben und den sich auch Branchen-Insider nicht vorstellen konnten. Rückblickend erscheint alles einen Sinn zu ergeben und der wirtschaftliche Erfolg sich wie von selbst einzustellen. Um Lenovos Leistung angemessen würdigen zu können, gilt es, auch ein Verständnis für die Verbindung von amerikanischer und chinesischer Unternehmenskultur zu entwickeln. So verkörpern das Unternehmen und die Marke Lenovo die Verknüpfung beider Welten: Ost und West, chinesische und US-amerikanische Unternehmenskultur. Dabei geht es nicht um die Abgrenzung und Gegenüberstellung östlicher und westlicher Kultur, sondern vielmehr um die Fähigkeit, unterschiedliche Denk- und Handlungsweisen in einem Unternehmen strategisch zu nutzen und so miteinander zu verbinden, dass sich das Beste aus beiden Welten wirksam entfalten kann. Mit dem Unternehmen Lenovo begegnet uns zugleich eine unternehmerische und eine gestalterische Kompetenz, die sich insbesondere darin äußert, aktuelle Marktsituationen zu beobachten, ihr Potenzial einzuschätzen und die sich bietenden Gelegenheiten wahrzunehmen.

Während unsere westliche Vorstellung von Management und Design häufig von einer Logik aus Mittel und Zweck bestimmt wird, scheint sich die besondere Wirkung von Design und Management bei Lenovo aus der Beobachtung von Marktsituationen und -bedingungen sowie aus den sich daraus bietenden Möglichkeiten zu ergeben. Durch die weltweite Präsenz von Lenovo erhält das Unternehmen detaillierte Einblicke in den Zustand und die unterschiedlichen Bedingungen der Regionen und Märkte, gerade auch in den Schwellenländern. Dadurch gelingt es, innovative und relevante Lösungen für unterschiedliche Märkte zu entwickeln.

There are no second chances when it comes to making a good first impression. Packaging that is both aesthetically appealing and well thought out creates a thrill of anticipation for the new smartphone.
Es gibt keine zweite Chance für einen ersten guten Eindruck. Eine ebenso ästhetische wie durchdachte Verpackung weckt die Vorfreude auf das neue Smartphone.

Outstanding design quality

Over the last few years, Lenovo has based its positioning largely on its product and design quality. In 2005, the company first took part in the Red Dot Award: Design Concept and immediately went on to win the highest award for its Yoga Concept. Further accolades followed. In the last two years alone, the international jury of the Red Dot Design Award has given Yao Yingjia and the Lenovo Design & User Experience Team 15 prizes. On two occasions they have included the highest accolade the competition has to bestow. The Lenovo ThinkCentre Edge 91Z personal computer and the IdeaPad U430s each won the Red Dot: Best of the Best. Both products are proof of Lenovo's Design & User Experience Team's ability to create technological, but also quality-focused products that match today's modern lifestyle and are simple and easy to use. Yao Yingjia and his team have managed to develop a design language that is understood around the globe.

This is also repeatedly confirmed by customers. Design is increasingly important to them and has therefore crystallised itself as a clear advantage on the Chinese and global markets. Before the takeover of the IBM computer division and the "Think" brand, Lenovo was known as a PC manufacturer mainly in its home country, China. More recently, it has been able continually to increase the quality of its design and the level of recognition in the market and win over consumers across the world through innovation, technology and outstanding design.

Ausgezeichnete Designqualität

Lenovo positioniert sich insbesondere in den letzten Jahren über seine Produkt- und Designqualität. Erstmals taucht das Unternehmen 2005 im Red Dot Award: Design Concept auf und gewinnt auf Anhieb die höchste Auszeichnung für sein Yoga Concept. Es folgen weitere Auszeichnungen. Allein in den letzten beiden Jahren verleiht die internationale Jury des Red Dot Design Award 15 Auszeichnungen an Yao Yingjia und das Lenovo Design & User Experience Team, darunter zwei Mal die höchste Auszeichnung des Wettbewerbs. Der Personal Computer Lenovo ThinkCentre Edge 91Z und das IdeaPad U430s werden jeweils mit dem Red Dot: Best of the Best prämiert. Lenovos Design & User Experience Team stellt mit beiden Produkten seine Kompetenz unter Beweis, technologische und zugleich qualitätsorientierte Produkte zu gestalten, die dem modernen Lebensstil entsprechen und einfach und angenehm zu benutzen sind. Yao Yingjia und seinem Team ist es gelungen, eine Designsprache zu entwickeln, die überall auf der Welt verstanden wird.

Dies wird auch durch die Kunden immer wieder bestätigt, für die gutes Design immer wichtiger wird und somit zum klaren Vorteil im chinesischen und im globalen Markt geworden ist. Vor der Übernahme der IBM-Computersparte und der Marke „Think" war das Unternehmen vor allem in seinem Heimatmarkt China als PC-Hersteller bekannt, konnte seine Designstärke und seine Markenbekanntheit dann in der jüngsten Vergangenheit kontinuierlich steigern und weltweit durch Innovationen, Technologie und herausragendes Design überzeugen.

»Being a designer means that I have to be insightful and farsighted, as well as able to predict trends.«
»Gestalter zu sein bedeutet, dass ich einfühlsam und weitsichtig sein und Trends vorhersagen können muss.«

Yao Yingjia, Vice President
Lenovo Design & User Experience

The Lenovo smartphone has been especially fashioned for mobile Internet use.
The smartphone can be operated entirely by touch thanks to its touchscreen display.
In addition, it also has an interface to which a QWERTY keypad may be attached.
Das Lenovo-Smartphone wurde speziell für die mobile Internetnutzung konzipiert.
Das Smartphone lässt sich dank seines Touchscreens vollständig über Berührungen steuern.
Zusätzlich besitzt es eine Schnittstelle, an die eine QWERTY-Tastatur angeschlossen werden kann.

The Lenovo IdeaCentre 720 is one of the slimmest all-in-one computer.
Its frameless display can be tilted backwards by 90 degrees to position the screen horizontally.
Das Lenovo IdeaCentre 720 ist einer der dünnsten All-in-one-Computer. Sein rahmenloses Display kann
um bis zu 90 Grad nach hinten geschwenkt werden, um den Bildschirm waagerecht zu positionieren.

En route to becoming global market leader

Lenovo is an exception among Chinese brands. In 2012, the company increased its turnover from US$ 21.6 billion to US$ 29.6 billion. It is not only the 37 per cent increase that is surprising, but also the fact that in 2012 Lenovo was already generating 58 per cent of its turnover from overseas markets. With respect to turnover from abroad, Lenovo clearly leads the field among China's most valued brands. The figures back up the company's "Protect & Attack" strategy and the stated aim of its CEO Yang Yuanqing to "become one of the leading technology companies in the world".

This goal is just within the company's grasp, as Lenovo is en route to becoming global market leader in its core business, the desktop and laptop segment. According to a report by the American market research institute IDC, an IT and telecommunications specialist, by the end of 2012, Lenovo occupied the second spot among the world's largest computer manufacturers, close on the heels of the market leader Hewlett-Packard. Another American market research institute, Gartner, even reckons that, at the end of 2012, Lenovo was ahead of Hewlett-Packard. Whichever of the two rivals leads the field, there is no doubt that Lenovo is the fastest growing PC manufacturer in the world. If nothing else, this growth is supported by the company's successful design development. As far back as June 2011, Lenovo entered into a joint venture with NEC and became No. 1 among Japanese PC manufacturers. In August 2011, Lenovo acquired European company Medion and became market leader in Germany.

If a company undergoes such growth in so short a time, it is proof of tremendous innovation and strategic vision. In this regard, the acquisitions have also paid off, as Lenovo today has customers in more than 160 countries. By acquiring IBM's computer division, Lenovo not only obtained the rights to the "Think" brand, but also gained access to its technology and to international markets. The garage legend has turned into a worldwide success story and today employs more than 30,000 people in over 60 countries. They include 3,000 technicians, researchers and scientists. Lenovo's research and development team have introduced numerous industry firsts as borne out by 2,000 patents and over 100 significant design awards, including 25 Red Dot Design Award prizes.

Auf dem Weg zum Weltmarktführer

Unter den chinesischen Marken ist Lenovo eine Ausnahmeerscheinung. Im Jahr 2012 konnte das Unternehmen seinen Umsatz von 21,6 Milliarden US-Dollar auf 29,6 Milliarden US-Dollar steigern. Erstaunlich daran ist nicht nur das Wachstum von 37 Prozent, sondern insbesondere die Tatsache, dass Lenovo im Jahr 2012 bereits 58 Prozent seines Umsatzes außerhalb Chinas generiert. Unter Chinas wertvollsten Marken hat Lenovo im Hinblick auf die im Ausland erzielten Umsätze klar die Nase vorn. Die Zahlen unterstreichen die „Protect & Attack"-Strategie des Unternehmens und den Anspruch ihres CEO, Yang Yuanqing, „eines der führenden Technologieunternehmen der Welt zu werden".

Das Ziel ist in greifbare Nähe gerückt, denn im Kerngeschäft, dem Segment für Desktop- und Laptop-Computer, ist Lenovo auf dem Weg zum Weltmarktführer. Nach einem Bericht des amerikanischen Marktforschungsinstituts IDC, das sich auf Informationstechnologie und Telekommunikation spezialisiert hat, ist Lenovo Ende 2012 weltweit die Nummer zwei unter den größten Computerherstellern, mit geringem Abstand zu Marktführer Hewlett-Packard. Das amerikanische Marktforschungsunternehmen Gartner sieht Lenovo gegen Ende des Jahres 2012 erstmals sogar vor Hewlett-Packard. Welcher der beiden Kontrahenten auch immer die Nase vorn haben mag, unstrittig bleibt, dass Lenovo einer der am schnellsten wachsenden PC-Hersteller ist und dieses Wachstum nicht zuletzt von der erfolgreichen Designentwicklung des Unternehmens getragen wird. Bereits im Juni 2011 bildet Lenovo ein Joint Venture mit NEC und wird so zur Nummer eins unter den PC-Herstellern in Japan. Im August 2011 übernimmt Lenovo in Europa die Firma Medion und ist jetzt Marktführer in Deutschland.

Wenn ein Unternehmen in nur wenigen Jahren eine derartige Entwicklung vorweisen kann, zeugt dies von einer enormen Innovationsstärke und von strategischer Weitsicht. Denn auch in diesem Punkt zahlen sich die Übernahmen aus. Lenovo hat heute Kunden in mehr als 160 Ländern. Mit dem Kauf der IBM-Computersparte hat Lenovo nicht nur das Recht an der Marke „Think" erworben, sondern auch den Zugang zur Technologie und zu den internationalen Märkten. Aus der Garagenlegende ist ein weltweit erfolgreicher Konzern mit mehr als 30.000 Mitarbeitern in über 60 Ländern geworden. Das Unternehmen beschäftigt heute mehr als 3.000 Techniker, Forscher und Wissenschaftler. Die Forschungs- und Entwicklungsteams von Lenovo haben viele Branchenneuheiten vorgestellt. Davon zeugen mehr als 2.000 Patente und über 100 wichtige Designauszeichnungen, darunter allein 25 Auszeichnungen im Red Dot Design Award.

»Lenovo is committed to becoming a global leader in the "PC+" era, the era of tablets, smartphones, smart TVs and PCs.«
»Lenovo hat sich zum Ziel gesetzt, weltweit führend im "PC+"-Zeitalter zu werden – der Ära der Tablets, Smartphones, Smart TVs und PCs.«

Yao Yingjia, Vice President
Lenovo Design & User Experience

The new lifestyle era "PC+"

Lenovo regards design as far more than a quality guarantee. Product design is an expression of the brand's value, as well as of the company's innovation capability. By combining technology and design, Lenovo has managed to emotionalise the brand and not only meet the expectations of Chinese companies and consumers, but also arouse interest and create demand on international markets. Despite or perhaps because of strong international competition in the computer sector, Lenovo's Design & User Experience Team is very focused on utilising its technical know-how and experience with computers to create innovative design concepts and solutions. This results in well thought-out and well-made products that meet modern lifestyle needs both in the business-to-business segment and the business-to-consumer sector.

However, Lenovo is much more than just a company that makes PCs. Its CEO Yang Yuanqing and his chief designer Yao Yingjia never tire of emphasising that Lenovo makes a wide range of computer technologies from smartphones to tablets, and smart TVs to PCs. In the smartphone sector, Lenovo is already the second largest supplier in China and is present in countries such as India, Indonesia, Russia, Vietnam and the Philippines. This is all part of what Lenovo calls the "PC+" world, in which people use PCs alongside other intelligent devices whose core technology may make them like PCs, but whose shape and use are very different from those of a traditional computer.

The "Lenovo way"

The development of worldwide solutions and innovative products that are part of the "PC+" era is unthinkable without the multinational Design & User Experience Team. More than 150 designers with different cultural backgrounds work for Lenovo. Undoubtedly, this requires a high degree of consensus and internal communication, but it is also invaluable for the development and design of new products. When speaking of the company culture, insiders at Lenovo like to refer to it as the "Lenovo way", their own way. That also comes out in the statement: "We do what we say and take responsibility for our actions". The Design & User Experience Team invests a good deal of time into tracking trends. By continuously monitoring and analysing cultural factors and their backgrounds, Yao Yingjia and his team are able to predict future consumer behaviour patterns and so create new design trends. "Lenovo is committed to becoming a global leader in the 'PC+' era, the era of tablets, smartphones, smart TVs and PCs," Yao Yingjia concludes.

---

[1] Karl E. Weick: The Social Psychology of Organizing. New York: McGraw-Hill, 1969
[2] Hans Joachim Fuchs: Die China AG. Munich: Finanzbuch Verlag, 2007
[3] Hans Joachim Fuchs: Die China AG, loc. cit.

Die neue Lifestyle-Ära „PC+"

Für Lenovo ist Design weitaus mehr als ein Qualitätsversprechen. Über das Design der Produkte werden auch der Wert der Marke sowie die Innovationsstärke des Unternehmens zum Ausdruck gebracht. Lenovo gelingt es, durch die Verknüpfung von Technologie und Design, die Marke zu emotionalisieren und nicht nur die Erwartungen der chinesischen Unternehmen und Konsumenten zu erfüllen, sondern auch Wünsche und Begehrlichkeiten auf internationalen Märkten zu wecken. Obwohl oder – besser gesagt – gerade weil die internationale Konkurrenz im Computersegment sehr hoch ist, arbeitet Lenovos Design & User Experience Team intensiv daran, das technologische Wissen und die Erfahrung im Umgang mit Computern für innovative Designstudien und Lösungen zu nutzen und sowohl im „Business-to-Business"-Bereich als auch im „Business-to-Consumer"-Bereich gut durchdachte und gut gemachte Produkte anzubieten, die dem modernen Lebensstil entsprechen.

Lenovo ist jedoch weit mehr als nur ein Unternehmen, das PCs herstellt. Lenovos CEO Yang Yuanqing und sein Designchef Yao Yingjia werden nicht müde zu betonen, dass Lenovo eine breite Palette von Computertechnologien entwickelt – von Smartphones über Tablets bis zu Smart TVs und PCs. Mit Blick auf das Marktsegment Smartphones ist Lenovo bereits der zweitgrößte Anbieter in China und in Ländern wie Indien, Indonesien, Russland, Vietnam und auf den Philippinen vertreten. Es ist alles Teil dessen, was Lenovo die „PC+"-Welt nennt, in der die Menschen PCs sowie verschiedene intelligente Geräte nutzen, die von ihrer Kerntechnologie her zwar PCs sind, aber mit Blick auf die Form und Anwendung nicht mehr an die traditionelle Form eines Computers erinnern.

Der „Lenovo-Weg"

Die Entwicklung von weltweiten Lösungen und innovativen Produkten der „PC+"-Ära ist ohne das weltumspannende Design & User Experience Team kaum vorstellbar. Über 150 Designer aus unterschiedlichen Kulturkreisen sind für Lenovo tätig. Dies erfordert ohne Zweifel ein hohes Maß an Abstimmung und interner Kommunikation und ist zugleich von unschätzbarem Nutzen für die Entwicklung und Gestaltung neuer Produkte. Mit Blick auf die Unternehmenskultur spricht man bei Lenovo daher auch gern vom eigenen Weg, dem „Lenovo-Weg", der sich auch in der Aussage „Wir tun, was wir sagen, und übernehmen Verantwortung für unser Tun" widerspiegelt. Das Design & User Experience Team investiert viel Zeit in das Aufspüren von Trends. Durch laufende Beobachtungen und die Analyse kultureller Faktoren und ihrer Hintergründe ist es Yao Yingjia und seinem Team möglich, künftige Entwicklungen im Verhalten der Verbraucher vorherzusagen und selbst neue Trends im Design zu setzen. „Lenovo hat sich zum Ziel gesetzt, weltweit führend im ‚PC+'-Zeitalter zu werden – der Ära der Tablets, Smartphones, Smart TVs und PCs", sagt Yao Yingjia.

---

[1] Karl E. Weick: Der Prozess des Organisierens. Frankfurt am Main: Suhrkamp Verlag, 1985
[2] Hans Joachim Fuchs: Die China AG. München: Finanzbuch Verlag, 2007
[3] Hans Joachim Fuchs: Die China AG, a. a. O.

»I do believe attitude is the key
to achieving anything in life.«
»Ich glaube, dass die Einstellung
der Schlüssel zum Erfolg ist.«

Yao Yingjia, Vice President
Lenovo Design & User Experience

Red Dot: Design Team of the Year 2013
Interview with Yao Yingjia, Chief Designer
Vice President of Lenovo Design & User Experience

The "Red Dot: Design Team of the Year" award this year goes to Yao Yingjia and the Lenovo Design & User Experience Team. Burkhard Jacob met Yao Yingjia for an interview in Milan.

Yao Yingjia, I assume you travel a lot.
What is your favourite city, aside from Beijing?

Yao Yingjia, I like cities like Milan, Barcelona and Munich. For me, as a designer, each of these cities has its own flair. But, I also like the climate and surroundings of Honolulu on Hawaii or the intercultural vibe of San Francisco. As a designer, I love the differences between all these cities.

How long have you been working for Lenovo?

Since 1996. I have only ever worked for Lenovo. That may sound unusual in a dynamic and fast changing country. But it is true because the company grew very fast, providing many new opportunities to challenge ourselves.

What are your qualifications?

Surprisingly, I come from the field of graphic design and then I started working for Lenovo. However, I also completed an executive MBA program, a postgraduate masters degree geared towards managers who have already gained a lot of practical experience. It is important not to lose sight of the connection between design, business and technology.

When did you first become aware of the significance of design?
Can you recall a particular situation?

Good question. Actually, I have loved pictures and illustrations ever since I was a child. When I was young I always wanted to be an artist. Later on I wanted to balance emotion and logic, to harmonise the power of thoughts and actions. As a designer, you have to combine all the elements of design and technology, passion and reason.

Did you have a mentor or someone you looked up to?

I have had lots of teachers in my life and I still have lots of people who set an example for me, like Yang Yuanqing, the chief executive officer of Lenovo.

What did you learn from these people?

Not to solely focus on the details, but to also keep an eye on interrelationships and the big picture. I occasionally also consider people like Yang Yuanqing to be designers. In this case, it is not about the design of a particular product or a user interface. Rather, it concerns the innovative solution to a problem, such as the organisation of a company. As a designer, I am also interested in successful entrepreneurs and extraordinary people with leadership qualities.

Die Ehrenauszeichnung „Red Dot: Design Team of the Year" geht in diesem Jahr an Yao Yingjia und das Lenovo Design & User Experience Team. Burkhard Jacob traf Yao Yingjia in Mailand zum Interview.

Yao Yingjia, ich nehme an, Sie reisen sehr viel.
Welche ist Ihre Lieblingsstadt? Beijing einmal ausgenommen.

Ich mag Städte wie Mailand, Barcelona und München. Für mich als Designer hat jede dieser Städte einen besonderen Reiz. Ich mag aber auch das Klima und die Umgebung in Honolulu auf Hawaii oder den kulturellen Mix in San Francisco. Als Designer liebe ich die Unterschiede zwischen diesen Städten.

Seit wann arbeiten Sie für Lenovo?

Seit 1996 habe ich ausschließlich für Lenovo gearbeitet. Das mag seltsam klingen in einem dynamischen Land, welches dabei ist, sich rasch zu verändern. Aber es trifft dennoch zu, da unser Unternehmen sehr schnell gewachsen ist und uns so viele neue Herausforderungen bot.

Welche Ausbildung haben Sie genossen?

Überraschenderweise komme ich aus dem Grafikdesign und fing dann an, für Lenovo zu arbeiten. Ich absolvierte aber auch ein Executive-MBA-Programm, einen postgradualen Masterstudiengang, der sich an Führungskräfte mit fortgeschrittener Berufserfahrung richtet. Es ist wichtig, die Verbindung zwischen Design, Business und Technologie im Auge zu haben.

Wann ist Ihnen die Bedeutung des Designs erstmals bewusst geworden? Erinnern Sie sich an eine besondere Situation?

Gute Frage. Eigentlich mag ich Bilder und Illustrationen schon seit meiner Kindheit. Und in frühen Jahren wollte ich immer Künstler werden. Später entdeckte ich dann den Wunsch, Emotion und Logik auszubalancieren, Handeln und Denken in Einklang zu bringen. Als Designer muss man Design und Technologie, Leidenschaft und Kalkül miteinander verbinden.

Hatten Sie einen Mentor? Oder ein Vorbild?

Ich hatte viele Lehrer in meinem Leben, und ich habe auch heute Menschen, die ein gutes Vorbild sind, wie zum Beispiel Yang Yuanqing, Lenovos geschäftsführenden Vorstand.

Was konnten Sie von diesen Menschen lernen?

Sich nicht nur auf die Details zu konzentrieren, sondern immer auch die Zusammenhänge und das große Ganze im Auge zu behalten. Manchmal betrachte ich auch Menschen wie Yang Yuanqing als Designer. Es geht dabei nicht um die Gestaltung eines Produktes oder das Design von Interfaces, sondern um die innovative Lösung für ein Problem, beispielsweise wenn es um die Organisation eines Unternehmens geht. Als Designer interessieren mich also auch erfolgreiche Unternehmer und außergewöhnliche Menschen mit Führungsqualitäten.

Yao Yingjia has loved illustrations ever since he was a child and wanted to be an artist. Later on as designer, he wanted to combine all the elements of design and technology, passion and reason.
Yao Yingjia mag Illustrationen schon seit seiner Kindheit und wollte immer Künstler werden. Als Designer entdeckte er dann später den Wunsch, Design und Technologie, Leidenschaft und Kalkül miteinander zu verbinden.

May I ask what motivates you and makes you happy?

I see this question against a backdrop of "work-life-balance". I am a relatively good basketball player and am very good at volleyball, almost at professional level. But, to get to the point: I like the element of team spirit in sport. I also like to work in a team. However, it isn't always easy to lead a team, although I am always very pleased if, after hammering it out together, we are able to reach a new or better solution. The most important results are achieved in the team. And, in design it is always possible to learn something or discover something new; not to underestimate the recognition that can be achieved when a design draft or a product is successful. That can be extremely rewarding.

You mention the Lenovo Design & User Experience Team. How do you see your role at Lenovo? Do you see yourself more as a designer within the team or as a manager of it?

I am passionate about working as a designer, but it is an undisputable fact that I need to consider more than just the design. When most people consider design, the first thing they think of is product design. I also have to consider other aspects such as management and issues like innovation and communication. These things operate on an entirely different level and are also entwined with economic and social issues.

What is your current assessment of the Chinese market and Chinese society at the moment?

The Chinese market is extremely dynamic and there is also a lot of change in society. Chinese consumers are on a path of self discovery. We are attempting to respond to this trend by using design, through technology, to facilitate a

Darf ich fragen, was Sie motiviert oder glücklich macht?

Ich sehe die Frage eher vor dem Hintergrund der „Work-Life-Balance". Ich bin ein relativ guter Basketball-Spieler und ziemlich gut in Volleyball – fast schon professionell. Um es auf den Punkt zu bringen: Ich mag den Teamgeist im Sport. Ich mag aber auch die Arbeit im Team. Natürlich ist es nicht immer einfach, ein Team zu führen, ich freue mich allerdings immer, wenn wir auch nach intensiven Diskussionen zu einer neuen oder besseren Lösung kommen. Die wichtigsten Ergebnisse werden im Team erzielt. Und im Design kann man immer etwas lernen oder etwas Neues entdecken; nicht zu vergessen die Anerkennung, die man erhält, wenn ein Entwurf oder ein Produkt gelungen ist. Das kann schon sehr beglückend sein.

Sie sprechen das Lenovo Design & User Experience Team an. In welcher Rolle sehen Sie sich bei Lenovo? Fühlen Sie sich mehr als Designer oder mehr als Manager eines Teams?

Ich mag die Arbeit als Designer sehr, aber es ist unbestritten, dass ich nicht nur über Design nachdenken kann. Viele denken beim Thema Design in erster Linie an Produktdesign. Ich verwende meine Zeit aber auch aufs Management und auf Themen wie Innovation und Kommunikation. Das ist eine ganz andere Ebene und berührt auch wirtschaftliche und gesellschaftliche Fragen.

Wie schätzen Sie den chinesischen Markt und auch die chinesische Gesellschaft augenblicklich ein?

Der chinesische Markt ist sehr dynamisch, und auch in der Gesellschaft ist viel in Bewegung. Die chinesischen Konsumenten möchten sich mehr und mehr selbst verwirklichen. Wir versuchen, auf diese Entwicklung zu reagieren und über die

Finger exercises. Yao Yingjia possesses both attention to detail and an overview of the whole picture.
Fingerübungen. Yao Yingjia mit der Aufmerksamkeit für Details und dem Blick für das Ganze.

whole new experience for the user in the general interaction with computer-based products. In the past, consumers have tended to buy the next generation of computers if only for the reason that the devices themselves were technically superior to their forerunners and could deliver a better performance. You simply bought the latest machine. Today, however, consumers associate the purchase of a new computer with a new experience. Devices today allow for a totally new interaction with computer-based technology. At the moment, Chinese manufacturers are realising just how important this development is. The focus is no longer on the manufacture of technical products but on new adventures and experiences which the consumer wants when using these devices. At Lenovo we try to create solutions that do not necessarily fit into the box. Rather, they serve to surprise the consumer with something new. Design today is about being able to present the look of tomorrow. And Lenovo does this extremely well.

Could you sum up Lenovo's design philosophy in just three words?

Lenovo design is "simple, unique, valuable". When our design team has created something new, we aim to make it easy to understand and simple to use. This also involves addressing the emotions we want to stimulate. Even if something has a simple design, it should still be distinctive and possess key elements that identify it as a typical Lenovo product; for example the colour, the material, the logo. Last but not least, a product should, and must, have a tangible benefit and value for the consumer. Only if customers identify value in our products will they want to have them, keep them and use them for the long-term.

Technologie hinaus Design zu nutzen, um für die Benutzer ein neues Erlebnis im Umgang mit computerbasierten Produkten zu ermöglichen. Früher haben die Konsumenten die nächste Computergeneration gekauft, weil die Geräte technisch besser waren und über eine höhere Leistung verfügten; man hat sich halt eine neue Maschine gekauft. Heute verbinden Konsumenten mit dem Neukauf eines Computers auch ein neues Erlebnis. Die Geräte von heute erlauben ganz neue Erfahrungen im Umgang mit computerbasierter Technologie. Augenblicklich lernen viele chinesische Hersteller, wie wichtig diese Entwicklung ist. Es geht nicht mehr nur um das Herstellen von technischen Produkten, sondern um neue Erlebnisse und Erfahrungen, die Konsumenten im Umgang mit Dingen haben möchten. Bei Lenovo versuchen wir, Lösungen zu entwickeln, die nicht nur die zeitgemäßen Erwartungen erfüllen, sondern die Konsumenten auch immer wieder aufs Neue überraschen. Im Design geht es heute darum, sich vorstellen zu können, wie die Produkte von morgen aussehen. Und das gelingt Lenovo sehr gut.

Können Sie versuchen, Lenovos Designphilosophie in drei Worten zu beschreiben?

Lenovo Design ist einfach, einzigartig und wertvoll: simple, unique, valuable. Wenn unser Designteam etwas Neues gestaltet, möchten wir, dass es leicht verständlich und einfach zu benutzen ist. Das schließt auch die Emotionen ein, die wir bei den Konsumenten ansprechen oder hervorrufen wollen. Auch wenn etwas einfach gestaltet ist, soll es dennoch unverwechselbar sein und über Schlüsselelemente verfügen, die es zu einem typischen Produkt der Marke Lenovo machen – zum Beispiel die Farbe, das Material, das Logo. Last, but not least soll und muss das Produkt auch einen konkreten Nutzen und einen Wert für den Konsumenten haben. Nur wenn Kunden einen Wert in unseren Produkten erkennen, möchten sie diese auch haben, behalten und langfristig nutzen.

Liu Chuanzhi, the founder of Lenovo, is an iconic figure in the Chinese computer industry.
He turned what was a workshop in a garage into a global company.
Liu Chuanzhi, Gründer von Lenovo, ist eine Ikone der chinesischen Computerindustrie.
Aus einer Garagenwerkstatt hat er einen Weltkonzern geformt.

Yang Yuanqing, CEO of the Lenovo Group, whose passion, vision and foresight
have transformed the company and brand into a global player in the IT industry.
Yang Yuanqing, CEO der Lenovo Group, hat das Unternehmen und die Marke
mit Leidenschaft und Weitsicht zum Global Player der IT-Industrie gemacht.

Is this understanding of design also reflected in the Lenovo brand?

The Lenovo brand is aimed at different target groups. On one side there are the consumers, who almost exclusively use our products for private purposes, and, on the other side, we have companies, which use our products in a professional capacity. That our products are easy to understand, user-friendly, useful and valuable must be true, at all costs, for both. This is also reflected in our brand.

Lenovo is a very young brand. Although the history of the company goes back to the 1980s, the brand itself only came into being in 2004.

That's very true. The Lenovo brand is still young and consequently very dynamic. We are passionate about developing many new ideas and products for the brand.

In 2005 Lenovo took over the Personal Computer division from IBM. At this stage, Lenovo was already the market leader in China. By acquiring the IBM personal computer business, it became a global player, virtually overnight.

We are on the way to becoming the global market leader. And that requires a new approach. In future, Lenovo will be organised into two divisions: the Lenovo Business Group and the Think Business Group. The Lenovo Business Group will be responsible for desktops, laptops and tablets as well as smartphones. The Think Business Group will improve the position of the "Think" brand and develop products for business customers. You see? We are already working on something new.

I recently spoke with Paul Lasewicz, who is in charge of the IBM corporate archive. He told me how especially proud IBM is of its design history.

I am too. IBM has made a huge contribution to the history of computer design and "Think" is without doubt a classic and, more importantly, a powerful brand, both in terms of design as well as its technical performance. I really like the products and the brand and I am really proud of them.

Spiegelt sich dieses Designverständnis auch in der Marke Lenovo wider?

Die Marke Lenovo spricht unterschiedliche Zielgruppen an. Auf der einen Seite stehen die Konsumenten, die unsere Produkte überwiegend privat nutzen, auf der anderen Seite finden wir Unternehmen, die unsere Geräte professionell nutzen. Für beide gilt aber gleichermaßen, dass unsere Produkte leicht verständlich, benutzerfreundlich, nützlich und wertvoll sind. Das spiegelt sich auch in unserer Marke wider.

Lenovo ist eine sehr junge Marke. Die Unternehmensgeschichte reicht zwar bis in die 1980er Jahre zurück, die Marke Lenovo existiert aber erst seit dem Jahr 2004.

In der Tat, die Marke Lenovo ist noch jung und damit auch sehr dynamisch. Wir haben die Leidenschaft, viele neue Ideen und Produkte unter dieser Marke zu entwickeln.

Im Jahr 2005 folgte die Übernahme der Personal Computer Division von IBM. Zu diesem Zeitpunkt war Lenovo bereits Marktführer in China. Mit dem Erwerb der IBM-Computersparte wurde Lenovo über Nacht zum Global Player.

Wir sind auf dem Weg zum Weltmarktführer. Und das verlangt einen neuen Ansatz. Künftig gliedert sich Lenovo in zwei Bereiche: die Lenovo Business Group und die Think Business Group. Die Lenovo Business Group wird sowohl für Desktops, Laptops und Tablets als auch für Smartphones zuständig sein. Die Think Business Group soll die Marke „Think" besser positionieren und Produkte für Geschäftskunden entwickeln. Das zeigt, dass wir bereits wieder an etwas Neuem arbeiten.

Vor Kurzem sprach ich mit Paul Lasewicz, der das IBM-Unternehmensarchiv leitet. Er erklärte mir, wie stolz IBM insbesondere auf seine Designgeschichte ist.

Das bin ich auch. IBM hat einen wertvollen Beitrag zur Geschichte des Computerdesigns geleistet, und „Think" ist ohne Zweifel eine klassische, aber auch nach wie vor eine sehr leistungsstarke Marke, sowohl mit Blick auf das Design als auch im Hinblick auf die technische Leistung. Ich mag die Produkte und die Marke sehr und bin ebenfalls sehr stolz darauf.

The global management team of the Lenovo Design & User Experience Centre.
Das globale Management-Team des Lenovo Design & User Experience Centre.

Is "Think" both a challenge and a heritage for Lenovo to bear?

This has a lot to do with our corporate culture and communication. With "Think", the Lenovo Think Business Group has become a premium brand for business customers and the high-end consumer segment. This also stimulates our international design team. We work with a huge variety of people from the USA, Europe and Asia, but we see ourselves as one team that is working on new possibilities for daily life and future products.

Can you give us any insights into what the future holds?
Lenovo often uses the slogan "PC+". What does this refer to?

The computer for many years now has been seen as a working tool. This was the age of the processor, networked devices and technical accessories. "PC+" emphasises the lifestyle aspect. This sees computers as much more than just a working tool and addresses the issue of which new experiences we can have with computers, such as the integration of computers in other products, like furniture or cars, or even architecture. Users do not perceive computers as just a calculator but as a means to explore new things in a playful fashion. Lenovo is committed to becoming a global leader in the "PC+" era, the era of tablets, smartphones, smart TVs and PCs. To achieve this objective, Lenovo's Design & User Experience Team will continue to apply our seasoned PC experience, entrepreneurial far-sightedness and excellent innovation design, to give our customers products that are reliable, functional, high-performing, easy-to-use, and of course, pleasing to the eye. Computers don't just impact other industries and our jobs, they are integral to our daily life. I think that Lenovo and the Lenovo Design & User Experience Team can make a huge contribution here, not just for the benefit of consumers and professional users but also for the development of design in the wider sense.

Yao Yingjia, we look forward to the surprises you have in store for us! Our hearty congratulations go to you and the Lenovo Design & User Experience Team on winning the award "Red Dot: Design Team of the Year 2013".

Ist „Think" zugleich Herausforderung und Verpflichtung für Lenovo?

Das hat viel mit Unternehmenskultur und Kommunikation zu tun. Die Lenovo Think Business Group wird mit „Think" zu einer Premiummarke für Geschäftskunden und im High-End-Bereich für Konsumenten. Dies ist zugleich auch Ansporn für unser internationales Designteam. Wir arbeiten mit den unterschiedlichsten Menschen aus den USA, Europa und Asien zusammen, verstehen uns aber als ein Team, das an neuen Möglichkeiten für unser tägliches Leben und an Zukunftsprodukten arbeitet.

Können Sie uns einen Vorgeschmack auf die Zukunft geben?
Lenovo verwendet häufig den Slogan „PC+". Was ist damit gemeint?

Der Computer wurde über viele Jahre als Arbeitsmittel verstanden. Das war die Zeit der Prozessoren, der vernetzten Geräte und der technischen Hilfsmittel. Mit „PC+" betonen wir den Lifestyle-Charakter. Das geht weit über den Computer als Arbeitsgerät hinaus und dreht sich um die Frage, welche neuen Erfahrungen wir im Umgang mit Computern machen können, zum Beispiel durch die Integration von Computern in anderen Produkten wie Möbel oder Automobile oder beispielsweise in der Architektur. Die Nutzer möchten den Computer nicht nur als Rechner wahrnehmen, sondern mit den Möglichkeiten des Computers spielerisch neue Erfahrungen sammeln. Lenovo hat sich zum Ziel gesetzt, weltweit führend im „PC+"-Zeitalter zu werden – der Ära der Tablets, Smartphones, Smart TVs und PCs. Um dies zu erreichen, wird das Lenovo Design & User Experience Team auch in Zukunft unsere bewährte PC-Erfahrung mit unternehmerischer Weitsicht und exzellentem, innovativem Design fortführen und so unseren Kunden Produkte bieten, die verlässlich, funktionell, leistungsfähig, leicht bedienbar und natürlich schön anzusehen sind. Der Computer beeinflusst also nicht nur andere Branchen und unsere Arbeit, sondern unser tägliches Leben. Ich glaube, dass Lenovo und damit auch das Lenovo Design & User Experience Team einen großartigen Beitrag dazu leisten kann – nicht nur für die Konsumenten und professionellen Anwender, auch für die Designentwicklung insgesamt.

Yao Yingjia, wir lassen uns gerne überraschen und gratulieren dem Lenovo Design & User Experience Team und Ihnen persönlich ganz herzlich zur Ehrenauszeichnung „Red Dot: Design Team of the Year 2013".

Red Dot: Best of the Best
The best designers of their category
Die besten Designer ihrer Kategorie

The designers of the Red Dot: Best of the Best
Only a few products in the Red Dot Design Award receive the "Red Dot: Best of the Best" accolade. In each category, the jury can assign this award to products of outstanding design quality and innovative achievement. Exploring new paths, these products are all exemplary in their design and oriented towards the future.

The following chapter introduces the people who have received one of these prestigious awards. It features the best designers and design teams of the year 2013 together with their products, revealing in interviews and statements what drives these designers and what design means to them.

Die Designer der Red Dot: Best of the Best
Nur sehr wenige Produkte im Red Dot Design Award erhalten die Auszeichnung „Red Dot: Best of the Best". Die Jury kann mit dieser Auszeichnung in jeder Kategorie Design von außerordentlicher Qualität und Innovationsleistung besonders hervorheben. In jeder Hinsicht vorbildlich gestaltet, beschreiten diese Produkte neue Wege und sind zukunftsweisend.

Das folgende Kapitel stellt die Menschen vor, die diese besondere Auszeichnung erhalten haben. Es zeigt die besten Designer und Designteams des Jahres 2013 zusammen mit ihren Produkten. In Interviews und Statements wird deutlich, was diese Designer bewegt und was ihnen Design bedeutet.

reddot design award
product design 2013

Designer portraits

Mikko Laakkonen
Aura

»I try to create objects that look simple and natural. Achieving that can be quite hard. Often it means that the manufacturing is very complex.«

»Ich versuche, Objekte zu entwerfen, die schlicht und natürlich aussehen. Das zu erreichen, kann schwer sein. Oft bedeutet es, dass die Herstellung sehr komplex ist.«

What do you like in particular about your own award-winning product?
The natural form that creates the space inside the sofa.

Is there a role model that inspired your work?
Human behaviour.

What are the greatest challenges that you currently see in your industry?
Make more from less.

Do you have something like a dream project?
To design a sailing boat. After the project is finished I'd sail the boat far, far away.

Was gefällt Ihnen an Ihrem eigenen, ausgezeichneten Produkt besonders gut?
Die natürliche Form, die den inneren Raum des Sofas bildet.

Gibt es ein Vorbild, das Ihre Arbeit inspiriert hat?
Das menschliche Verhalten.

Worin bestehen für Sie aktuell die größten Herausforderungen in Ihrer Branche?
Aus weniger mehr zu machen.

Haben Sie so etwas wie ein Wunschprojekt?
Ein Segelboot zu entwerfen. Sobald das Projekt beendet ist, würde ich damit weit, weit weg segeln.

Red Dot: Best of the Best
Aura
Sofa
See page 62

reddot design award
best of the best 2013

Canon Electronic Business Machines (H.K.) Co., Ltd.
X Mark II

»It is important to have a clear understanding
of our brand vision and design direction.«
»Es ist wichtig, ein klares Verständnis unserer
Markenvision und Gestaltungsrichtung zu haben.«

What do you like in particular about
your own award-winning product?
This unique and subtle form blends comfortably into any environment. The size
is minimised, yet it features the largest
possible keys, while taking ergonomics
into account.

How would you define
the term "quality"?
Quality is a level of achievement with
a definite target that has to be reached
or exceeded. Quality should be judged
objectively – not only from the designers' perspective, but also by gaining
recognition from the public.

Was gefällt Ihnen an Ihrem
eigenen, ausgezeichneten Produkt
besonders gut?
Diese einzigartige und subtile Form passt
sich bequem jeder Umgebung an. Die
Größe ist reduziert, weist unter Berück-
sichtigung der Ergonomie aber dennoch
die größtmöglichen Tasten auf.

Wie würden Sie den Begriff
der Qualität beschreiben?
Qualität ist ein Leistungsniveau mit defi-
niertem Ziel, das erreicht oder übertroffen
werden muss. Die Qualität muss objektiv
beurteilt werden – nicht nur aus der Perspektive der Gestalter, sondern sie sollte
auch Bestätigung von der Öffentlichkeit
erfahren.

Red Dot: Best of the Best
X Mark II
Calculator
Taschenrechner
See page 108

reddot design award
best of the best 2013

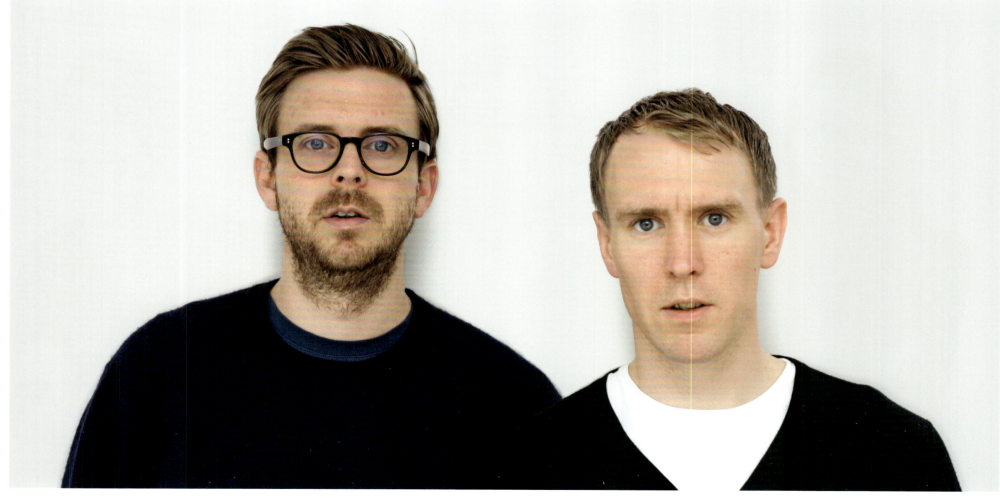

Designer portraits

Form Us With Love, Grønlund Design, StokkeAustad
RBM Noor

»Design quality is when the cleverness of the product
and its method of manufacture push the industry forward.«
»Designqualität liegt vor, wenn die Klugheit des Produkts
und seiner Herstellungsmethode die Industrie vorantreibt.«

Is there a specific approach that is
of significance to you and your work?
StokkeAustad: Dialogue is the most
important tool in the office. We use it
to analyse the task at hand and really
find out exactly what it is we are solving.

Which social, cultural or economic
developments have a particularly
strong influence on the design of
today?
StokkeAustad: The explosion in social
media has led to a democratisation
that is set to continue. The Internet has
levelled the playing field, allowing talent
to be exposed regardless of precon-
ceived notions of merit.

Gibt es einen bestimmten Ansatz,
der für Sie und Ihre Arbeit von
Bedeutung ist?
StokkeAustad: Das allerwichtigste Instrument im Büro ist der Dialog. Er dient uns
dazu, die anstehende Aufgabe zu analysieren und genau herauszufinden, was
wir zu lösen haben.

Welche gesellschaftlichen, kulturellen
oder wirtschaftlichen Entwicklungen
beeinflussen das Design augenblicklich
besonders stark?
StokkeAustad: Das explosionsartige Auf-
kommen sozialer Medien hat zu einer
anhaltenden Demokratisierung geführt.
Das Internet hat das Spielfeld geebnet,
sodass sich Talent trotz vorgefasster
Meinungen dazu, was Qualität sei, vorur-
teilsfrei zeigen kann.

Red Dot: Best of the Best
RBM Noor
Canteen and Conference Chair
Kantinen- und Konferenzstuhl
See page 66

reddot design award
best of the best 2013

Designer portraits

Sam Baskar, Chris Shook
Black & Decker Gyro Driver™

»We must continue to be innovative and creat ve,
and to think outside the box when designing products
that aid everyone in daily life.«

»Wir müssen weiterhin innovativ und kreativ sein und
über den Tellerrand hinausschauen, wenn wir Produkte
gestalten, die den Menschen im Alltag helfen.«

Red Dot: Best of the Best
Black & Decker Gyro Driver™
Motion Sensing Screwdriver
Bewegungsgesteuerter Akku-Schrauber
See page 116

reddot design award
best of the best 2013

Designer portraits

Jörg Peschel
NEBV

»In an industrial context, products have to be intuitive and easy to use. So giving them a 'clear' and functional design certainly is advantageous.«

»Im industriellen Kontext müssen Produkte für den Nutzer intuitiv bedienbar sein. Da hilft es natürlich, wenn die Gestaltung ,aufgeräumt' und funktional ist.«

What is, in your opinion, the significance of design quality in your industry?
In the B2B industry, there is definitely still potential for design. Studies give attest to this and it also shows at trade fairs. Festo is an innovator in the field of automation, but awareness of good design is growing.

How would you define the term "design quality"?
On the one hand, design quality means good usability and sustainability. However, products that are "fun" also have a quality, one that often is intrinsically linked to the design. It is exciting to observe the merging of these different aspects in the relatively new field of "experience design".

Wie schätzen Sie den Stellenwert des Designs in Ihrer Branche ein?
In der B2B-Branche ist definitiv noch Potenzial für Design. Das belegen Studien und man kann es auf Messen sehen. Festo ist im Bereich der Automation ein Vorreiter. Aber das Bewusstsein für gute Gestaltung wächst.

Wie würden Sie den Begriff der Designqualität beschreiben?
Einerseits ist Designqualität gute Bedienbarkeit und Nachhaltigkeit. Aber auch Produkte, die „Spaß machen", haben eine Qualität, die meist eng mit dem Design verknüpft ist. Ich beobachte mit Spannung die Verschmelzung dieser unterschiedlichen Attribute in dem noch jungen Feld „Experience Design".

Red Dot: Best of the Best
NEBV
Connecting Cables
Verbindungsleitungen
See page 182

reddot design award
best of the best 2013

Designer portraits

Davide Lamparelli
Wärtsilä Propulsion Control System

»I really enjoy designing equipment for modern, professional applications and I am constantly seeking to improve my knowledge and experience in this field.«

»Ausrüstungen für moderne, professionelle Anwendungen zu gestalten, bereitet mir wirklich Freude, und ich versuche stets, mir noch mehr Wissen und Erfahrungen auf diesem Gebiet anzueignen.«

Is there a role model that inspired your work?
The project was inspired by the body language of sea captains. I was impressed by how they were able to mimic actions by using finger and head movements. With great attention to detail, and by using their whole body, these captains were able to describe what the balance of the vessel feels like.

Do you have a favourite quotation?
"Life is like riding a bicycle. To keep your balance, you must keep moving."
(Albert Einstein)

Gibt es ein Vorbild, das Ihre Arbeit inspiriert hat?
Inspiriert ist das Projekt von der Körpersprache, die Schiffskapitäne benutzen. Die Art, wie sie Vorgänge mit Kopf- und Fingerbewegungen kommunizieren, hat mich beeindruckt. Mit viel Sorgfalt fürs Detail und unter Einsatz des gesamten Körpers können diese Kapitäne beispielsweise beschreiben, wie ihr Schiff im Wasser liegt.

Haben Sie ein Lieblingszitat?
„Das Leben ist wie ein Fahrrad. Man muss sich vorwärts bewegen, um das Gleichgewicht nicht zu verlieren."
(Albert Einstein)

Red Dot: Best of the Best
Wärtsilä Propulsion Control System
Wärtsilä Antriebssteuerungssystem
See page 158

Designer portraits

Stephan Niehaus
Hilti TE-CD / TE-YD

»The philosophy of our innovations is exemplified by our hollow drill bits: starting with real-life requirements, the creative solution delivers an entire system that is perfectly matched in its most minute detail and which thus creates real added value.«
»Die Philosophie unserer Innovationen zeigt sich in dem Hohl-bohrer beispielhaft: von den realen Bedürfnissen über die kreative Lösung zu einem bis in die Details aufeinander abgestimmten Gesamtsystem, das tatsächlichen Mehrwert schafft.«

What do you like in particular about your own award-winning product?
While the design of our appliances puts the technical requirements and semantic implementation of power and endurance centre-stage, the astoundingly different design principle of this hollow drill fascinates me. It turns the status quo on its head and breaks away from the familiar in favour of added-value for users.

Is there a particular preceding model that inspired your work?
The drill represents the core of a series of diverse, interdependent product innovations. Our design consideration did not centre on optimising a traditional drill, but on eliminating dust emissions while drilling.

Was gefällt Ihnen an Ihrem eigenen, ausgezeichneten Produkt besonders gut?
Während beim Design unserer Geräte der formale Charakter und die semantische Umsetzung von Kraft und Ausdauer im Vordergrund stehen, fasziniert mich an diesen Hohlbohrern die verblüffende Andersartigkeit des Designprinzips, das den Status quo auf den Kopf stellt und zugunsten eines Mehrwerts für den Kunden mit Gewohntem bricht.

Gibt es ein bestimmtes Vorgänger-modell, das Ihre Arbeit inspiriert hat?
Der Bohrer ist das Herzstück einer Reihe unterschiedlicher, ineinandergreifender Produktinnovationen. Im Zentrum der Überlegung stand nicht die Optimierung eines traditionellen Bohrers, sondern das Ziel, ohne Staubentwicklung bohren zu können.

Red Dot: Best of the Best
Hilti TE-CD / TE-YD
Hollow Drill Bits
Hohlbohrer
See page 128

reddot design award
best of the best 2013

Ruben Beijer, Alexander Flebbe
PV Premium

»The goal was to integrate a photovoltaic system seamlessly into a roof surface without compromising on function.«
»Das Ziel war, eine Solaranlage möglichst unauffällig in eine Dachfläche zu integrieren, ohne dabei die Funktion einzuschränken.«

How would you define the term "design quality"?
"Design quality" means that the design must meet the minimum technical requirements. In addition, design quality is determined by the reaction of the users.

What are the greatest challenges that you currently see in your industry?
The biggest challenge in the photovoltaic market is to ensure that the design and technical features of the products achieve or exceed the standard requirements of the building industry.

Wie würden Sie den Begriff der Designqualität beschreiben?
Designqualität bedeutet, dass die Gestaltung den technischen Mindestanforderungen genügen muss. Darüber hinaus bestimmt sich Designqualität über die Resonanz der Nutzer.

Worin bestehen für Sie aktuell die größten Herausforderungen in Ihrer Branche?
Die größte Herausforderung im Solaranlagenmarkt ist es sicherzustellen, dass die Gestaltung und die technischen Merkmale eines Produkts den Standardanforderungen in der Gebäudeindustrie genügen oder sie gar übertreffen.

Red Dot: Best of the Best
PV Premium
Photovoltaic In-Roof System
Photovoltaik Indach-System
See page 150

Alexander Müller, Sebastian Maier
Dräger Perseus A500

»Our idea of good design is products that even
after years of use still have a right to exist.«
»Unsere Vorstellung von gutem Design sind Produkte,
die auch nach Jahren ihre Daseinsberechtigung haben.«

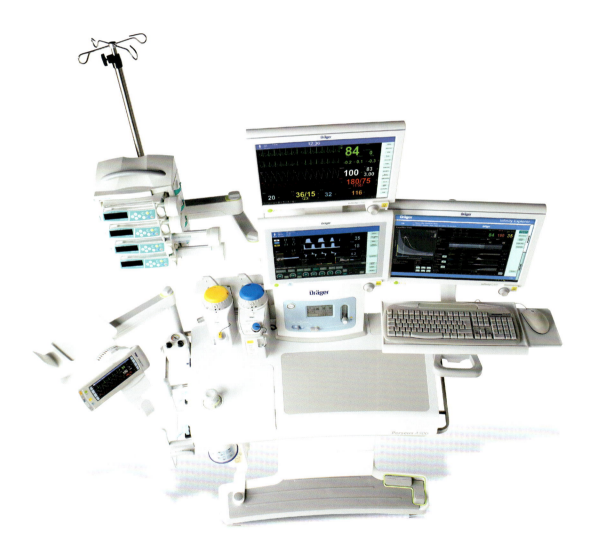

What do you like in particular about your own award-winning product?
That we managed to comply with the requirements of an anaesthetic work environment and transform it into a workplace that is truly of high value.

What are the greatest challenges that you currently see in your industry?
In our industry, medical and laboratory technology design is still often misunderstood and mistaken for styling. The biggest challenge is not only to create good design but also to communicate it, in order to improve its image in the mid-term.

Was gefällt Ihnen an Ihrem eigenen, ausgezeichneten Produkt besonders gut?
Dass wir es geschafft haben, aus der Anforderung eines Anästhesiearbeitsplatzes im wahrsten Sinne des Wortes auch einen solchen vollwertigen Arbeitsplatz zu gestalten.

Worin bestehen für Sie aktuell die größten Herausforderungen in Ihrer Branche?
In unserem Bereich, der Medizin- und Labortechnik, wird Design immer noch häufig verkannt und mit Styling verwechselt. Die größte Herausforderung ist nicht nur, gutes Design zu machen, sondern es auch zu kommunizieren, um das Image mittelfristig zu verbessern.

Red Dot: Best of the Best
Dräger Perseus A500
Anaesthesia Workstation
Anästhesiearbeitsplatz
See page 210

reddot design award
best of the best 2013

Designer portraits

Nathan Pollock, Lee Rodezno
WiTouch Pro

»Our aim was to design a product that was both visually beautiful and simple so that the form intuitively suggests how the product is to be used.«

»Unser Ziel war es, ein Produkt zu entwerfen, das optisch schön und schlicht zugleich ist, sodass sich die Verwendung des Produkts intuitiv über dessen Form erklärt.«

Is there a specific approach that is of significance to your work?
As designers we see our key role as developing solutions to client's problems from the user's perspective. Of course, we work closely with our clients because teamwork is essential, but it is crucial that we represent the user and innovate on their behalf to improve their experience.

What are the greatest challenges that you currently see in your industry?
The biggest challenge is fighting the public's demand for ridiculously cheap products. This has created a throw-away culture which is unsustainable.

Gibt es einen bestimmten Ansatz, der für Ihre Arbeit von Bedeutung ist?
Als Gestalter sehen wir unsere Hauptrolle darin, aus dem Blickwinkel der Anwender heraus Lösungen für die Probleme unserer Klienten zu entwickeln. Natürlich arbeiten wir eng mit unseren Klienten zusammen, denn Teamwork ist wichtig; das Wichtigste aber ist, dass wir die Anwender vertreten und ihnen durch unsere innovative Tätigkeit neue Erfahrungen bieten.

Worin bestehen für Sie aktuell die größten Herausforderungen in Ihrer Branche?
Die größte Herausforderung ist der Kampf gegen die allgemeine Nachfrage nach irrwitzig billigen Produkten. Dies hat zu einer Wegwerfkultur geführt, die nicht nachhaltig ist.

Red Dot: Best of the Best
WiTouch Pro
Wireless TENS Unit
Kabellose TENS-Einheit
See page 296

Designer portraits

Jens Kaschlik
CyFlow Cube 6

»A product proves to be of high design quality when people like to have it close, when they enjoy looking at it, when they simply cannot stop touching and using it.«

»Ein Produkt ist dann von hoher Designqualität, wenn man sich gern damit umgibt, wenn man es gern betrachtet, wenn man es einfach nicht schafft, seine Finger davon zu lassen.«

What do you like in particular about your own award-winning product?
The CyFlow Cube 6 is like a magic cube, highly complex in its inner workings, yet entirely logical, easy to understand and distinctive in appearance.

Is there a specific approach that is of significance to your work?
It is crucial to interpret technology within its own context. It is a prerequisite for a truly great product to be able to emerge.

Was gefällt Ihnen an Ihrem eigenen, ausgezeichneten Produkt besonders gut?
Der CyFlow Cube 6 ist wie ein Zauberwürfel, im Inneren hochgradig komplex und doch absolut logisch, leicht zu begreifen und signifikant in der Erscheinung.

Gibt es einen bestimmten Ansatz, der für Ihre Arbeit von Bedeutung ist?
Das Wichtigste ist, die Technik in ihrem Umfeld zu verstehen. Nur so kann anschließend ein wirklich tolles Produkt entstehen.

Red Dot: Best of the Best
CyFlow Cube 6
Flow Cytometry System
Durchflusszytometrie-System
See page 252

Designer portraits

Eung Seok Kim, Sang Hoon Lee
5aver

»When it comes to design, our biggest challenge
is to set a goal and go beyond it.«
»Beim Design ist unsere größte Herausforderung
uns ein Ziel zu setzen und dieses noch zu übertreffen.«

What do you like in particular about your own award-winning product?
A lot of effort was put into the manufacture of the product "5aver" in order to ensure portability, so that the product could be stored or carried around regardless of the circumstances. Furthermore, our product strives to achieve a simple and aesthetic design.

Were there specific challenges when designing the product which turned out to be really difficult?
The process of creating a completely new design was a huge challenge. Since it is a product that deals with life, decreasing the size of the product while maintaining the standardised stability required in every country, was the greatest difficulty.

Was gefällt Ihnen an Ihrem eigenen, ausgezeichneten Produkt besonders gut?
In die Herstellung des Produkts „5aver" wurde viel Arbeit gesteckt, um die Transportfähigkeit und somit das Verstauen und Mitführen des Geräts unter allen Umständen zu ermöglichen. Zudem strebt unser Produkt eine schlichte und ästhetische Gestaltung an.

Gab es besondere Herausforderungen bei der Produktgestaltung, die sich als besonders schwierig erwiesen haben?
Der Prozess, eine komplett neue Gestaltung zu kreieren, war eine große Herausforderung. Da sich das Produkt um das Leben dreht, bestand die größte Schwierigkeit darin, seine Größe zu reduzieren und gleichzeitig die in jedem Land standardisierte Stabilität einzuhalten.

Red Dot: Best of the Best
5aver
Emergency Flashlight and Mask
Notfall-Taschenlampe und Maske
See page 250

Roland Heiler
Porsche Design P'9981

»I agree with Leonardo da Vinci that 'simplicity is the ultimate sophistication'.«
»Ich halte es mit Leonardo da Vinci: ‚Einfachheit ist die höchste Stufe der Vollendung'.«

Were there specific challenges when designing the product, which turned out to be really difficult?
It was a true challenge for the engineers to make the keyboard extend to the very edges while protecting it from potential damage at the same time. The leather trim on the back of the case was also technically demanding, which is why every single device is individually crafted by hand.

Is there a specific approach that is of significance to your work?
Functional design is in our DNA. The aesthetics always have to be in equilibrium with the function, according to our credo of "form equals function".

Gab es besondere Herausforderungen bei der Produktgestaltung, die sich als besonders schwierig erwiesen haben?
Zum einen war es für die Ingenieure eine Herausforderung, die Tasten wirklich bis zur Außenkante zu führen und sie dennoch gegen unbeabsichtigte Beschädigung zu schützen. Auch die Belederung der Rückseite ist handwerklich anspruchsvoll und wird deshalb bei jedem einzelnen Gerät von Hand ausgeführt.

Gibt es einen bestimmten Ansatz, der für Ihre Arbeit von Bedeutung ist?
Funktionales Design ist in unserer DNA. Die Ästhetik muss mit der Funktion immer im Gleichgewicht stehen, gemäß unserem Credo „form equals function".

Red Dot: Best of the Best
Porsche Design P'9981
Smartphone from BlackBerry
See page 304

Designer portraits

reddot design award
best of the best 2013

Designer portraits

Yingjia Yao
IdeaPad U430s

»I do believe attitude is the key to achieving anything in life, so I try to face everything in my life with a positive attitude.«
»Ich glaube, dass die Einstellung der Schlüssel zum Erfolg ist, und versuche deshalb, alles in meinem Leben mit einer positiven Einstellung anzugehen.«

What do you like in particular about your own award-winning product?
The IdeaPad U430s is simply designed. It exhibits the design language of a dynamic journal, as well as the human-centred design that makes products appealing. The biggest challenge was how to implement a large screen yet allow the product to remain portable and practical, while at the same time retaining a desirable design quality.

Is there a specific approach that is of significance to your work?
Being a designer means that I have to be insightful and farsighted, as well as able to predict trends.

Was gefällt Ihnen an Ihrem eigenen, ausgezeichneten Produkt besonders gut?
Das IdeaPad U430s ist schlicht gestaltet, trägt die Formensprache eines dynamischen Journals sowie eine menschengerechte Gestaltung, die Produkte anziehend macht. Die größte Herausforderung war, wie wir ein großes Display implementieren und das Produkt dennoch mobil und anwendungspraktisch gestalten könnten und dabei gleichzeitig eine begehrenswerte gestalterische Qualität beibehalten.

Gibt es einen bestimmten Ansatz, der für Ihre Arbeit von Bedeutung ist?
Gestalter zu sein bedeutet, dass ich einfühlsam und weitsichtig sein und Trends vorhersagen können muss.

Red Dot: Best of the Best
IdeaPad U430s
Laptop
See page 374

Designer portraits

Mylene Tjin
HP Envy 120 e-All-in-One

»Design quality is the intersection of form, function, materials, finishes and craftsmanship. At this point they become greater as a whole than the sum of their parts.«

»Designqualität ist die Schnittstelle von Form, Funktion, Materialien, Veredelungen und Handwerkskunst, an der sie zusammen als Ganzes größer sind als die Summe der Teile.«

Is there a role model that inspired your work?
Architecture inspires me greatly. I draw energy from the refreshing ways in which space, material and light interplay and have the ability to evoke different senses in the observer.

Were there specific challenges when designing the product, which turned out to be really tough?
The glass scanner surface proved to be a technical challenge, as the right amount of transparency was required to actually check documents on the scanner, while trying to achieve an opaque, unobtrusive effect when not in use. I wanted to create the effect of a privacy filter screen that muted the design, because the design objective was to achieve an elegant presence.

Gibt es ein Vorbild, das Ihre Arbeit inspiriert hat?
Architektur inspiriert mich sehr. Aus dem erfrischenden Zusammenspiel von Raum, Materialien und Licht sowie der Wirkungsmacht, im Betrachter verschiedene Sinnesempfindungen auszulösen, beziehe ich Kraft.

Gab es besondere Herausforderungen bei der Produktgestaltung, die sich als besonders schwierig erwiesen haben?
Die gläserne Scan-Oberfläche erwies sich als Herausforderung, da sie zum Einlesen von Dokumenten die exakte Transparenz aufweisen muss, im Ruhezustand aber eine undurchsichtige, unauffällige Wirkung erzielen sollte. Ich wollte den Effekt einer Sichtschutzscheibe kreieren, die die Gestaltung dämpft, denn das Ziel war es, dem Scanner eine elegante Ausstrahlung zu verleihen.

Red Dot: Best of the Best
HP Envy 120 e-All-in-One
Inkjet All-in-One Printer
All-in-One-Tintenstrahldrucker
See page 394

Offices
Büro

Office furniture, furniture for reception halls and waiting rooms,
office systems, office equipment, accessories
Büromöbel, Möbel für Empfangs- und Wartebereiche, Bürosysteme,
Büroausstattung, Accessoires

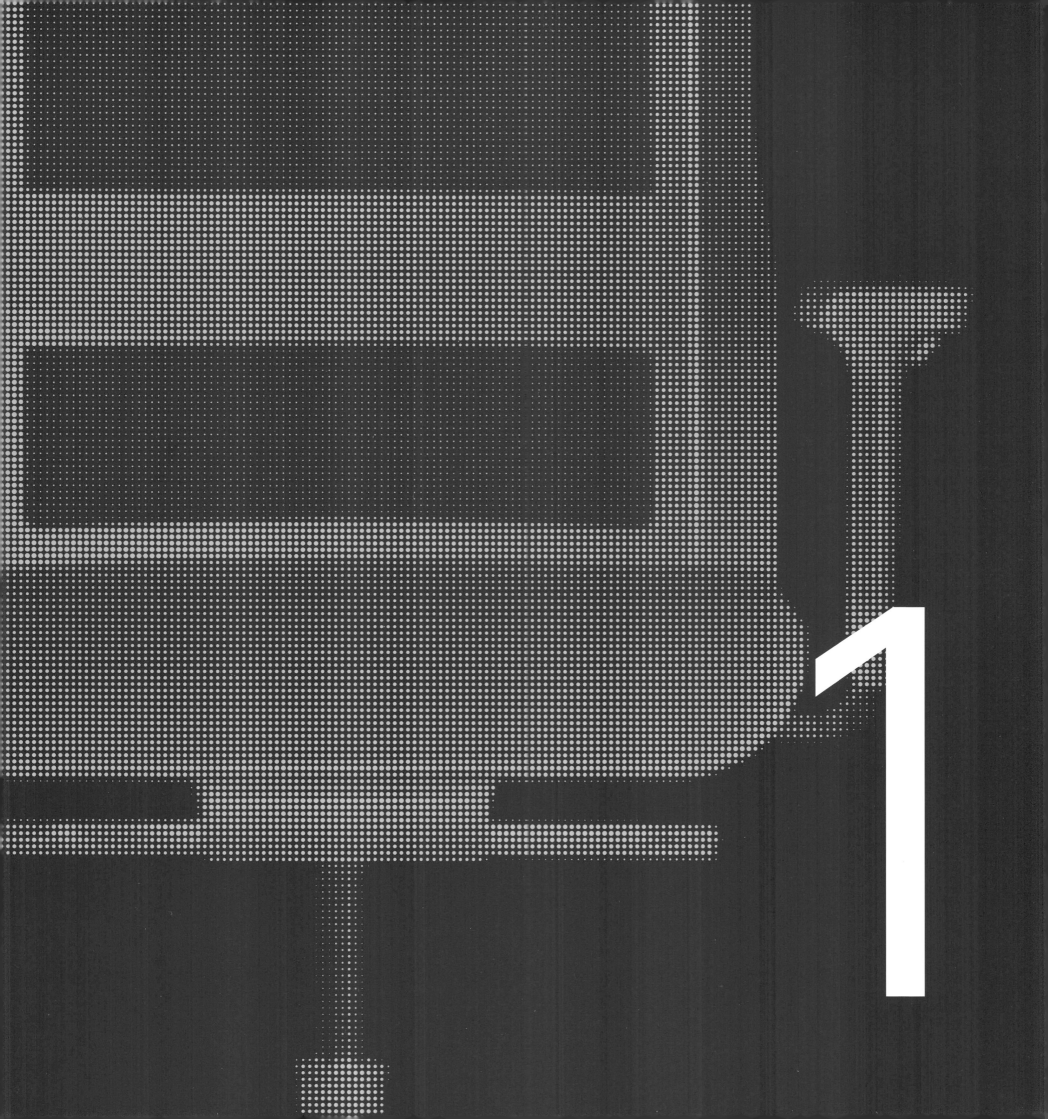

Aura
Sofa

Manufacturer
Inno Interior Oy,
Espoo, Finland

Design
Studio Mikko Laakkonen,
Helsinki, Finland

Web
www.inno.fi
www.mikkolaakkonen.com

reddot design award
best of the best 2013

Invitingly generous

The definition of rooms plays an important role in the large-scale architecture of today. The aim is to find a clever way to create pools of privacy and peace even in open plan spaces. The Aura sofa puts these requirements at the heart of its sensitive design. An organically opening shape gives it its identity. The high back is particularly noticeable though it flows harmoniously into the lines of the other elements. This high back has a room-defining role. Through it, the sofa becomes an aesthetic room divider that can be used in many different ways. The Aura sofa is well thought out ergonomically and is very comfortable to sit in. To increase its potential for use in interior design, the Aura concept includes inviting armchairs. Their backs are slightly lower, which gives the person sitting in them a better view of their surroundings. Grouped together, the Aura sofa and armchairs create a contemporary suite for open spaces and modern living. The appealing design form of the Aura sofa enriches any interior. It also opens up new possibilities for arranging rooms and, in doing so, creates high-quality living and working environments.

Einladende Offenheit

In der zeitgemäß großzügigen Architektur spielt die Definition von Räumen eine wichtige Rolle. Es geht darum, auf geschickte Weise auch in Großräumen Bereiche von Privatheit und Geborgenheit entstehen zu lassen. Die feinsinnige Gestaltung des Sofas Aura stellt diese Bedürfnisse in den Mittelpunkt. Eine sich organisch öffnende Form verleiht ihm seine Identität. Auffällig ist dabei sein hoch gestalteter Rücken, der sich harmonisch in die Linienführung der übrigen Elemente einfügt. Der hohe Rücken hat eine raumbildende Funktion: Durch ihn wird dieses Sofa zu einem ästhetischen Raumteiler, der vielfältig eingesetzt werden kann. Das Sofa Aura ist ergonomisch gut durchdacht und bietet dem Sitzenden viel Komfort. Um weitere Möglichkeiten der Raumgestaltung zu bieten, integriert das Konzept von Aura auch einladend anmutende Loungesessel. Deren Rückenlehne ist ein wenig niedriger gestaltet und bietet dem Sitzenden so eine gute Übersicht. In der Gruppe formen das Aura Sofa und die Sessel zeitgemäße Sitzgruppen für Open-Space-Lösungen und ein modernes Wohnen. Mit seiner sympathisch anmutenden Formensprache bereichert das Sofa Aura das Interieur. Es eröffnet neue Möglichkeiten für die Strukturierung von Räumen und schafft dabei ein hohes Maß an Wohn- und Arbeitsqualität.

Statement by the jury

The innovative proportions of the Aura sofa are enough to persuade anyone of its appeal. The minimalist design has an iconographic quality. It is obviously based on extensive studies of the current significance of seating furniture. This sofa with its high back perfectly addresses the requirements for a contemporary room divider.

Begründung der Jury

Das Sofa Aura begeistert mit seinen innovativen Proportionen. Es zeigt einen Minimalismus von ikonografischer Qualität. Es wird deutlich, dass seine Gestaltung auf ausführlichen Studien über die aktuelle Bedeutung eines Sitzmöbels basiert. Dieses Sofa erfüllt mit seinem hohen Rücken auf perfekte Weise die Ansprüche für den Einsatz als ein zeitgemäßer Raumteiler.

Furniture for reception halls and waiting rooms

plot
Modular Loungescape

Manufacturer
Brunner GmbH,
Rheirau, Germany
Design
Osko+Deichmann
(Oliver Deichmann, Blasius Osko),
Berlin, Germany
Web
www.brunner-group.com
www.oskodeichmann.com

With its multilevel seating options, this lounge furniture by the name of plot invites people to interpret seating in a new way, again and again. Its modular elements allow individual loungescapes to be created, which use the three cushioned levels flexibly as backrest, table, couch, shelf or seat. It is available in monochrome colour combinations and in seven different colour sets, where the top surfaces differ in colour and material from the sides. With these numerous colour and modular combinations, the plot furniture adapts to a wide variety of room situations and can, for example, be used to partition communication areas.

Mit Sitzflächen auf verschiedenen Ebenen lädt plot dazu ein, das Sitzen immer wieder neu zu interpretieren. Mithilfe der modularen Elemente lässt sich eine individuelle Wohnlandschaft gestalten, wobei die drei gepolsterten Ebenen flexibel als Lehne, Tisch, Ablage, Liege oder Sitz genutzt werden können. Insgesamt stehen monochrome Farbzusammenstellungen sowie sieben verschiedene Farbkanons zur Verfügung, die sich auf den oberen Seiten in Farbe und Material von den seitlichen Flächen unterscheiden. Mit zahlreichen Farb- und Modulkombinationen lässt sich plot an vielfältige Raumsituationen anpassen oder zum Beispiel in unterschiedliche Kommunikationsbereiche unterteilen.

LUC
Lounge Chair
Loungesessel

Manufacturer
Rossin GmbH,
Neumarkt-Laag (Bozen), Italy
Design
Lorenz*Kaz
(Steffen Kaz, Catharina Lorenz),
Milan, Italy
Web
www.rossin.it
www.lorenz-kaz.com

The Luc lounge chair features a smoothly and harmoniously curved polyurethane shell that rests on slim feet. Its spacious seat encloses the seated individual and provides a high level of comfort and ease. The generous dimensions of the lounge chair encourage relaxation and foster both openness and personal space for retreat. Different colours and bases turn this model into an individualistic piece of furniture.

Der Loungesessel Luc zeigt eine sanft harmonisch gebogene Kunststoffschale, die auf schlanken Füßen platziert wurde. Ihre ausladende Sitzfläche umschließt den Sitzenden und bietet ihm dabei einen angenehmen Komfort. Die großzügigen Maße des Loungesessels laden zum Verweilen ein und ermöglichen gleichermaßen Rückzug wie Offenheit. Verschiedene Farben und Untergestelle machen das Modell zum individuell einsetzbaren Sitzmöbel.

RBM Noor
Canteen and Conference Chair
Kantinen- und Konferenzstuhl

Manufacturer
Scandinavian Business Seating
AS, Oslo, Norway

In-house design
Scandinavian Business Seating
Design Team

Design
StokkeAustad (Øystein Austad, Jonas Ravlo Stokke),
Oslo, Norway
Form Us With Love (Petrus Palmér, John Löfgren),
Stockholm, Sweden
Grønlund Design (Susanne Grønlund),
Aarhus, Denmark

Web
www.rbmfurniture.de
www.stokkeaustad.com
www.formuswithlove.se
www.gronlunddesign.com

reddot design award
best of the best 2013

Cutting a dash
Conferences and canteen meals are important communal moments in office life. The RBM Noor canteen and conference chair has been designed to improve the quality of seating even in these locations. This chair is very comfortable and its curved, flexible backrest adapts well to the user's body shape. Its innovative aesthetics are based on a clear division of seat and backrest. Viewed from the side, this chair's sweeping forms are arresting and make it appear friendly and dynamic. The RBM Noor product range offers eight different versions of the chair. As these all come with different shells made of plastic or with a 3D veneer, they all have their own individual touch. The collection also offers options for multifunctional use. There are versions with a five-star-pedestal, with wooden, stainless steel or wire legs and numerous different colours, so the chairs can be combined in a large variety of ways. Thanks to its curvy silhouette, the RBM Noor canteen and conference chair creates an informal atmosphere in every type of environment. It shows its qualities wherever people meet.

Mit neuem Schwung
Im Büro sind Konferenzen wie auch das Essen in der Kantine wichtige Momente der Gemeinsamkeit. Der Kantinen- und Konferenzstuhl RBM Noor wurde mit der Maxime gestaltet, die Qualität des Sitzens auch an diesen Orten zu verbessern. Dieser Stuhl bietet viel Komfort, und mit seiner geschwungenen und flexiblen Rückenlehne passt er sich gut den Konturen des Sitzenden an. Seine innovative Ästhetik basiert dabei auf einer visuell deutlichen Trennung von Sitz und Rückenlehne. Aus der Seitenansicht begeistert dieser Stuhl durch seine schwungvolle Gestaltung, die ihm Freundlichkeit und Dynamik verleiht. Die Produktfamilie des RBM Noor bietet acht unterschiedliche Varianten. Da diese mit unterschiedlichen Schalen aus Kunststoff oder mit einem 3D-Furnier gestaltet sind, besitzen sie eine jeweils eigene Anmutung. Die Kollektion bietet zudem viele Variationen für einen multifunktionalen Einsatz. Erhältlich sind Versionen mit einem Fußkreuz, mit Beinen aus Holz, Stahlrohr oder Draht. Da sich zudem unterschiedliche Farben anbieten, lassen sich diese Stühle vielfältig kombinieren. Mit seiner geschwungenen Silhouette schafft der Kantinen- und Konferenzstuhl RBM Noor dabei eine freundliche Atmosphäre in jeder Art von Umgebung – er zeigt seine Qualitäten überall da, wo Menschen sich treffen.

Statement by the jury
The RBM Noor canteen and conference chair has created a new generation of office furniture. Its curved, minimalist form brings movement into day-to-day life. The design of the seat shell is based on well-engineered technologies and is comfortable to sit on even over longer periods of time. This chair shows authentic character.

Begründung der Jury
Mit dem Kantinen- und Konferenzstuhl RBM Noor entstand eine neue Typologie von Büromöbeln. Seine minimalistische und schwungvolle Formensprache bringt Aktion in den Alltag. Die Gestaltung der Sitzschale basiert auf ausgereiften Technologien und bietet viel Sitzkomfort auch bei längerem Sitzen. Dieser Stuhl hat Charakter und ist authentisch.

Furniture for reception halls and waiting rooms

RBM Noor
Canteen and Conference Chair
Kantinen- und Konferenzstuhl

Manufacturer
Scandinavian Business Seating AS,
Oslo, Norway
In-house design
Scandinavian Business Seating Design Team
Design
StokkeAustad
(Øystein Austad, Jonas Ravlo Stokke),
Oslo, Norway
Form Us With Love
(Petrus Palmér, John Löfgren),
Stockholm, Sweden
Grønlund Design
(Susanne Grønlund),
Aarhus, Denmark
Web
www.rbmfurniture.de
www.stokkeaustad.com
www.formuswithlove.se
www.gronlunddesign.com

The RBM Noor canteen and conference chair series was designed with the vision of making the world a better place to sit in and, accordingly, with the goal of extending the comfort standard in this product segment. The curved, ergonomic shape of the chair and its flexible backrest perfectly accommodate the user's body, offering comfort and stability. It is available in eight different versions, for instance the version made of 3D veneer with wooden legs, which features high-quality aesthetics. Its curved silhouette and selected materials illustrate the demands made on quality by this distinctive chair.

Die Kantinen- und Konferenzstuhlserie RBM Noor wurde mit der Vision gestaltet, die Welt zu einem besseren Ort des Sitzens zu machen und so das Komfortniveau in diesem Segment zu erweitern. Die geschwungene, an den Körper angepasste Formgebung des Stuhls und eine flexible Rückenlehne nehmen den Körper des Nutzers optimal auf und geben ihm Stabilität. Der Stuhl ist in acht Varianten erhältlich. Das aus 3D-Furnier hergestellte Modell mit Beinen aus Holz besitzt mit seiner geschwungene Silhouette eine hochwertige Ästhetik, die zusammen mit der Materialwahl den Qualitätsanspruch dieses Stuhls verdeutlicht.

LUC
Chair
Stuhl

Manufacturer
Rossin GmbH,
Neumarkt-Laag (Bozen), Italy
Design
Lorenz*Kaz
(Steffen Kaz, Catharina Lorenz),
Milan, Italy
Web
www.rossin.it
www.lorenz-kaz.com

The design of this chair is defined by a characteristic and reduced polyurethane shell with high lateral lines. Its harmonious proportions, generous dimensions and the slightly flexible back provide a high level of seating comfort. A variety of colours and bases allow an almost limitless array of combinations. The Luc thus becomes a personalised piece of seating furniture for any room, whether public or private.

Die Gestaltung dieses Stuhls ist durch eine charakteristisch reduzierte Kunststoffschale mit hoher Seitenlinie gekennzeichnet. Seine harmonischen Proportionen, die großzügigen Maße und der leicht wippende Rücken bieten einen hohen Sitzkomfort. Eine Vielzahl an Farben und verschiedenen Untergestellen ermöglichen schier unendliche Kombinationen. Luc wird dadurch zum personalisierten Sitzmöbel für jeden Raum, ob öffentlich oder privat.

Statement by the jury
The Luc chair stands out with its distinctively simple seating shell, which is produced in a single moulding operation and lends the chair its unique appearance.

Begründung der Jury
Luc besticht durch seine ausgeprägt schlichte Sitzschale, die aus einem Guss gefertigt wurde und seinem Erscheinungsbild eine unverkennbare Note verleiht.

MESAMI 2
Visitor Chair
Besucherstuhl

Manufacturer
LÖFFLER GmbH,
Reichenschwand, Germany
In-house design
Friedrich Močnik
Web
www.loeffler.de.com
Honourable Mention

Characterised by well-balanced proportions and premium workmanship, the MESAMI 2 visitor chair is suitable for use in conference halls, in restaurants and at home. Its adhesive-free cushioned seat pan promises comfortable posture, while a leather strap at the backrest and the matching hand-laced leather armrest lend the chair a distinctive appearance. The chair is available with different frame options.

Statement by the jury
A clear, reduced design and full-surface padding turn the MESAMI 2 into a piece of furniture that is highly versatile in use.

Der durch ausgewogene Proportionen und eine hochwertige Verarbeitung gekennzeichnete Besucherstuhl MESAMI 2 eignet sich ebenso für den Konferenzsaal wie für die Gastronomie und das Zuhause. Seine verklebungsfrei gepolsterte Sitzschale verspricht eine bequeme Haltung. Die Lederschlaufe an der Rückenlehne sowie die per Hand parallel geschnürte Kernlederarmlehne verleihen dem Stuhl Ausdruck. MESAMI 2 ist mit verschiedenen Gestellvarianten erhältlich.

Begründung der Jury
Eine klare, zurückhaltende Gestaltung und eine durchgehende Polsterung machen den MESAMI 2 zum vielseitig einsetzbaren Sitzmöbel.

NOA
Chair
Stuhl

Manufacturer
Pedrali Spa,
Mornico al Serio, Italy
Design
Marc Sadler,
Milan, Italy
Web
www.pedrali.it
www.marcsadler.it

Clear proportions and an elegant design are the outstanding characteristics of the stackable Noa chair. An innovative production technique has created a highly comfortable chair with an upholstered seat that is situated in a glossy polycarbonate shell. The upholstery is available as fabric, leather or synthetic leather in different colours, while the frame comes in chrome-plated or powder-coated steel.

Statement by the jury
The Noa is defined by classic design vocabulary, which, along with the wide variety of available colours and materials, makes it flexible to use.

Klare Proportionen und eine elegante Gestaltung sind die hervorstechenden Merkmale des stapelbaren Noa. Eine innovative Produktionstechnik ermöglicht den hohen Komfort dieses Stuhls, dessen Sitz gepolstert ist und sich in einer Schale aus Polycarbonat mit glänzender Oberfläche befindet. Als Bezug steht Stoff, Leder oder Kunstleder in verschiedenen Farben zur Wahl und das Gestell aus Edelstahl ist verchromt oder lackiert erhältlich.

Begründung der Jury
Noa ist durch eine klassische Formensprache geprägt, die ihn auch dank der großen Farb- und Materialauswahl vielseitig einsetzbar macht.

Bay Chair
Chair
Stuhl

Manufacturer
Bene AG,
Waidhofen an der Ybbs, Austria
Design
PearsonLloyd
(Luke Pearson, Tom Lloyd),
London, GB
Web
www.bene.com
www.pearsonlloyd.com

The Bay Chair is a hybrid between a swivel chair and an informal seat for meetings. It accommodates flexible work strategies by adapting to the individual and by being easy to use. A soft, homely design facilitates relaxed seating when working at a laptop, reading or making phone calls away from the desk and brings a human touch into office spaces. In addition, its simple and distinctive form makes a strong visual statement.

Der Bay Chair ist eine Mischung aus Drehstuhl und informeller Sitzgelegenheit für Meetings. Er kommt flexiblen Arbeitsstrategien entgegen, indem er sich an den Nutzer anpasst und einfach zu bedienen ist. Durch sein weiches, wohnliches Design ermöglicht er entspanntes Sitzen beim Arbeiten mit dem Laptop, beim Lesen oder beim Telefonieren abseits des Schreibtischs und bringt so Human Touch ins Büro. Seine einfache markante Form setzt zudem ein visuelles Statement.

Statement by the jury
This office swivel chair conveys a sense of individuality and comfort and thus stands out in its category with an expressive appearance.

Begründung der Jury
Dieser Bürodrehstuhl vermittelt Individualität und Komfort und hebt sich dadurch von vergleichbaren Produkten des Segments ausdrucksstark ab.

CBS
Chair Programme
Stuhlprogramm

Manufacturer
Sidiz, Inc.,
Seoul, South Korea
In-house design
Tae-Jin Ko, Tae-Hee Ryu
Design
Claudio Bellini Design+Design
(Claudio Bellini, Giovanni Ingignoli),
Milan, Italy
Web
www.sidiz.com
www.claudiobellini.com

The multifunctional CBS chair programme embodies a modern vision of a chair series conceived to satisfy the expectations of a broad target group. The elegant design underscores the chair's comfort with a wide, enveloping back, which is based on an extremely thin one-piece nylon shell. Finely cut in the lumbar area, the back provides enhanced freedom of movement during sitting. The CBS chair is available with a black or white shell, each with or without coloured pads. There is also an optional metal frame, allowing for several different configurations.

Das multifunktionale Stuhlprogramm CBS versteht sich als die moderne Vision einer Stuhlserie, die den Erwartungen einer breiten Zielgruppe gerecht werden kann. Die elegante Gestaltung unterstreicht den Komfort mit einer breiten, kuvertierten Lehne, die auf einem sehr dünnen Stück Nylon Shell basiert. Im Lendenbereich ist die Lehne fein geschnitten und bietet dadurch mehr Bewegungsfreiheit beim Sitzen. Der Stuhl wird in Schwarz und in Weiß angeboten, jeweils mit und ohne farbige Polsterung sowie optionalem Metall-rahmen, und kann zudem in verschiedenen Variationen konfiguriert werden.

Statement by the jury
The CBS chair programme is striking with its original, sophisticated proportions and a modern colour and material concept.

Begründung der Jury
Das Stuhlprogramm CBS sticht durch seine originellen Proportionen sowie ein zeitgemäßes Farb- und Materialkonzept ins Auge.

Comforto 29
Office Chair
Bürostuhlserie

Manufacturer
Haworth GmbH,
Ahlen, Germany
In-house design
Haworth Design Studio
Design
Steve Nemeth
Web
www.haworth-europe.com

Comforto 29 is a family of seats comprised of swivel chairs with and without armrests that allows for ergonomic seating in the workplace. The swivel chair features an automatic weight-control mechanism, so that different users need not spend any time adjusting the chair. Comforto 29 was specifically designed as an affordable product. And thanks to carefully selected materials, the chair is almost completely recyclable.

Comforto 29 ist eine Stuhlfamilie, die aus Arbeitsdrehstühlen mit und ohne Armlehnen besteht und ergonomisches Sitzen am Arbeitsplatz ermöglicht. Der Drehstuhl ist mit einer automatischen Gewichtsmechanik ausgestattet, so dass sich wechselnde Nutzer nicht mehr mit der Anpassung des Sitzes an ihren Körper beschäftigen müssen. Comforto 29 ist ein bewusst erschwinglich gestaltetes Produkt. Und dank sorgfältig ausgesuchter Materialien kann der Stuhl nach seiner Nutzungsdauer zu einem Großteil dem Wertstoffkreislauf wieder zugeführt werden.

FLO
Office Chair
Bürostuhl

Manufacturer
PATRA,
Ansan City, Gyeonggi Province, South Korea
In-house design
Changgon Lee
Web
www.patrainc.com

The twisted form of the simply and elegantly designed FLO swivel chair is inspired by the leaves of the Sansevieria plant and meets both functional and aesthetic demands. Especially eye-catching are design features like its flowing lines and the rounded edges and surfaces. The chair's twisted structure guarantees a secure backrest and tilting capacity. The curved armrest provides stability and lends the swivel chair a unique appearance. Its lightweight tilt was developed using a new technique and produced without die-casting. FLO is available in two different heights, three frame colours and as a cantilever type.

Der schlicht und elegant gestaltete Dreh-stuhl FLO ahmt mit seiner gedrehten Struktur den Wuchs der Sansevieria-Blätter nach und erfüllt neben funktionalen auch ästhetische Ansprüche. Besonders augen-fällig sind die geschwungenen Linien sowie die abgerundeten Kanten und Oberflächen. Die gedrehte Struktur garantiert eine sichere Rückenstütze und Neigungsfähigkeit. Die geschwungene Armlehne verleiht dem Drehstuhl Stabilität und ein unverwechselbares Erscheinungsbild. Seine leichte Neigung wurde mit einem neuentwickelten Verfahren ohne Druckguss erzeugt. FLO ist in zwei verschiedenen Höhen, drei Rahmenfarben sowie als Freischwinger erhältlich.

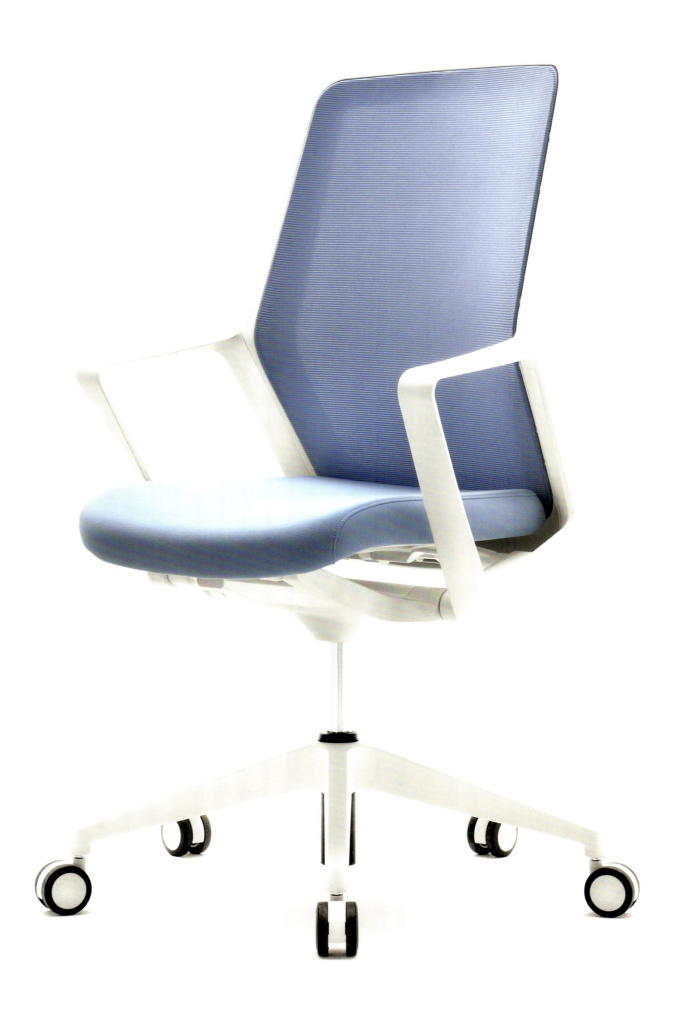

Office furniture

Tola
Office Chair
Bürostuhl

Manufacturer
Koleksiyon.
Istanbul, Turkey
Design
f/p cesign
(Fritz Frenkler, Anette Ponholzer),
Munich, Germany
Web
www.koleksiyon.com.tr
www.f-p-design.com

The Tola conference and office chair is designed for work environments where distinctions between executives, management and work teams are fading. The modular design with its overlapping back shells offers a wide variety of applications, ranging from office chair with low backrest to a comfortable conference chair with synchronous mechanics, a high back and a neck rest. Tola also convinces with a timeless language of form: it is modern and functional, but not technical, and integrates into working and living environments alike.

Der Konferenz- und Arbeitsstuhl Tola ist prädestiniert für eine Arbeitsumgebung, in der die Trennung zwischen Chef, Management und Arbeitsgruppen durchlässiger wird. Das modulare Design mit seinen überlappenden Rückenschalen bietet eine breite Palette an Anwendungsmöglichkeiten, vom Arbeitsstuhl mit niedriger Rückenlehne bis hin zum Konferenzsessel mit Synchronmechanik, hohem Rücken und Nackenstütze. Tola überzeugt auch durch seine zeitlose Formensprache, ist modern und funktional, aber nicht technisch, und lässt sich sowohl in das Arbeits- wie auch das Wohnumfeld integrieren.

Statement by the jury
With a seamless connection of arm and backrest, the Tola office chair promises enhanced comfort. Thanks to different modular elements for head and back support, it moreover encourages versatile use.

Begründung der Jury
Mit der nahtlosen Verbindung von Arm- und Rückenlehne zeigt sich der Bürostuhl Tola sehr komfortabel. Verschiedene modulare Elemente für Kopf und Rücken prädestinieren ihn darüber hinaus für einen flexiblen Einsatz.

Diagon
Swivel Chair
Bürodrehstuhl

Manufacturer
Girsberger Holding AG,
Bützberg, Switzerland
Design
Burkhard Vogtherr,
Mulhouse, France
Web
www.girsberger.com
www.vogtherr.com

The Diagon swivel chair stands out through its aluminium backrest support. While it appears to be a decorative design feature, the support also has an ergonomic function. It is based on the idea of developing a comfortable backrest by mounting it on two rubber cushions to create a movable lumbar area. The high comfort of this swivel chair is especially evident in the dynamic sitting position. The flexibly mounted backrest maximises the opening angle between seat and backrest when leaning back, thereby ensuring a pleasant motion experience. The backrest nevertheless features a slim design.

Der Bürodrehstuhl Diagon fällt durch seinen Rückenlehnenträger aus Aluminium ins Auge. Dieser wirkt wie ein dekoratives Designmerkmal, ist jedoch funktional begründet. Er geht auf die Idee zurück, eine bequeme Rückenlehne zu entwerfen und sie im Lendenwirbelbereich beweglich an zwei Gummipuffern aufzuhängen. Der hohe Komfort des Bürostuhls zeigt sich insbesondere beim dynamischen Sitzen. Die beweglich gelagerte Rückenlehne maximiert den Öffnungswinkel von Sitz und Lehne beim Zurücklehren und sorgt dabei für einen sehr angenehmen Bewegungsablauf. Dabei fällt die Form der Rückenlehne dennoch schlank aus.

Statement by the jury
The design of the Diagon swivel chair combines stability with a comfortable seating experience. This turns the chair into a high-grade product in its segment.

Begründung der Jury
Die Gestaltung des Bürodrehstuhls Diagon kombiniert Stabilität mit einem komfortablen Sitzgefühl. Das macht ihn zu einem hochwertigen Produkt in seinem Segment.

poi
Swivel Chair
Drehstuhl

Manufacturer
Wiesner-Hager Möbel GmbH,
Altheim, Austria
Design
neunzig° design
(Barbara Funck, Rainer Weckenmann),
Werdlingen, Germany
Web
www.wiesner-hager.com
www.neunzig-grad.com

poi represents a new generation of swivel chairs: The successful combination of aesthetics, comfort and sophisticated ergonomics makes it unique and economically competitive. The elegant monocoque design is the characteristic feature of poi. The shell has an enveloping effect and conveys a feeling of security and protection. The soft core is inviting and comfortable. The colour variations of the seat upholstery add a certain touch to the office and make poi versatile – stylish, classy, or young and fresh.

poi repräsentiert eine neue Generation von Drehstühlen: Die gelungene Kombination von Ästhetik, Komfort und ausgereifter Ergonomie macht ihn einzigartig und auch wirtschaftlich attraktiv. Charakteristisch für poi ist die elegante Schalenbauweise. Die Schale wirkt einhüllend und vermittelt Geborgenheit und Schutz. Der weiche Kern fühlt sich einladend und komfortabel an. Die Farbvariationen des Sitzpolsters setzen Akzente und machen poi wandlungsfähig – stylisch, edel oder jung und frisch.

W70
Conference Chair
Konferenzstuhl

Manufacturer
Wagner –
Die Wohlfühlmarke der Topstar GmbH,
Langenneufnach, Germany
Design
Wagner Design GmbH
(Peter Wagner)
Web
www.wagner-wellness.de

The focal point of the W70 conference chair is the upper part made of solid plastic. Its backrest and seat surface were cast as a whole in one single operation, lending the chair a high degree of stability and its distinctive contouring. Thanks to welting technology, the frame of the backrest can be covered by a wide variety of fabrics, from leather or felt to a breathable mesh cover. The Dondola®3 technology is integrated invisibly and enables sitting dynamically, which actively prevents back strain by constantly facilitating micro movements.

Basis des Konferenzstuhls W70 ist das Oberteil aus stabilem Kunststoff, dessen Lehne und Sitzfläche als Ganzes in einem Arbeitsgang gegossen werden. Dies verleiht ihm eine hohe Stabilität und ermöglicht die markante Konturführung. Der Rahmen der Rückenlehne wird mithilfe der Kedertechnik bespannt, sodass zahlreiche Stoffe von Leder über Filz bis zum atmungsaktiven Netzbezug eingesetzt werden können. Die unsichtbar integrierte Dondola®3-Technik sorgt für ein dynamisches Sitzverhalten, durch dessen ständige Mikrobewegungen Rückenschäden aktiv vorbeugt wird.

DOCKLANDS
Individual Workstation
Einzelarbeitsplatz

Manufacturer
Bene AG,
Waidhofen an der Ybbs, Austria
Design
PearsonLloyd
(Luke Pearson, Tom Lloyd),
Londor, GB
Web
www.bene.com
www.pearsonlloyd.com

Docklands, a work place concept for temporary activities, is conceived as an anchor point for employees and visitors. Its modular elements structure the office space, foster acoustic effectiveness and support focused work. The furniture features a high-quality design and is fully equipped with power supply and network access. Its Dock-In Bays are compact desk units, which were designed to offer working space with more privacy and also to allow for individual configuration.

Docklands, ein Arbeitsplatzkonzept für temporäre Tätigkeiten, versteht sich als Ankerpunkt für Mitarbeiter und Besucher. Seine Elemente gliedern den Raum, sind akustisch wirksam und unterstützen konzentriertes Arbeiten. Die hochwertig gestalteten Möbel sind mit Strom und Netzwerkzugang voll ausgestattet. Ihre Dock-In Bays sind kompakte Schreibtischbuchten, die für den ungestörten Aufenthalt entworfen wurden und unterschiedlich konfiguriert werden können.

CUBE_S
Modular Workplace
Modularer Arbeitsplatz

Manufacturer
Bene AG,
Waidhofen an der Ybbs, Austria
In-house design
Christian Horner
Web
www.bene.com

The Cube_S workplace programme is both a functional and an aesthetic response to new work culture. To meet individual requirements, modular elements allow versatile combinations of working space and storage areas in open office settings. An efficient use of floor space, comfort for employees and flexibility are all placed centre stage. For instance, the storage compartment at the side of the desk can serve as shelf, cabinet or room partition.

Das Arbeitsplatzprogramm Cube_S stellt eine funktionsgerechte wie ästhetische Antwort auf die neue Arbeitskultur dar. Dank modularer Bausteine können Arbeitsplatz und Stauraum im Open Office vielfältig miteinander kombiniert und so den individuellen Anforderungen der Nutzer gerecht werden. Im Vordergrund stehen Flächeneffizienz, Flexibilität und der Komfort der Mitarbeiter. So kann etwa der seitlich angebundene Stauraum als Schrank, Regal, Ablage oder Raumteiler fungieren.

Statement by the jury
Cube_S promotes the flexible use of office space and moreover presents itself as a simple furniture system with a high-quality design.

Begründung der Jury
Cube_S ermöglicht eine flexible Raumnutzung und zeigt sich zudem als ein schlicht und hochwertig gestaltetes Möbelsystem.

e-motion
Office Furniture System
Büromöbelsystem

Manufacturer
Tuna Ofis ve Ev Mobilyaları,
Istanbul, Turkey
Design
Ozan Sinan Tığlıoğlu
Web
www.tunaofis.com

e-motion was designed as a simple modular solution for today's dynamic office culture and enables flexible configurations for individual office situations. The furniture system builds on the creative design of the base frame, which transforms a basic profile into an aesthetic form thanks to a simple manufacturing process. e-motion presents both a sturdy office system and a modular organising unit, which can be freely rearranged or extended. It therefore adapts to new trends and workplace concepts.

e-motion wurde als einfache modulare Lösung für die dynamische Bürokultur von heute entworfen und ermöglicht flexible Konfigurationen für individuelle Bürosituationen. Das Möbelsystem baut auf der kreativen Gestaltung des Gestells auf, dessen einfacher Produktionsprozess ein elementares Profil in eine ästhetische Form überführt. e-motion stellt ein stabiles Bürosystem sowie eine modulare Organisationseinheit dar, die sich bequem und einfach umgestalten und erweitern lässt. Sie wird so den neuen Trends und Arbeitsplatzkonzepten gerecht.

Statement by the jury
The functional e-motion office system is characterised by a likeable, no-frills design. It can be changed effortlessly and thus responds to current and future requirements.

Begründung der Jury
Das funktionale Bürosystem e-motion ist durch ein schnörkelloses, sympathisches Design gekennzeichnet. Es lässt sich spielend leicht verändern und ist so auf die Anforderungen der Zeit abgestimmt.

Office furniture systems

TYPUS
Table
Tisch

Manufacturer
Wilde+Spieth Designmöbel GmbH & Co. KG,
Esslingen, Germany
Design
Heidi Edelhoff, Alexander Nettesheim,
Aachen, Germany
Web
www.wilde-spieth.com
www.alexander-nettesheim.com

The Typus table catches the eye with the angular design of its legs, inspired by the supporting construction of bridges. The legs are constructed with flat, coated rectangular tubes, which are arranged in such a way that the table braces itself. In spite of their filigree appearance, these legs provide a high degree of stability. They lend the table an elegant and dynamic form, distribute loads evenly below the table top and guarantee a firm stand. The table top has bevelled edges to enhance the lightweight appearance.

Der Tisch Typus fällt durch die winklige Ausstellung seiner Beine ins Auge, die von den Tragwerkkonstruktionen bei Brücken inspiriert wurde. Die Tischbeine wurden aus flachem, beschichtetem Vierkantrohr konstruiert und so angeordnet, dass sich der Tisch selbst ausste ft. Trotz ihrer filigranen Anmutung sind diese Beine sehr stabil. Sie bilden eine elegante dynamische Form, verteilen die aufkommenden Kräfte gut unter der Tischplatte und gewährleisten einen sicheren Stand. Die Tischplatte ist an den Kanten abgeschrägt, um die leichte Optik zu verstärken.

Dreyfuss
Meetingtable
Besprechungstisch

Manufacturer
Neudoerfler Office Systems GmbH,
Neudörfl, Austria
In-house design
Manfred Neubauer
Design
Thomas Hribar,
Baden bei Wien, Austria
Web
www.neudoerfler.com
www.thomashribar.com

The Dreyfuss product line, designed for use in meetings, exhibits a modern lightweight character and an appealing design language. The table legs – a contemporary interpretation of pedestals with cantilever legs – are made from two interlaced steel sheets that stabilise each other. The resulting branching brackets constitute the formal centre of the table, lending it an unmistakable character.

Die für Besprechungen entwickelte Produktlinie Dreyfuss präsentiert sich mit moderner Leichtigkeit und einer ansprechenden Formensprache. Die Tischfüße, eine zeitgemäße Interpretation von Säulenfüßen mit Bodenausleger, bestehen jeweils aus zwei Stahlblechen, die sich mittels Verschränkung gegenseitig stabilisieren. Die so entstehenden Verzweigungen bilden den formalen Mittelpunkt des Tisches und geben ihm seine unverwechselbare Charakteristik.

Statement by the jury
The Dreyfuss table stands out through the sophisticated construction of its legs. They contrast with the simple glass top and emphasise the table's unique appearance.

Begründung der Jury
Der Tisch Dreyfuss besticht durch die raffinierte Konstruktion seiner Beine. Sie kontrastieren die schlichte Glasplatte und betonen den originellen Gesamteindruck.

Metronome
Conference Table
Konferenztisch

Manufacturer
Nienkämper,
Toronto, Canada
Design
FigForty
(Lee David Fletcher, Terence Woodside),
Toronto, Canada
Web
www.nienkamper.com
www.fig40.com

The design of the Metronome conference table imparts aesthetics and lightness – the key elements of this collection are simplicity and sophistication. The aluminium frame is inspired by sculptural structures found in large buildings, where mass and gravity are optimally balanced. This approach lends the table its characteristic appeal. The materials were selected to ensure that all elements are recyclable. The meticulously manufactured table legs are available in clear anodised aluminium or fully polished aluminium, while the tabletop may be selected from a wide range of wood veneers and laminates. The centre of the table opens up to expose a brightly coloured compartment, which conceals all power connectivity, enabling this table to be used flexibly.

Die Gestaltung des Konferenztisches Metronome vermittelt Ästhetik und Leichtigkeit – die Kernthemen dieser Kollektion sind Einfachheit und Raffinesse. Das Aluminiumgestell ist inspiriert von den skulpturalen Strukturen großer Gebäude, bei denen Masse und Schwerkraft optimal ausbalanciert sind. Dieser Ansatz verleiht dem Tisch seine charakteristische Anmutung. Die Materialien wurden so gewählt, dass sämtliche Elemente später recycelt werden können. Die sorgfältig gefertigten Tischbeine sind in klarem anodisiertem Aluminium mit polierten Details oder in vollständig poliertem Aluminium erhältlich, und für die Tischplatte stehen verschiedene Holzfurniere sowie Laminate zur Auswahl. Die Mitte der Tischplatte enthüllt beim Öffnen eine bunte Mulde in leuchtenden Farben, in der die gesamten Stromanschlüsse verborgen sind, sodass eine flexible Nutzung des Tischs möglich ist.

Statement by the jury
Metronome convinces with a clear, elegant design and well-balanced proportions. In addition, the intelligently hidden power connections in the table centre make it a highly contemporary piece of office furniture.

Begründung der Jury
Metronome überzeugt durch ein klares, elegantes Design und ausgewogene Proportionen. Die intelligent in die Mitte des Tischs versenkten Stromanschlüsse machen ihn außerdem zu einem überaus zeitgemäßen Büromobiliar.

Office furniture

KINETICis5
Communication Furniture
Kommunikationsmöbel

Manufacturer
Interstuhl GmbH & Co. KG,
Meßstetten-Tieringen, Germany
Design
Phoenix Design GmbH + Co. KG,
Stuttgart, Germany
Web
www.interstuhl.de
www.phoenixdesign.com

Dynamics and flexibility determine communication forms in the offices of today and the future. For situations where classic sitting positions are considered no longer up-to-date and where workplaces are becoming more and more mobile, the KINETICis5 product family is a suitable solution. With its unique design of elegant standing seats and ascetically reduced tables, it integrates perfectly into modern interiors. All individual elements are easily moved, so that the series provides a high degree of flexibility. The system can be quickly set up for relaxed and spontaneous meetings, with a laptop stand in close reach at all times.

Dynamik und Flexibilität prägen die Kommunikationsformen im Büro von heute und morgen. Wo klassisches Sitzen nicht mehr als zeitgemäß empfunden wird und Arbeitsplätze immer mobiler werden, stellt die Produktfamilie KINETICis5 eine passende Lösung dar. Mit ihrer eigenständigen Gestaltung von eleganten Stehsitzen und ebenso asketisch reduzierten Tischen fügt sie sich in moderne Interieurs ein. Ihre einzelnen Elemente lassen sich schnell und leicht bewegen, sodass diese Serie ein hohes Maß an Flexibilität bietet. Ein kurzfristig einberufenes Meeting lässt sich hier entspannt und ohne großen Aufwand abhalten, wobei das Laptop-Tablar stets parat ist.

Statement by the jury
The design of this product family stands out through the strong elegance that is fostered by the interplay of striking dynamic lines and high-quality materials.

Begründung der Jury
Die Gestaltung dieser Produktfamilie besticht durch die hohe Eleganz, zu der sich markante dynamische Linien und hochwertige Materialien miteinander verbinden.

Patis
Desktop Partition
Schreibtischtrennwand

Manufacturer
Morita Aluminum Industry, Inc.,
Osaka, Japan
In-house design
Kazuaki Sagara, Kentaro Uno
Design
hers design inc.
(Prof. Chiaki Murata),
Osaka, Japan
Web
www.moritaalumi.co.jp
www.hers.co.jp

The Patis desktop partition was designed in response to the need to find a balance between personal productivity and group communication in large office spaces. It is easy to install on flat worktop surfaces and is stabilised by an elegant aluminium base. Reusable adhesive pads on the underside ensure that the partition remains upright and also allow them to be easily rearranged. The height was chosen to guarantee both work privacy and unobstructed communication with colleagues. Available in two lengths, the partition boards are horizontally slidable, allowing flexible adjustments for different room configurations.

Die Schreibtischtrennwand Patis trägt der Notwendigkeit Rechnung, in einem weiträumigen Büroraum eine Balance zwischen persönlicher Produktivität und Gruppenkommunikation zu finden. Sie ist leicht auf glatten Arbeitsflächen anzubringen und wird durch einen eleganten Aluminiumfuß stabilisiert. Wiederverwendbare Klebepolster an der Unterseite sorgen dafür, dass die Trennwand nicht kippt, und ermöglichen zugleich ein einfaches Versetzen. Die Höhe wurde so gewählt, dass ungestörtes Arbeiten wie auch ungehinderte Kommunikation mit Kollegen gewährleistet werden. Die in zwei Längen erhältlichen Trennwandplatten lassen sich horizontal verschieben, sodass sie flexibel an unterschiedliche Raumkonfigurationen angepasst werden können.

Statement by the jury
The Patis presents a simple solution to partitioning work places. In addition, its simple, reduced design turns it into an aesthetically appealing eye-catcher.

Begründung der Jury
Patis bietet eine einfache Lösung zur Trennung von Arbeitsplätzen. Ihre schlichte reduzierte Gestaltung macht sie zudem zu einem ästhetischen Blickfang.

terri tory
Modular Storage Range
Modulares Stauraumprogramm

Manufacturer
Sedus Systems GmbH,
Geseke, Germany
Design
Formwelt Industriedesign
(Henriette Deking, Luca Cianfanelli),
Murich, Germany
Web
www.sedus.de
www.formwelt.com

The terri tory is a modular storage range with multiple benefits. As a space-defining piece of furniture, it combines three functions in one: storage area, spatial division and privacy screen. The easy-to-handle modular system with its wide range of accessories provides a high degree of individuality. With access from one or both sides, the storage space is freely configurable, and the 400 mm grid enables the easy planning of configurations and a high level of flexibility. The boxes are clamped to the function rail by means of a specially developed, centred connection fitting. The included assembly tool allows users to create individualised set-ups.

terri tory ist ein modulares Stauraumprogramm mit Mehrfachnutzen. Als raumbildendes Möbel vereint es die Funktionen Stauraum, Raumzonierung und Abschirmung. Der einfach zu handhabende Baukasten mit vielfältigem Zubehör bietet ein hohes Maß an Individualität. Der einseitige und wechselseitige Stauraum ist frei konfigurierbar und erlaubt durch das 400er Raster eine einfache Planbarkeit und hohe Flexibilität. Die Boxen werden über einen eigens entwickelten, mittig sitzenden Verbindungsbeschlag mit der Funktionsschiene verspannt. Das mitgelieferte Montagetool erlaubt es dem Nutzer, individuelle Aufbauten zu schaffen.

Statement by the jury
The terry tori storage space range reflects a symbiosis of clearly conceived design and enhanced individuality in terms of both its construction and use.

Begründung der Jury
Das Stauraumprogramm terry tori stellt eine Symbiose aus klarer, durchdachter Gestaltung und hoher Individualität im Hinblick auf seine Konstruktion und Nutzung dar.

VARICOLOR
Drawer Box
Schubladenbox

Manufacturer
Durable Hunke & Jochheim GmbH & Co. KG,
Iserlohn, Germany
Design
GBO DESIGN,
Helmond, Netherlands
Web
www.durable.de
www.gbo.nl

Varicolor is a box with colourful drawers made of premium-quality plastics. When the drawer fronts are closed, slender coloured lines – in the same hue that highlights the inside of the respective drawer – serve as organisation and orientation aids. The silent and smooth-running drawers with pull-out stops hold documents in formats up to DIN C4, including folio and letter size. The boxes are stackable with non-slip plastic feet and can be arranged in rows inside all standard office cabinets and sideboards. The drawers come with exchangeable labels that are inserted into transparent windows and can be marked either by hand or computer.

Varicolor ist eine Box mit farbiger Schubladen in hochwertiger Kunststoffqualität. Bei geschlossener Schubladenfront dienen feine Farblinien, deren Farbe jeweils auch den Innenraum der Schublade ausfüllt, als Organisations- und Orientierungshilfe. Ausgestattet mit geräuschfreiem Schubladenleichtlauf samt Auszugssperre, eignet sich die Schubladenbox für Formate bis DIN C4, inklusive Folio- und Letter-Size-Format. Sie ist mithilfe rutschfester Kunststoff-Füßchen stapelbar und für den Einsatz in Reihenaufstellung in allen Standard-Büroschränken und -Sideboards geeignet. Die auswechselbaren Einsteck-etiketten für die transparenten Beschriftungsfenster sind per Computer oder von Hand beschreibbar.

Statement by the jury
The Varicolor drawer box presents itself as a simply designed, rigid storage system which, with its delicate coloured lines, integrates harmoniously into any interior.

Begründung der Jury
Die Schubladenbox Varicolor präsentiert sich als schlicht gestaltetes, stabiles Aufbewahrungssystem, das sich mit seinen zarten Farbstreifen in jedes Ambiente einfügt.

D-Wings
Cord Organiser
Kabelführungssystem

Manufacturer
Feng Chun Rubber Industrial Co.,
Yingko, Taipei Hsien, Taiwan
Design
Rubto International, Inc.,
UT Wire (Kevin Chen),
Anaheim, USA
Web
www.rubbointernational.com
www.ut-wire.com

D-Wings is an easy-to-install cord organiser which holds and cleanly guides cords across any flat surface. Thanks to the self-adhesive backing, the system can be installed on glass, metal, concrete or other surfaces without any drilling or stapling. In addition, it guides cords around curves, corners or edges. The product features an innovative clamp shape with a slip at the top, which is opened by applying gentle pressure on two sides. This allows cords to be quickly removed and reinserted whenever needed. D-Wings has a paintable body so that it can be mixed and matched with any decor and cord colour.

D-Wings ist ein montagefreundliches Kabelführungssystem, mit dem Kabel auf ebenen Oberflächen sauber verlegt werden können. Dank der selbstklebenden Rückseite kann das System ohne Bohren oder Klammern auf Glas, Metall, Beton und anderen Oberflächen befestigt werden. Darüber hinaus führt es Kabel auch um Rundungen, Ecken und Kanten. Das Produkt verfügt über eine innovative Bügelform mit einem Schlitz an der Oberseite, der sich durch leichtes beidseitiges Drücken öffnen lässt. Durch diese Öffnung können Kabel im Handumdrehen eingeführt oder entfernt werden. D-Wings ist überstreichbar und kann so problemlos an jedes Dekor und jede Kabelfarbe angepasst werden.

Statement by the jury
D-Wings impresses with a clear, reduced design, a carefully conceived concept, and use that is simple and flexible in equal measure.

Begründung der Jury
D-Wings beeindruckt durch eine klare, reduzierte Gestaltung, eine durchdachte Konzeption und schließlich eine Nutzung, die so flexibel wie einfach ist.

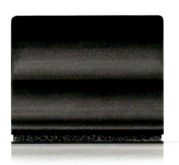

Mobile Self-Adjustable Sorter – Elba For Business
Sorter
Ordnungsmappe

Manufacturer
Elba, Hamelin Group,
Hérouville Saint Clair, France
In-house design
Vincent Lemaistre, Keyne Dupont
Web
www.elba.com
www.hamelinbrands.com
Honourable Mention

The Elba for Business sorter was designed for managers and executives who require sophisticated organisation of their documents. It adapts automatically to the volume of its content, since the cover is flexible and the adjustable mechanism can hold up to 400 sheets of paper. The dividers are arranged like an accordion, allowing for efficient organisation. The sorter is light and consists of rewritable polypropylene material with a soft-touch surface.

Statement by the jury
This appealingly designed sorter meets the demands of its target group with an efficient organisational structure.

Die Ordnungsmappe Elba for Business wurde für Führungskräfte entwickelt, die eine intelligente Organisation ihrer Unterlagen benötigen. Sie passt sich automatisch dem Volumen des Inhalts an, da das Cover beweglich ist und ihr justierbarer Mechanismus bis zu 400 Blatt Papier fasst. Die Fächer sind wie ein Akkordeon angeordnet, was eine effektive Ablage erlaubt. Die Mappe ist leicht und besteht aus wiederbeschreibbarem Polyprophylen mit Soft-Touch-Oberfläche.

Begründung der Jury
Die ansprechend gestaltete Mappe wird dem Anspruch ihrer Zielgruppe an eine wirksame Ordnungsstruktur gerecht.

Eco Clip
Reusable
Sketchbook Maker
Clip für Mehrweg-Skizzenbücher

Manufacturer
Igloo Design Strategy,
Kfar Saba, Israel
In-house design
Arik Yuval, Yariv Sade
Design
Michal Granit, David Sade
Web
www.igloo-design.com

The Eco Clip is used to make sketchbooks from old drafts and leftover paper. It can turn 15 sheets of A4-sized paper into a 60-page A5 sketchbook, simply by folding the A4 papers in half and binding them with the product. This innovative concept of reusing draft papers saves on both costs and trees. The Eco Clip is suitable for use in offices and at home and is available in six colours.

Statement by the jury
The Eco Clip presents an unconventional method for saving paper. Intuitive to use and offering a simple design, it is an easy way to protect the environment.

Mit dem Eco Clip lassen sich Skizzenbücher aus übrig gebliebenem und schon gebrauchtem Papier erstellen. So kann man 15 Blatt A4-Papier in ein A5-Sketchbook mit 60 Seiten verwandeln, indem man das A4 Papier in der Hälfte faltet und mit dem Produkt bindet. Dieses innovative Konzept, um Papier wiederzuverwenden, spart Kosten und schont den Baumbestand. Der Eco Clip ist für zuhause wie für das Büro geeignet und in sechs Farben erhältlich.

Begründung der Jury
Der Eco Clip bietet eine unkonventionelle Möglichkeit Papier zu sparen. Schlicht gestaltet und intuitiv zu handhaben, macht er es einem leicht, die Umwelt auf einfache Art zu schonen.

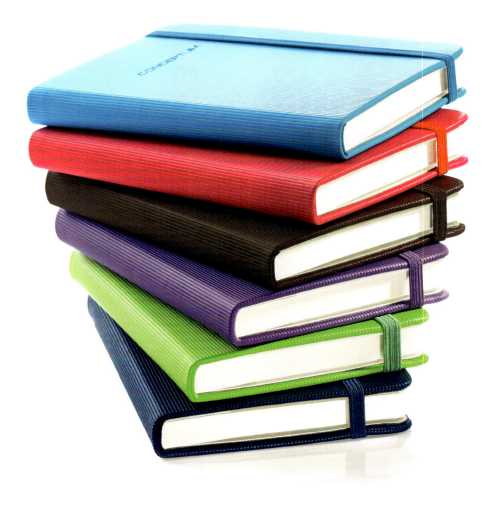

CONCEPTUM
Notebook
Notizbuch

Manufacturer
Sigel GmbH,
Mertingen, Germany
In-house design
Werner Bögl
Web
www.sigel.de

The Colour collection of the Conceptum notebook range is designed to provide space for thoughts. With vividly fresh or muted colours, the notebooks clearly stand out and feature, besides a hard cover and a premium Softwave surface, a number of functional details: pen loop, elastic fastener, page numbering, a directory that makes it easy to find notes, and perforated note pages to tear out. Three filing compartments – a quick pocket, an archive pocket and a card slot – round off the notebook features. Here, aesthetics and function meet to encompass a wide range of products: the Colour collection comprises 20 books in six different colours and two formats, as well as a choice of lined or squared page ruling.

Die Kollektion Colour der Notizbuchserie Conceptum ist dafür gemacht, Gedanken Raum zu geben. Mit leuchtend frischen sowie gedeckten Farben setzt sie Akzente und ist neben dem Hardcover mit hochwertiger Softwave-Oberfläche mit weiteren Details sinnvoll ausgestattet: Stiftschlaufe, Gummibandverschluss, Seitennummerierung, Inhaltsverzeichnis zum leichten Auffinden von Notizen, perforierte Notizseiten zum Heraustrennen und die drei Aufbewahrungsfächer Quickpocket, Archivtasche und Kartenfach vervollständigen die Ausstattung. Ästhetik trifft auf Funktion – samt großer Bandbreite. Denn Colour besteht aus 20 Büchern in sechs Farben und zwei Formaten und ist in linierter wie karierter Lineatur erhältlich.

Statement by the jury
This notebook range stands out through a clear design. Displaying a simple, convenient form that is complemented by contemporary colours and many useful details, these notebooks embody high practical value.

Begründung der Jury
Diese Notizbuchserie fällt durch eine klare Gestaltung auf. Schlicht und handlich in der Form sorgen zeitgemäße Farben und viele nützliche Details für einen hohen Gebrauchswert.

senseBook
Notebook
Notizbuch

Manufacturer
Holtz Office Support,
Wiesbaden, Germany
In-house design
Tobias Liliencron,
Christopher Holtz-Kathan
Web
www.sensebook.de

senseBook is a notebook that touches the senses. Its smooth cover is hand-sewn and made of premium calf leather. The fine, cream-coloured paper is firm and perfect for writing down thoughts, memories, sketches and project records. Details like page numbers or perforated pages for the simple removal of sheets make collecting thoughts easy and clear. The notebooks come in three different sizes, with each available in three page styles: ruled, squared or blank. Moreover, the leather cover is offered in two versions: the Flap model with a cord for secreting away information and the Red Rubber version with a rubber strap for recording sudden inspirations.

senseBook ist ein Notizbuch, das die Sinne berührt. Der glatte Einband ist handgenäht und besteht aus edlem Rinderleder. Das feine Papier ist stabil und cremefarben und eignet sich für Gedanken, Erinnerungen, Skizzen oder Projektaufzeichnungen. Details wie Seitenzahlen oder perforierte Seiten zum Heraustrennen machen das Gedankensammeln einfach und übersichtlich. Die Notizbücher gibt es in drei Größen, jeweils liniert, kariert oder blanko. Zudem sind sie mit zwei verschiedenen Einbänden erhältlich: Flap zum Schnüren, um Geheimnisse zu bewahren, und Red Rubber mit Klippband zum schnellen Aufzeichnen von Gedankenblitzen.

Statement by the jury
This notebook range fascinates with its premium leather finish and a design that pays great attention to detail.

Begründung der Jury
Die Notizbuch-Reihe begeistert durch ihre edle Ausführung in Leder und eine mit viel Liebe zum Detail entworfene Gestaltung.

LAMY 2000 metal
LAMY 2000 Metall
Writing Instruments
Schreibgeräte

Manufacturer
C. Josef Lamy GmbH,
Heidelberg, Germany
In-house design
Wilhelm Berberich
Design
Gerd A. Müller
Web
www.lamy.de

The Lamy 2000 is fascinating with its seamless, matte-brushed finish and smooth transitions of its product segments. All visible parts are made of solid stainless steel. The combination of design, material and surface structure results in sublime elegance and a weight that promotes a comfortable writing experience. Characteristic, contrasting details, such as individual chamfers and lower surfaces, have a polished finish. In all cases, the clip is spring-mounted and displays a high-gloss finish on the back. The fountain pen features a gold nib and a piston mechanism, the ballpoint pen the usual push mechanism, while the roller ball, like the fountain pen, has a cap. The mechanical pencil features a push mechanism with integrated eraser tip and comes with a 0.7 mm lead refill.

Statement by the jury
The Lamy 2000 writing instrument series merges a combination of premium materials with a minimalist design to foster an outstanding appearance that is classic and elegant in equal turn.

Lamy 2000 fasziniert durch die nahtlose strichmattierte Oberfläche und den ansatzlosen Übergang der Produktsegmente. Alle sichtbaren Komponenten sind massiv aus Edelstahl gefertigt. Aus der Kombination von Design, Material und Oberflächenstruktur ergibt sich eine erhabene Eleganz und ein ausbalanciertes Gewicht, das angenehmes Schreiben ermöglicht. Charakteristische, kontrastierende Details wie einzelne Fasen und tiefer liegende Oberflächen wurden poliert. Der Clip ist jeweils fremdgefedert und der Clip rücken hochglanzpoliert. Der Füllhalter wurde mit Goldfeder und Kolbenmechanik ausgestattet, der Kugelschreiber mit der gewohnten Druckmechanik, und der Tintenroller ist wie der Füllhalter als Kappenmodell ausgeführt. Der Druckbleistift besitzt eine Druckmechanik mit Radiertip und schreibt mit Bleiminen der Strichstärke 0.7 mm.

Begründung der Jury
Die Schreibgeräteserie Lamy 2000 verbindet die Kombination hochwertiger Materialien mit einer minimalistischen Gestaltung zu einem überaus klassisch eleganten Erscheinungsbild.

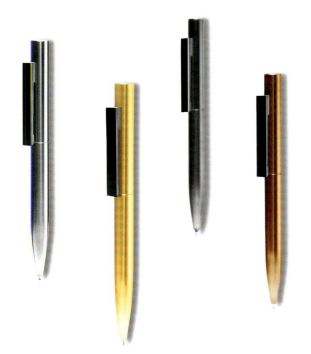

SIGNER LINER
Ballpoint Pen
Kugelschreiber

Manufacturer
SENATOR GmbH & Co. KGaA,
Groß-Bieberau, Germany
Design
White Studios
(Fabian Fischer),
Berlin, Germany
Web
www.senatorglobal.com
www.white-studios.net

The Signer Liner ballpoint pen presents a clear, purist design. With a material combination of high-quality stainless steel and plastics, as well as four elegant metallic barrel colours and nine clip colours, the pen provides optimal conditions for tailored brand communication. Barrel and clip can be individually adjusted to customer requirements and corporate identity specifications.

Statement by the jury
A classic, reduced design and selected materials turn the Signer Liner into a product that communicates expressiveness and elegance.

Der Kugelschreiber Signer Liner zeigt ein puristisches, klares Erscheinungsbild. Mit einer Materialkombination aus hochwertigem Edelstahl und Kunststoff sowie vier verschiedenen, elegant metallischen Schaft- und neun Clipfarben bietet er optimale Voraussetzungen für eine maßgeschneiderte Markenkommunikation. Schaft und Clip können so individuell an Kundenwünsche und Corporate-Identity-Vorgaben angepasst werden.

Begründung der Jury
Eine klassisch reduzierte Gestaltung und ausgewählte Materialien machen den Signer Liner zu einem Produkt, das Ausdruckskraft und Eleganz vermittelt.

MARKSMAN TRIGON
Stylus Ballpoint
Stylus Kugelschreiber

Manufacturer
PF Concept International BV,
Netherlands
In-house design
Joeri van der Leeden
Web
www.pfconcept.com
Honourable Mention

The Marksman Trigon pen is a simply designed writing instrument with a twist action ballpoint and a stylus touchscreen tip. Designed specifically for the promotional gift market, it provides a solution that works equally well with or without the addition of customer branding. The pen features a brass body and matte-black finish and is presented in a black gift box.

Statement by the jury
A well-proportioned design with tapered tip and end define this ballpoint pen as a timeless companion for both conventional writing and electronic interaction.

Der Marksman Trigon Kugelschreiber ist ein schlicht gestaltetes Schreibgerät mit Drehmechanismus und einer Stylus Touchscreen-Spitze. Speziell für das Segment der Werbegeschenke konzipiert, bietet er sich als eine Lösung an, die ebenso mit oder ohne den Zusatz der betreffenden Marke funktioniert. Er verfügt über eine Messing- und eine mattschwarze Lackierung und wird in einer schwarzen Geschenkbox präsentiert.

Begründung der Jury
Seine ebenmäßige Gestaltung mit sich verjüngender Spitze und Stiftabschluss definiert diesen Kugelschreiber als einen zeitlosen Begleiter für konventionelles Schreiben und elektronische Interaktion.

th.INK
Writing Instruments
Schreibgeräte

Manufacturer
Pelikan Vertriebsgesellschaft mbH & Co. KG,
Hannover, Germany
In-house design
Web
www.pelikan.com
Honourable Mention

th.INK is a series of writing instruments designed in trendy colours. It comprises fountain pens and ballpoint pens, which are suited for continuous use at school, university or at home. Both models have a soft, ergonomic grip profile for long hours of relaxed writing. With elements in black or violet and a combination of high-gloss black and matte soft elements, these pens are true eye-catchers.

Statement by the jury
The writing instruments of the th.INK series stand out in particular through their contrasting surfaces and colour-accentuated elements.

th.INK ist eine in modischen Farben entworfene Schreibgeräteserie. Sie enthält Füllhalter und Kugelschreiber, die sich für den Dauereinsatz in der Schule, der Uni oder zu Hause eignen. Beide Modelle besitzen ein weiches, ergonomisches Griffprofil für langes und unverkrampftes Schreiben. Mit Elementen in Schwarz oder Violett sowie der Kombination aus hochglänzendem Schwarz und matten Soft-Elementen sind sie echte Hingucker.

Begründung der Jury
Die Schreibgeräte der Serie th.INK ragen insbesondere durch die Kontraste ihrer Oberflächen sowie die verschiedenfarbig abgesetzten Elemente heraus.

LAMY scala
Writing Instruments
Schreibgeräte

Manufacturer
C. Josef Lamy GmbH,
Heidelberg, Germany
Design
sieger design GmbH & Co. KG
(Michael Sieger),
Sassenberg, Germany
Web
www.lamy.de
www.sieger-design.com
Honourable Mention

The scala writing instruments lend elegance and emotion to the two core shapes of Lamy's design language: the cylinder and the square block. The form of the pen's body is strictly cylindrical, while the clip is square. The contour of the body is interrupted by high-gloss trim plated in chrome. The clip breaks up the geometrical austerity, creating a flowing connection between the two surfaces. Such contrasts in form and surface make this series of writing instruments especially fascinating.

Statement by the jury
The scala merges contrasting surfaces with basic geometric shapes into one harmonious unit.

Die Schreibgeräte scala verleihen den beiden Grundtypen der Lamy Gestaltungssprache, dem Zylinder und dem Quader, Eleganz und Emotion. Die Gehäuseform ist streng zylindrisch, während der Clip eine rechteckige Form hat. Die Gehäuselinie wird durch die hochglanzverchromten Abschlüsse unterbrochen, wohingegen der Clip die geometrische Strenge durchbricht. Er schafft eine fließende Verbindung zwischen den beiden Oberflächen. Diese Kontraste in Form und Oberfläche machen den besonderen Reiz der Modellreihe aus.

Begründung der Jury
In scala verschmelzen kontrastierende Oberflächen mit geometrischen Grundformen zu einer harmonischen Einheit.

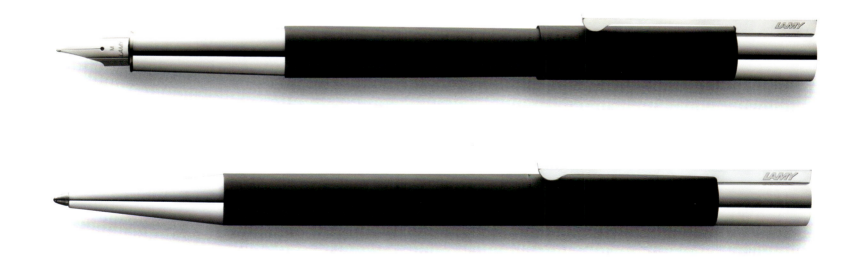

Power
Ballpoint Pen
Kugelschreiber

Manufacturer
Pelikan Vertriebsgesellschaft mbH & Co. KG,
Hannover, Germany
In-house design
Web
www.pelikan.com

The Power ballpoint pen comes in a reduced design and new colour combinations. The different variants are all equipped with a blue refill and, thanks to the slightly concave design, rest optimally in the hand. The writing instruments feature a new colour concept that turns the pens into vivid, colourful eye-catchers. The Pool colour version is a two-tone combination of light, cool grey and luminous blue.

Statement by the jury
The outstanding characteristic of this ballpoint pen is its dynamic design, expressed in both the form and the colour concept.

Der Kugelschreiber Power erscheint in einer reduzierten Gestaltung und neuen Farbkombinationen. Die verschiedenen Ausführungen enthalten eine blaue Ersatzmine und lassen sich dank der leicht konkaven Formgebung optimal in der Hand halten. Neu ist ein Farbkonzept, das die Schreibgeräte zum farbenfrohen Hingucker macht. Die Variante Pool ist eine zweifarbige Verbindung aus hellem kühlem Grau und einem leuchtenden Blau.

Begründung der Jury
Seine dynamische Gestaltung, die in der Form wie auch im Farbkonzept zum Ausdruck kommt, ist das hervorstechende Kennzeichen dieses Kugelschreibers.

Twist
Writing Instrument
Schreibgerät

Manufacturer
Pelikan Vertriebsgesellschaft mbH & Co. KG,
Hannover, Germany
In-house design
Web
www.pelikan.com
Honourable Mention

Twist is a new trendy writing instrument. It is available as a fountain pen and rollerball pen and is eye-catching with its twisted look. Thanks to this ergonomic form, the writing instruments lie well in the hand and fit perfectly for both left- and right-handed users. Both models are available in four bright and vivid colour combinations. They also guarantee that writing with the Twist is a lot of fun.

Statement by the jury
In addition to its distinctive colouring, the Twist stands out through its twisted form. It serves an ergonomic purpose and ensures an optimal grip.

Twist ist ein neues Trend-Schreibgerät. Erhältlich als Füllhalter und Tintenroller macht es durch seinen gedrehten Look auf sich aufmerksam. Diese ergonomische Formgebung sorgt dafür, dass das Schreibgerät gut in der Hand liegt und universell für Rechts- wie für Linkshänder geeignet ist. Beide Modelle gibt es in jeweils vier leuchtenden Farbkombinationen. Nicht nur sie tragen dazu bei, dass das Schreiben mit dem Twist große Freude bereitet.

Begründung der Jury
Twist fällt neben seiner Farbigkeit durch die gedrehte Form auf. Sie ist ergonomisch begründet und gewährleistet einen optimalen Griff.

Job
Highlighter
Textmarker

Manufacturer
Schneider Schreibgeräte GmbH,
Schramberg, Germany
Design
z.B. Designers
(Nicola Harrison, James Harrison),
Darmstadt, Germany
Web
www.schneiderpen.com
www.zb-designers.de

Highlighters are indispensable tools for marking important text passages. The Job highlighter reflects a contemporary design and is overall flatter than previous models, without departing from the familiar design language of highlighters. To improve both its aesthetic appeal and its usability, a gentle elevation was added to the clip surface, which serves as a pressure point for pushing off the cap with the thumb. The marker is held intuitively at the circular recesses on both sides of the body and, thanks to its rounded ends, can also be placed in a shirt pocket with ease.

Um wichtige Textpassagen farbig hervorzuheben, sind Textmarker ein unverzichtbares Arbeitsmittel. Job zeigt sich in einer zeitgemäßen Gestaltung und insgesamt flacher als bisherige Modelle, ohne dabei die gewohnte Formensprache für Textmarker zu verlassen. Um seine ästhetische Anmutung wie auch die Benutzerfreundlichkeit zu verbessern, erhielt die Clipoberfläche eine sanfte Erhebung, sodass sich die Kappe mit dem Daumen leicht abschieben lässt. An den kreisrunden Mulden auf beiden Seiten des Gehäuses greift man den Marker intuitiv. Aufgrund seiner flacheren Formgebung mit abgerundeten Enden lässt er sich zudem einfacher in Hemdtaschen oder Schlaufen einstecken.

Statement by the jury
Various ergonomic details, together with a simple, clear design, ensure that the Job highlighter rests comfortably in the hand.

Begründung der Jury
Verschiedene ergonomische Details sowie eine schlichte, klare Gestaltung führen dazu, dass der Textmarker Job besonders gut in der Hand liegt.

my.pen
Eraser
Radierer

Manufacturer
Herlitz PBS AG,
Berlin, Germany
In-house design
Juliane Rogoll
Web
www.herlitz.de

In addition to its colouring, the my.pen eraser displays a surprisingly organic and flowing shape, which was designed to meet the requirements of daily school life. Due to this shape and its ergonomic grooves, it fits perfectly in the hand, offers a good grip and enables accurate erasing. With a rounded and tapered eraser surface, both large areas and fine lines can be made to disappear in next to no time. The eraser is small and fits easily into any pencil case.

Statement by the jury
The particular design of the my.pen convinces aesthetically through many trendy colours and functionally through its ideal usability.

Neben seiner Farbgebung überrascht der my.pen Radierer durch seine organische fließende Form, die auf die Bedürfnisse im Schulalltag abgestimmt ist. Aufgrund dieser Form und den ergonomischen Rillen liegt er gut in der Hand, bietet Halt und ermöglicht zielgenaues Radieren. Mit einer abgerundeten und einer zugespitzten Radierfläche lässt er sowohl größere Flächen als auch feine Linien im Handumdrehen verschwinden und passt platzsparend in jedes Federmäppchen.

Begründung der Jury
Die besondere Formgebung des my.pen Radierers überzeugt ästhetisch mit vielen Trendfarben und funktional durch ideale Gebrauchsmöglichkeiten.

Staple-free Stapler
Klammerloser Hefter

Manufacturer
PLUS Corporation,
Tokyo, Japan
Design
Human Code Japan,
Tokyo, Japan
Web
www.plus.co.jp/en
www.hcj.co.jp

This simply and compactly designed staple-free stapler is a product of Japanese technology. It staples up to five sheets of paper without the need for staples and is especially eco-friendly since more than 90 per cent of its materials are made from recycled primary products. Stapled sheets can directly go through a shredder, since no metal needs to be removed from the paper first.

Statement by the jury
This staple-free stapler surprises with its inventive functionality. Due to its compact design, it is also comfortable to use.

Der schlicht und handlich gestaltete klammerlose Hefter ist ein Produkt japanischer Technologie. Er heftet bis zu fünf Blätter ohne den Gebrauch üblicher Heftklammern und ist besonders umweltfreundlich, da mehr als 90 Prozent seiner Materialien aus recycelten Vorprodukten bestehen. Geheftete Blätter können später direkt dem Aktenvernichter zugeführt werden, da das Papier nicht erst entklammert werden muss.

Begründung der Jury
Der klammerlose Hefter überrascht durch seine originelle Funktionalität. Aufgrund seiner kompakten Gestaltung liegt er zudem gut in der Hand.

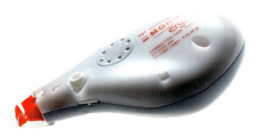

MONO ergo
Correction Tape
Korrekturroller

Manufacturer
Tombow Pencil Co., Ltd.,
Tokyo, Japan
In-house design
Hiromi Kuronuma
Design
Kanazawa University
(Prof. Katsuyuki Shibata),
Kanazawa, Japan
Web
www.tombow.com
www.kanazawa-u.ac.jp/e

With its ergonomic shape, this refillable correction tape roller guarantees comfortable handling. Its flexible head allows for the precise application of correction tape, even on uneven paper surfaces. A recess in the grip area shows exactly where to position the thumb to ensure ideal handling. The integrated cap protects the tape from dirt when not in use.

Statement by the jury
The Mono ergo correction tape roller stands out due to its smoothly contoured shape, which makes it very convenient and practical to use.

Der nachfüllbare Korrekturroller gewährt aufgrund seiner ergonomischen Form eine angenehme Handhabung. Sein flexibel gelagerter Kopf ermöglicht, dass das Korrekturband auch auf unebener Papieroberfläche exakt aufgetragen werden kann. Eine Mulde in der Griffzone zeigt an, wo der Daumen liegen soll, um eine ideale Anwendung zu gewährleisten. Die integrierte Schutzkappe schützt das Band vor Verschmutzung, wenn es nicht in Gebrauch ist.

Begründung der Jury
Der Korrekturroller Mono ergo fällt durch seine abgerundete Formgebung auf. Dadurch liegt er gut in der Hand – die Grundlage für einen funktionsgerechten Gebrauch.

Convex Scissor
Konvex Schere

Manufacturer
PLUS Corporation,
Tokyo, Japan
Design
Human Code Japan,
Tokyo, Japan
Web
www.plus.co.jp/en
www.hcj.co.jp

The Convex Scissors promise a sharper cut than conventional scissors, setting a new standard in terms of performance. Since the scissor blades are curved according to Bernoulli's principle, the angle of the blades is kept at a constant optimal cutting angle of 30 degrees. Therefore, only a third of the usual force is required, and the scissors effortlessly cut even hard or thin materials such as cardboard or vinyl sheets.

Die Konvex Schere verspricht, schärfer als herkömmliche Scheren zu schreiden, und setzt damit einen neuen Leistungsstandard. Da die Scherenblätter nach dem Bernoulli-Prinzip geschwungen wurden, stehen die Klingen am Schnittpunkt stets optimal im Schnittwinkel von 30 Grad. Dadurch wird nur ein Drittel des üblichen Kraftaufwands benötigt und selbst besonders harte oder dünne Materialien wie Pappe oder Folien lassen sich mühelos schneiden.

Statement by the jury
The Convex Scissors stand out through their high practical value: the blades are both sharp and uniquely curved, which reduces the necessary force and effort.

Begründung der Jury
Die Konvex Schere zeichnet sich durch einen hohen Gebrauchswert aus: Gleichzeitig scharf und besonders geschwungen, verringern die Klingen den benötigten Kraftaufwand.

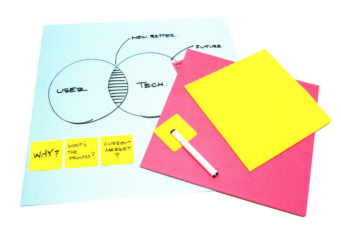

Post-it® Big Pad
Stationery Product
Büroprodukt

Manufacturer
3M Company,
St. Paul, USA
In-house design
Kristopher Clover
Web
www.3m.com/innovation

The new Post-it® Big Pad was designed as a tool for project teams to cluster and think through ideas in a flexible and efficient way. It changes the traditional use of flip charts by bringing the conversations to the table, where ideas can be sketched out and discussed in a dynamic face-to-face dialogue. Once these ideas come to fruition, the sheets can be mounted on the wall.

Statement by the jury
Big Pad responds to the requirements of today's work and office environments by presenting an inventive solution for the ideas and project outlines developed during teamwork.

Das neue Post-it® Big Pad wurde als Hilfsmittel für Projektteams entwickelt, um Ideen auf flexible und effiziente Weise zu sammeln und zu durchdenken. Es verändert die Verwendung von Flipcharts, indem es die Konversation an den Tisch holt, wo die Ideen skizziert und in einem dynamischen Face-to-Face-Dialog diskutiert werden können. Sobald es an die Verwirklichung dieser Ideen geht, können die Blätter an die Wand geheftet werden.

Begründung der Jury
Big Pad reagiert auf die Anforderungen der heutigen Arbeits- und Bürokultur und stellt eine originelle Lösung für im Team entwickelte Ideen oder Projektskizzen dar.

eN
Tape Dispenser
Klebefilmspender

Manufacturer
minimalife inc.,
Tokyo, Japan
Design
PLANE Co., Ltd.
(Hiroaki Watanabe),
Tokyo, Japan
Web
www.minimalife.jp
www.plane-id.co.jp
Honourable Mention

Unlike common tape dispensers, which hold a roll of tape with a reel, the eN has a rotating mechanism with two small built-in rollers. Volume and footprint are reduced to a minimum, lending this tape dispenser a clear, minimalist appearance. Another visually appealing feature is the integration of the tape roll, which is exposed by two thirds. With a weight of 900 grams, the eN proves to be a solid, effortless-to-use desk accessory.

Statement by the jury
This tape dispenser integrates basic geometrical forms in an austere and clear design, which simultaneously underlines the user-friendliness of the device.

Anders als übliche Klebefilmspender, bei denen das Klebeband von einer Achse gehalten wird, verfügt eN über einen Rotationsmechanismus mit zwei eingebauten kleinen Rädchen. Das Volumen und die Auflagefläche des Produkts wurden auf ein Minimum reduziert und verleihen diesem Abroller eine klare, minimalistische Anmutung. Optisch reizvoll ist zudem die Integration der Klebefilmrolle, die zu zwei Dritteln freiliegt. Der eN besitzt ein Gewicht von 900 Gramm und erweist sich so als stabiles, mühelos zu bedienendes Schreibtisch-Accessoire.

Begründung der Jury
Der Klebefilmspender integriert die geometrischen Grundformen in eine strenge, klare Gestaltung, die zugleich die Nutzerfreundlichkeit des Gerätes unterstreicht.

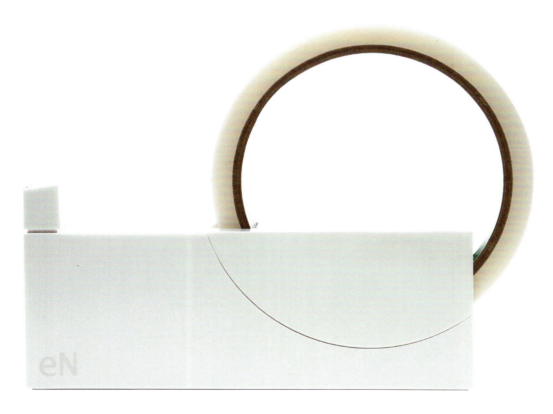

Post-it® Full Adhesive Roll

Stationery Product
Büroprodukt

Manufacturer
3M Company,
St. Paul, USA
In-house design
Kristopher Clover
Web
www.3m.com/innovation

Post-it® Full Adhesive Roll delivers the simplicity of a Post-it® Note in the form of a roll with a completely coated adhesive backside. The dispenser embodies a minimalist design that was inspired by the shape of a speech bubble. It lies smoothly in the hand and features a clean-cutting tear surface. The wide version utilises an innovative spindle, which enables the use of three roll widths in various configurations.

Die Post-it®-Kleberolle bietet die Einfachheit eines Post-its® in Form einer Rolle, deren Rückseite komplett mit Klebstoff beschichtet ist. Sie ist als minimalistischer Spender gestaltet, der von der Form einer Sprechblase inspiriert wurde. Dieser Spender liegt geschmeidig in der Hand und besitzt eine sauber schneidende Abreißfläche. Die breite Version nutzt eine innovative Spindel, die den Gebrauch von drei Rollenbreiten in verschiedenen Konfigurationen ermöglicht.

Magnetic Glass Board
Glas-Magnetboard

Manufacturer
Sigel GmbH,
Mertingen, Germany
In-house design
Werner Bögl
Web
www.sigel.de

The artverum magnetic glass board facilitates presentation and organisation at a high level. In terms of function, the reduced, appealing board is also an elegant alternative to classic whiteboards. Available in different sizes, these planning and organisation boards are magnetic, suitable for being written on by hand, and arranged either as single elements or a wall ensemble. The TÜV-approved safety mount without brackets emphasises the elegance of the frameless magnetic boards, which are perfectly suited to offices, meeting areas and conference rooms.

Das Glas-Magnetboard artverum erlaubt Präsentation und Organisation auf hohem Niveau. Das reduzierte, formschöne Board bietet auch funktional eine stilvolle Alternative zum klassischen Whiteboard. Denn die in verschiedenen Maßen erhältlichen Plan- und Organisationstafeln sind magnetisch und beschriftbar und lassen sich ebenso als Einzelelement wie auch als Wandensemble arrangieren. Die TÜV-geprüfte Sicherheitsaufhängung ohne Halter unterstützt die Eleganz der rahmenlosen Magnetboards, die sich ideal für Büros, Besprechungszimmer und Konferenzzimmer eignen.

Statement by the jury
The artverum magnetic glass board merges elegant aesthetics and versatile use to form an expressive object in the room.

Begründung der Jury
In dem Glas-Magnetboard artverum verschmelzen edle Ästhetik und variabler Gebrauch zu einem ausdrucksstarken Objekt im Raum.

artverum projection
Magnetic Glass Board
Glas-Magnetboard

Manufacturer
Sigel GmbH,
Mertingen, Germany
In-house design
Werner Bögl
Web
www.sigel.de

The magnetic glass board called artverum projection unites design and function. Its simple, linear style and white satin safety glass turn this board into an elegant eye-catcher. It can be used as a projection screen or a writing board and features magnetic qualities, thus providing an ideal basis for presentations of any kind. The meticulously manufactured glass surface on the front side, together with a special pigment coating on the back, ensures good quality and contrast for images projected by a beamer. Thanks to the TUV-approved safety mounting, which does not require holders or brackets, the frameless board appears to float on the wall.

Das Glas-Magnetboard artverum projection verbindet Design mit Funktion. Aufgrund seiner schlichten, geradlinigen Formgebung und dem weißen, seidenmatten Sicherheitsglas wird das Board zu einem edlen Blickfang. Es ist projektionsfähig, beschreibbar und magnetisch und bietet Präsentationen verschiedenster Art dadurch eine optimale Grundlage. Die aufwendig bearbeitete Glasoberfläche auf der Vorderseite und die Rückseitenbeschichtung aus speziellen Farbpigmenten ermöglichen eine Beamer-Projektion mit guter Bild- und Kontrastqualität. Dank halterloser, TÜV-geprüfter Sicherheitsaufhängung scheint das rahmenlose Magnetboard an der Wand zu schweben.

Statement by the jury
artverum projection impresses with a simple, high-quality design that is also functional. It is suitable for video presentations and can also be written on by hand.

Begründung der Jury
artverum projection besticht durch eine schlichte, hochwertige Gestaltung, die auch funktional punktet. Das Magnetboard eignet sich ebenso für Video-Präsentationen wie für die Beschriftung von Hand.

X Mark II
Calculator
Taschenrechner

Manufacturer
Canon Electronic Business
Machines (H.K.) Co., Ltd.,
Hong Kong

In-house design
Chak Yun Hei, Rex Hung Hoi

Web
www.canon-ebm.com.hk

reddot design award
best of the best 2013

Clearly defined
The first calculator came onto the market in the 1960s. Since then, no office is complete without these compact calculating aids. The X Mark II is a successful interpretation of the classic calculator. It is very contemporary and extremely clearly laid out. The conceptual aim was to combine environmental benefits and design in a new way. The minimalist casing of the calculator is therefore made exclusively of recycled materials. The X Mark II is pleasant to handle, very slim and its harmonious lines and curved surfaces are impressive. The keypad is user friendly as the individual keys are clearly defined and are pleasant to touch. The laser-engraved control keys are made to last as the symbols do not wear off with time. The calculator is entirely solar-powered and does not require any batteries, which makes it self-sufficient in use. The elegant design language of the X Mark II is an expression of the intense deliberation on how to link ecology and design. Its minimalist design gives calculators a new aesthetic appeal.

Klar definiert
Die ersten Taschenrechner kamen in den 1960er Jahren auf den Markt. Seitdem sind diese kompakten Rechenhilfen unverzichtbar für jedes Büro. Der X Mark II stellt eine gelungene Interpretation des Taschenrechners dar, denn er ist zeitgemäß und außerordentlich klar in seinem Ausdruck. Das Ziel seiner Gestaltung war es, Ökologie und Design auf neue Weise miteinander in Einklang zu bringen. Das minimalistische Gehäuse dieses Taschenrechners besteht deshalb vollständig aus recycelten Materialien. Der X Mark II liegt gut in der Hand, er ist sehr flach und beeindruckt durch seine harmonische Linienführung und subtilen Radien. Seine Tastatur ist nutzerfreundlich, da die einzelnen Tasten gut definiert sind und auch haptisch angenehm bedient werden können. Da die Symbole zudem per Laser in die Funktionstasten eingraviert werden, sind sie langlebig und nutzen sich im Gebrauch nicht ab. Dieser Taschenrechner bezieht seine Energie vollständig aus der Solartechnologie. Er arbeitet ohne Batterien und ist dadurch unabhängig einsetzbar. Die elegante Formensprache des X Mark II ist Ausdruck eines intensiven Nachdenkens über das Zusammenspiel von Ökologie und Design. Seine minimalistische Gestaltung verleiht dem Taschenrechner eine neue Ästhetik.

Statement by the jury
The calculator here undergoes a fascinating re-interpretation. The slim lines of the casing of the X Mark II, as well as the combination of materials turn it into a stylish accessory. All the elements have been harmoniously combined and the user interface is pleasant to touch. The ecological direction taken by this product, as well as its durability are groundbreaking.

Begründung der Jury
Der Taschenrechner erfährt hier eine faszinierend neue Interpretation. Die flache Linie des Gehäuses des X Mark II sowie die Kombination seiner Materialien definieren ihn als einen stilvollen Begleiter. Sehr harmonisch sind alle Elemente miteinander vereint, die Nutzeroberfläche ist haptisch angenehm. Die ökologische Ausrichtung dieses Produktes sowie seine Langlebigkeit sind wegweisend.

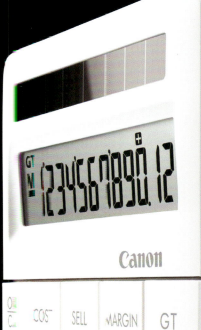

Power Slate
Portable Power
Tragbare Batterie

Manufacturer
Nexiom Company Limited,
Hong Kong
In-house design
Lei Zheng, Vincent Lau
Web
www.nexiom.cc

Power Slate is a mobile backup battery in letter paper size. At a height of only 5.2 mm, it is even thinner than standard USB connections. With ring binder holes on the side, the unit is convenient to store and carry in folders. Moreover, with a capacity of 13,500 mAh / 50 Wh, it guarantees extended battery operation for laptops, tablets, smartphones and MP3 players. A multitude of connectors and cord accessories ensure compatibility with any mobile device.

Statement by the jury
Power Slate represents an intelligent electronic tool, which – integrated in files and folders – allows a broad range of mobile devices to be charged at any given location.

Power Slate ist eine mobile Backup-Batterie im Letter-Format und mit nur 5,2 mm Höhe sogar dünner als ein Standard-USB-Anschluss. Mit Ringbuchlochung an der Seite lässt sie sich einfach und bequem in Ordnern mitführen und gewährleistet mit einer Kapazität von 13.500 mAh/50 Wh einen langen Betrieb von Laptops, Tablets, Smartphones oder MP3-Playern. Eine Vielzahl von Schnittstellen und Anschlusszubehör macht sie kompatibel mit allen mobilen Geräten.

Begründung der Jury
Power Slate stellt ein intelligentes Elektron ktool dar, mit dem sich – integriert in die Bürounterlagen – die verschiedensten Mobilgeräte stets und überall aufladen lassen.

Leitz Complete
Multicharger
Multi-Ladestation

Manufacturer
Esselte Leitz GmbH & Co KG,
Stuttgart, Germany
In-house design
Jason Mc Grath
Web
www.esselteleitz.de

Up to four mobile devices can be connected to the Leitz Complete multicharger at the same time. It provides a perfect set-up angle, which allows comfortable writing and reading on the device while it is charging. The multicharger features four USB connections, among others, and also non-slip rubber elements which ensure that the mobile devices are securely positioned and that the charger stands firmly on the table.

Statement by the jury
Besides its high functionality, this multicharger features a design inspired by the reduced, elegant look of the connected mobile devices.

An die Leitz Complete Multi-Ladestation können bis zu vier Mobilgeräte gleichzeitig angeschlossen werden. Sie bietet einen idealen Aufstellwinkel, sodass während des Ladevorgangs bequem an den Geräten geschrieben oder gelesen werden kann. Die Ladestation ist u. a. mit vier USB-Anschlüssen ausgestattet und verfügt über rutschfeste Gummielemente, die für den sicheren Halt der Mobilgeräte sowie einen stab len Stand auf dem Tisch sorgen.

Begründung der Jury
Neben ihrer hohen Funktionalität besitzt diese Multi-Ladestation eine Gestaltung, die sich an die reduzierte wie elegante Linie der angeschlossenen Mobilgeräte anlehnt.

File-it
Filer and USB-Flash
Hefter und USB-Stick

Manufacturer
xlyne GmbH,
Neuenrade, Germany
Design
Emamidesign
(Arman Emami),
Berlin, Germany
Web
www.xlyne.com
www.emamidesign.de

File-it is a latch for file folders with an integrated USB memory stick. It allows digital documents to be archived directly into a physical folder next to paperwork. By activating a black button, the integrated USB memory stick slides out to be connected to any USB port. Suitable for archiving documents, job application papers, business correspondence and much more, File-it stands out with an intuitive design.

Statement by the jury
File-it combines analogue and digital data archives in an innovative manner, and its functionality is underlined by a simple, friendly design.

File-it ist ein Hefter mit integriertem USB-Speicher. So bietet er die Möglichkeit, digitale Unterlagen zusammen mit Papierunterlagen direkt in dem jeweiligen Aktenordner zu archivieren. Mit der schwarzen Taste lässt sich der integrierte Speicherstick herausschieben, um mit einem USB-Anschluss verbunden zu werden. Zur Archivierung von Dokumenten, Bewerbungsunterlagen, Geschäftspapieren und vielem mehr geeignet, macht File-it durch eine intuitive Gestaltung auf sich aufmerksam.

Begründung der Jury
File-it verbindet die Sammlung analoger und digitaler Daten auf innovative Weise, unterstrichen durch eine schlichte, sympathische Gestaltung.

PLUG POT
Power Strip
with Cable Organiser
Kabelmanagement-Box

Manufacturer
ninebridge,
Anyang City, Gyeonggi Province,
South Korea
Design
we'd design
(Kim Jongwon, Cho Kyunghoon),
Seongnam City, Gyeonggi Province,
South Korea
Web
www.nine-bridge.co.kr
www.wed-design.com

Plug Pot is a wire box that unites the functions of a covered power strip and a utility box for organising and managing wires. Together with a separate socket for mobile devices, individual power buttons have been installed on the top of the box, allowing for the easy power shut-down of electronic devices. Inspired by the role model of a flower pot, the design fits into different interiors and has a likeable appeal.

Statement by the jury
The aesthetically designed Plug Pot wire box not only integrates many cables but also allows individual devices to be switched on and off.

Plug Pot ist eine Kabelbox, die die Funktionen einer Mehrfachsteckdose mit Deckel und einer Utility Box für das Kabelmanagement miteinander verbindet. Zusammen mit einem separaten Anschluss für Mobilgeräte wurden einzelne Schalter an der Oberseite angebracht, um gerade nicht benutzte Elektronikgeräte einfach abschalten zu können. Gestaltet nach dem Vorbild eines Blumenkastens, passt sich das Produkt verschiedenen Interieurs an und besitzt eine sympathische Ausstrahlung.

Begründung der Jury
Die ästhetisch gestaltete Kabelbox Plug Pot integriert nicht nur die vielen Kabel, sondern ermöglicht auch das An- und Abschalten einzelner Geräte.

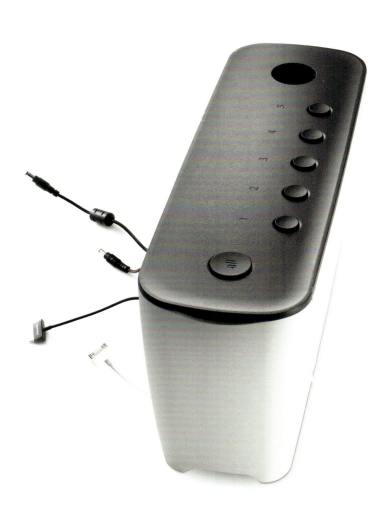

trystrams SPREAD/
Stretchable Cases
Small Cases
Etuis

Manufacturer
Kokuyo S&T Co., Ltd.,
Tokyo, Japan
In-house design
Kiyoshi Sakurai
Design
Design Studio S
(Fumie Shibata),
Tokyo, Japan
Web
www.kokuyo-st.co.jp
www.design-ss.com

The trystrams Spread cases are made of a lightweight, stretchable nylon with a solid core layer. The extremely flat Pen Board fits into any shirt pocket and is especially suitable for carrying pens, sticky notes and paper clips. The Pocketable Pouch comes with a strap for attaching it to clothes and thus is always at hand. Available in many vivid colours, the cases immediately catch the eye and present themselves as a friendly sheath for pens, mobile phones and much more. This makes the trystrams Spread attractive to people who are creative and frequently on the move, but also to many others as well.

trystrams Spread sind Etuis aus leichtem, elastischem Nylonmaterial mit einer stabilen Kernschicht. Das besonders flache Pen Board passt in jede Brusttasche und eignet sich besonders für Stifte, Klebezettel und Büroklammern. Das Pocketable Pouch ist mit einer Schlaufe versehen, mit der man es stets griffbereit direkt an der Kleidung befestigen kann. In vielen bunten Farben erhältlich, machen die Etuis sofort auf sich aufmerksam und dienen als sympathische Hülle für Stifte, Handy und Vieles mehr. Das macht trystrams Spread nicht nur für Menschen attraktiv, die kreativ tätig und viel unterwegs sind.

Statement by the jury
The colourful trystrams Spread cases prove to be stylish accessories that are especially practical and comfortable when on the move.

Begründung der Jury
Die farbenfrohen Etuis trystrams Spread präsentieren sich als stilvolle Accessoires, die sich gerade unterwegs als praktisch und komfortabel erweisen.

MONDO Loupe
Magnifier
Lupe

Manufacturer
Moto Design Co., Ltd.
Tokyo, Japan
Design
Miyake Design
(Kazushige Miyake),
Tokyo, Japan
Web
www.motodesign.co.jp
www.mondo-web.com

Loupe is a magnifying glass designed in the form of a rectangle that conforms to the shape and size of books and newspapers. With a highly simple and wide grip, which extends sideways at a 45-degree angle and thus allows for more comfortable reading, the magnifier is suitable for both right- and left-handed users. The unpretentious design promises steadiness and turns the product into an elegant, appealing daily companion.

Mit Loupe wurde ein Vergrößerungsglas in Form eines Rechtecks entworfen, um es damit an die Gestalt und die Größe von Büchern und Zeitungen anzupassen. Mit einem sehr schlichten, breiten Griff, der im 45-Grad-Winkel seitlich absteht und dadurch angenehmes Lesen ermöglicht, eignet sich die Lupe ebenso für Links- wie für Rechtshänder. Die unprätentiöse Gestaltung verspricht Stabilität und macht das Produkt zu einem formschönen Begleiter im Alltag.

Statement by the jury
Loupe convinces through its classic, reduced design, which distinguishes this magnifier as an everyday product of simple beauty.

Begründung der Jury
Loupe überzeugt durch ihr klassisches, reduziertes Design, das sie als ein Gebrauchsgut von schlichter Schönheit kennzeichnet.

Industry and crafts
Industrie und Handwerk

Tools, machines, industrial plant and equipment, components, robots, technology, measuring and testing equipment, cash dispenser, switches, timers, security technology
Werkzeuge, Maschinen, Anlagen, Komponenten, Roboter, Betriebstechnik, Mess- und Prüftechnik, Geldautomaten, Schalter, Zeitsysteme, Sicherheitstechnik

Black & Decker Gyro Driver™
Motion Sensing Screwdriver
Bewegungsgesteuerter Akku-Schrauber

Manufacturer
Black & Decker,
Shenzhen, China

In-house design
Sam Baskar

Web
www.blackanddecker.com

reddot design award
best of the best 2013

Perfect twist

Fastening screws requires skill and a steady hand
because it is not always easy to adjust the direction
and speed ideally for each task. Therefore, the
design of the Black & Decker Gyro Driver offers
an exciting concept of interaction: the battery-
powered screwdriver senses the movement of the
wrist. It thus allows users to easily control and adjust
the motor's direction and speed. Users simply have
to grip the trigger to activate and twist right for
forward and twist left for reverse. The motor speed
is controlled by the movement in the wrist, so that
users can vary the speed by simply rotating the wrist
between 0 to 30 degrees. This gyroscopic motion-
sensing coordination system makes fastening and
unfastening screws an easy task. An LED work light,
which provides additional lighting for the work space,
lets users work in dark areas or difficult to access
corners. This battery-powered screwdriver rests
comfortably in the hand, is well-balanced and has a
pleasant feel. Intuitive and self-explanatory in use,
it allows all screwing movements to be controlled.
The Black & Decker Gyro Driver motion sensing
screwdriver responds in a highly direct way to the
user's movements. It thus turns into a friendly and
also reliable assistant.

Der perfekte Dreh

Das Eindrehen von Schrauben erfordert Geschick,
denn es ist nicht immer leicht, die Drehrichtung und
die Geschwindigkeit der jeweiligen Situation anzupassen. Die Gestaltung des Black & Decker Gyro Driver
bietet hier ein spannendes Konzept der Interaktion:
Der Akku-Schrauber reagiert auf Handbewegungen.
Der Nutzer kann auf diese Weise die Drehrichtung und
die Geschwindigkeit bestimmen. Will er eine Schraube
einschrauben, dreht er die Hand nach rechts. Zum
Ausschrauben dreht er die Hand nach links. Die Geschwindigkeit wird durch die Stärke der Handdrehung
bestimmt, denn der Nutzer kann sie mit leichten Dre-
hungen zwischen 0 und 30 Grad variieren. Durch diese
bewegungsgesteuerte Koordination können Schrauben leicht ein- und ausgedreht werden. Die Arbeit in
dunklen Bereichen oder an schwer zugänglichen Ecken
ermöglicht dabei ein LED-Arbeitslicht, welches das Ar-
beitsfeld zusätzlich ausleuchtet. Dieser Akku-Schrauber
liegt gut ausbalanciert in der Hand und ist haptisch
angenehm. Intuitiv und selbsterklärend lassen sich mit
ihm alle Schraubbewegungen steuern. Der bewegungsgesteuerte Akku-Schrauber Black & Decker Gyro Driver
reagiert sehr direkt auf die Handlungen des Nutzers.
Auf diese Weise wird er zu einem freundlichen und auch
zuverlässigen Helfer.

Statement by the jury

This battery-powered motion sensing screwdriver
impresses in particular with its innovative right-
and left-twist feature. It possesses innovative
qualities that makes fastening and unfastening
screws fun. The Black & Decker Gyro Driver is
intuitive to use and rests well in the hand. It is
designed in an emotionalising form language,
which lends it a unique identity.

Begründung der Jury

Dieser bewegungsgesteuerte Akku-Schrauber beeindruckt insbesondere durch die innovative Funktion
des Rechts- und Linksdrehers. Er besitzt interaktive
Qualitäten, mit denen das Schrauben Spaß macht.
Der Black & Decker Gyro Driver kann intuitiv bedient
werden und liegt gut in der Hand. Gestaltet ist er
mit einer emotionalisierenden Formensprache, die
ihn unverwechselbar macht.

Rotary Hammer
Bohrhammer

Manufacturer
Robert Bosch Ltda,
Campinas, Brazil
In-house design
Julian Bergmann
Design
Farné Design e Comunicação Ltda
(Alfredo Farné),
Embu, Brazil
Web
www.skil.com.br/es
www.farnedesign.com

This handy, high-performance rotary hammer masters a wide variety of tasks when working with wood, bricks, concrete and metal. Thanks to the quick-change chuck, tips can be exchanged quickly and effortlessly. The integrated pneumatic impact drill function with high blow rate ensures quick and effective drilling, even in hard materials. The variable speed selection feature allows the drilling force to be precisely adjusted to the work at hand. Rotation can be set with one hand via a selector switch, thus ensuring easy screwing and unscrewing. The compact design and ergonomic grip contribute to smooth and safe handling.

Der handliche und leistungsstarke Bohrhammer bewältigt ein breites Spektrum an Aufgaben rund ums Holz, Bausteine, Beton und Metall. Dank des Schnellspannbohrfutters können die Bohrer schnell und mühelos gewechselt werden. Die zuschaltbare pneumatische Schlagbohrfunktion mit hoher Schlagzahl sorgt für schnelles und effektives Bohren auch in harten Materialien. Die Drehzahlvorwahl ermöglicht eine präzise Anpassung der Bohrkraft an die jeweilige Arbeit. Die Laufrichtung kann über einen Wahlschalter einhändig eingestellt werden, was einen leichten Wechsel zwischen Ein- und Ausdrehen von Schrauben gewährleistet. Das kompakte Design und der ergonomisch geformte Griff tragen zur einfachen und sicheren Bedienung bei.

Statement by the jury
The precision and performance of the rotary hammer is expressed in a sovereign design language with particularly powerful and dynamic features.

Begründung der Jury
Die Präzision und Leistungsfähigkeit des Bohrhammers wird souverän in eine Formensprache überführt, die besonders kraftvoll und dynamisch anmutet.

Impact Drill 16 mm
Schlagbohrmaschine

Manufacturer
Robert Bosch Ltda,
Campinas, Brazil
In-house design
Julian Bergmann
Design
Farné Design e Comunicação Ltda
(Alfredo Farné),
Embu, Brazil
Web
www.skil.com.br/es
www.farnedesign.com

This 750 watt impact drill with a 16 mm drill chuck embodies a very modern brand design that delivers functionality and robustness with a composition of straight and organic lines. The tool thus represents reliability and efficiency in the professional skilled-trade sector. Practical additional features, such as variable speeds and a rubber-coated grip, allow for extreme outdoor usage with both comfort and safety.

Statement by the jury
Distinctive colour and material accents give the impact drill an unmistakable look and facilitate intuitive handling as well.

Diese 750 Watt starke Schlagbohrmaschine mit 16-mm-Bohrfutter verkörpert ein sehr modernes Markendesign, das mit seiner Komposition aus geraden und organischen Linien sowohl Funktionalität als auch Robustheit zum Ausdruck bringt. Praktische Zusatzfunktionen wie die Geschwindigkeitsregulierung und der gummierte Griff erlauben ein sicheres und bequemes Arbeiten auch unter extremen Außenbedingungen.

Begründung der Jury
Markante Farb- und Materialakzente verleihen der Schlagbohrmaschine ein unverwechselbares Äußeres und unterstützen die intuitive Bedienung.

Impact Drill 10 mm
Schlagbohrmaschine

Manufacturer
Robert Bosch Ltda,
Campinas, Brazil
In-house design
Julian Bergmann
Design
Farné Design e Comunicação Ltda
(Alfredo Farné),
Embu, Brazil
Web
www.skil.com.br/es
www.farnedesign.com

The compact 10 mm impact drill is ideal for all around applications. The variable speed selection enables optimised control of the drill performance, while the switch lock facilitates continuous operation. The organically formed lines and the comfortable soft grip, in combination with the ergonomic grip size, provide pleasant haptics and a secure hold.

Statement by the jury
The stylish design of the impact drill is characterised by a high degree of functionality and efficiency. It is particularly light and handy.

Die kompakte 10-mm-Schlagbohrmaschine ist ideal für den Allzweckeinsatz geeignet. Die Drehzahlvorwahl sorgt für eine optimale Kontrolle der Bohrleistung, während die Schalterarretierung den Dauerbetrieb erleichtert. Die organisch geformten Linien und der bequeme Softgrip in Kombination mit der ergonomischen Griffgröße sorgen für eine angenehme Haptik und einen sicheren Halt.

Begründung der Jury
Die stilsicher gestaltete Schlagbohrmaschine zeichnet sich durch eine hohe Funktionalität und Effizienz aus. Sie ist besonders leicht und handlich.

TE 30-A36
Cordless Combihammer
Akku-Kombihammer

Manufacturer
Hilti Corporation,
Schaan, Liechtenstein
In-house Design
Design
Proform Design,
Winnenden, Germany
Web
www.hilti.com
www.proform-design.de

The TE 30-A36 is a battery-powered combihammer that combines cordless mobility with the full power of a comparable corded tool – while offering a battery capacity that lasts for a whole working day. Thanks to an additional chiselling function, the device is suitable for a wide range of areas. The patented automatic ATC power cut-out system (active torque control) ensures maximum safety in repetitive drilling applications on concrete, even when encountering difficult surfaces. The AVR technology (active vibration reduction) reduces vibrational forces. Furthermore, the tool convinces with an optimal power-to-weight ratio. Its centre of gravity is positioned between the two grips, ensuring perfectly balanced and easy handling.

Der Akku-Kombihammer TE 30-A36 verbindet die volle Leistung eines vergleichbaren Kabelgeräts mit der Autonomie eines Akku-Geräts. Dabei reicht die Akkukapazität für einen ganzen Arbeitstag. Durch die zusätzliche Meißelfunktion bietet das Gerät ein breites Anwendungsspektrum. Bei Serienbohrungen von Dübellöchern in Beton sorgt die patentierte automatische Schnellabschaltung ATC (Aktive-Drehmoment-Kontrolle) für maximale Sicherheit auch bei schwierigen Untergründen. Die AVR-Technologie (Aktive Vibrationsreduzierung) verringert die Vibrationsbelastung. Das Werkzeug überzeugt zudem durch ein optimales Gewicht-Leistungs-Verhältnis. Der Geräteschwerpunkt im Zentrum beider Griffe ermöglicht eine optimale Balance und Handhabung.

TE 2-A18 & TE 2-A22
Cordless Rotary Hammer Drill
Akku-Bohrhammer

Manufacturer
Hilti Corporation,
Schaan, Liechtenstein
In-house Design
Design
Matuschek Design & Management GmbH,
Aalen, Germany
Web
www.hilti.com
www.matuschekdesign.de

The TE 2-A18 & TE 2-A22 cordless rotary hammer drills were developed for light-duty drilling in concrete and masonry and are perfect tools for professional use in interior finishing and metalwork. The drills achieve the same performance as a corded tool. The built-in LED work light provides better visibility in poorly lit spaces. The sophisticated air-circulation concept for cooling results in an especially compact design that reduces the weight of the tool and also facilitates access in tight corners.

Die Akku-Bohrhammer TE 2-A18 & TE 2-A22 wurden für leichte Bohrungen in Beton und Mauerwerk entwickelt und stellen das ideale Arbeitsgerät für den professionellen Innenausbau und Metallarbeiten dar. Sie bieten dieselbe Leistung wie ein Netzgerät. Das integrierte LED-Arbeitslicht verhilft zu einer besseren Sicht in schlecht beleuchteten Räumen. Aufgrund des ausgeklügelten Kühlluft-Zirkulation-Konzepts konnte darüber hinaus in der Gestaltung eine sehr kompakte Bauform realisiert werden, die nicht nur das in der Hand liegende Gewicht reduziert, sondern auch der Eckenzugänglichkeit der Geräte zu Gute kommt.

DD 150-U & DD 160
Diamond Core Drilling Systems
Diamant-Kernbohrsysteme

Manufacturer
Hilti Corporation,
Schaan, Liechtenstein
In-house Design
Design
Busse Design + Engineering,
Elchingen, Germany
Web
www.hilti.com
www.busse-design.com

This diamond core drilling system, available in two versions, represents an optimal solution for a wide range of concrete and masonry application areas. It facilitates hand-guided or rig-based wet drilling on concrete and hand-guided dry drilling in masonry with equal ease. The 360-degree adjustable side handle with integrated dust and water management functionality guarantees enhanced working comfort. The LED performance indicator of the powerful three-speed transmission enables controlled drilling to be carried out. Moreover, it helps inexperienced users achieve the optimum rate of drilling progress and maximises the life cycle of the core bits.

Das Diamant-Kernbohrsystem ist in zwei Varianten erhältlich und stellt eine optimale Lösung für vielfältige Anwendungen in Beton und Mauerwerk dar. Hand- oder ständergeführte Nassbohrungen in Beton können mit derselben Leichtigkeit durchgeführt werden wie handgeführte Trockenbohrungen in Mauerwerk. Der um 360 Grad justierbare Seitengriff mit integriertem Staub- und Wassermanagement garantiert höchsten Arbeitskomfort. Die LED-Leistungsanzeige des kraftvollen 3-Gang-Getriebes ermöglicht kontrollierte Bohrleistungen und hilft auch ungeübten Anwendern, die optimale Bohrgeschwindigkeit und maximale Bohrkronen-Lebensdauer zu erreichen.

HDE 500-A18 & HDM 500

Dispensers
Auspressgeräte

Manufacturer
Hilti Corporation,
Schaan, Liechtenstein
In-house Design
Design
Proform Design,
Winnenden, Germany
Web
www.hilti.com
www.proform-design.de

The HDE and HDM dispensers allow for fast mortar injections without great physical effort at construction sites. They are used together with the pollutant-free Hilti injectable mortar and firestop foams. Both dispensers feature volume control for accurate hole filling and an automatic pressure-release mechanism that reduces mortar consumption and wastage. The ergonomic design ensures effortless dispensing even in deep anchor holes. Particular attention has also been placed on the safe operation of the devices. Self-explanatory colour coding virtually rules out the risk of confusing system components.

Mit den Auspressgeräten HDE und HDM gelingen Mörtelinjektionen bei Baustellenarbeiten schnell und ohne großen Kraftaufwand. Die Geräte werden zusammen mit dem schadstofffreien Hilti Injektionsmörtel oder Brandschutzschäumen verwendet. Sie verfügen über eine Mengenregulierung für die präzise Bohrlochfüllung sowie einen automatischen Entspannmechanismus für geringen Mörtelverbrauch. Die ergonomische Form erlaubt ein müheloses Auspressen auch bei großen Verankerungstiefen. Besonderes Augenmerk wurde auf eine sichere Anwendung der Auspressgeräte gelegt. So ist eine Verwechslungsgefahr der Systemkomponenten durch die eindeutige Farbcodierung praktisch ausgeschlossen.

Statement by the jury
The comfortable size of the grip and the sophisticated mechanics of the dispensers facilitate controlled mortar injection and also ensure fatigue-free working.

Begründung der Jury
Das angenehme Greifvolumen und die ausgeklügelte Mechanik der Auspressgeräte erlauben die kontrollierte Mörtelinjektion und ermüdungsfreies Arbeiten.

GARANT High-Performance Milling Systems
GARANT Hochleistungsfrässysteme

Manufacturer
Hoffmann GmbH
Qualitätswerkzeuge,
Munich, Germany
Design
Böhler CID
Corporate Industrial Design
(Katja Lautenbach,
Melchior von Wallenberg-Pachaly),
Fürth, Germany
Web
www.hoffmann-group.com
www.boehler-design.com

The Garant high-performance milling systems are comprised of quality tools that have been combined in a useful way. They are arranged in a clear manner in a perfectly formed foam inlay and safely stored in the system box. Due to the wide variety of milling tools, a flexible storage system is required for holding them safely and precisely. To address this need, the boxes come in three different base sizes and in two heights each. With its symmetrical design, each box segment can be used as a top or a bottom.

Die Garant Hochleistungsfrässysteme bestehen aus sinnvoll kombinierten Qualitätswerkzeugen, die in einer perfekt abgestimmten Schaumeinlage übersichtlich angeordnet und in der Systembox sicher verpackt sind. Durch die große Vielfalt an Fräsern wird ein flexibles Aufbewahrungssystem benötigt, welches diese sicher und präzise aufnimmt. Das gelingt durch drei verschieden große Grundflächen der Boxen und jeweils zwei unterschiedliche Schalenhöhen. Aufgrund ihrer Symmetrie kann jede Schale als Ober- und Unterteil verwendet werden.

Statement by the jury
The structure of the high-performance milling systems has been cleverly thought-out down to the smallest detail. It provides both visual order and safe storage.

Begründung der Jury
Der bis ins kleinste Detail durchdachte funktionale Aufbau der Hochleistungsfrässysteme sorgt für eine aufgeräumte Optik und sichere Aufbewahrung.

Exact 220E, 280E, 360E
Pipe Cutting System
Rohrschnittsystem

Manufacturer
Exact Tools Oy,
Helsinki, Finland
In-house design
Mika Priha
Web
www.exacttools.com

With an adjustable user interface, the Exact pipe cutting system solves the challenges of on-site pipe cutting while providing an exact cut with all pipe materials. The names of the different models – 220E, 280E and 360E – refer to the maximum cutting capacity, with the letter E standing for electronic control. All saws are portable and exceptionally safe to use since the cutting blade is fully protected. It is nevertheless easy to cut the pipe to the exact length using the cutting reference on the lower blade guard. All components in yellow indicate important functions to the user, while the other colours, grey and black, ensure that the tools do not appear worn even after extended periods of use.

Mit einer anpassbaren Benutzerschnittstelle löst das Rohrschnittsystem Exact die Probleme beim Rohrschnitt vor Ort und bietet zudem einen exakten Schnitt mit allen Rohrmaterialien. Die Bezeichnungen der Modelle – 220E, 280E und 360E – beziehen sich auf die maximale Schnittkapazität, und der Buchstabe E steht für elektronische Steuerung. Alle Rohrsägen sind tragbar und für den Benutzer außerordentlich sicher in der Anwendung, da das Sägeblatt vollständig geschützt ist. Dennoch ist es mithilfe der Schnittreferenz auf dem unteren Sägeblattschutz einfach, das Rohr auf die exakte Länge zuzuschneiden. Alle gelben Komponenten kennzeichnen wichtige Funktionen für den Benutzer, während die restliche Farbgebung in Grau und Schwarz dafür sorgt, dass die Werkzeuge auch nach längerer Verwendung nicht abgenutzt aussehen.

AccuPocket
Power Source
Schweißstromquelle

Manufacturer
Fronius International GmbH,
Wels-Thalheim, Austria
In-house design
Julia Huemer
Design
Designbüro Formquadrat
(Mario Zeppetzauer),
Linz, Austria
Web
www.fronius.com
www.formquadrat.com

The AccuPocket power source for welding applications is one of the first devices to utilise the enormous power of lithium-ion technology. The combination of welding and battery technology enables welding to be carried out completely independently of the mains and without any cables. The outer shape is also completely geared to meet the harsh requirements of everyday welding environments. An intelligent mix of metal and plastic combines mobility with robustness.

Statement by the jury
AccuPocket is an extraordinarily practical device. Its compactness is highlighted by the red edging on its front and back.

Die Schweißstromquelle AccuPocket setzt als eines der ersten Geräte auf die geballte Kraft der Lithium-Ionen-Technologie. Die Verbindung der Schweiß- und Akkutechnologie ermöglicht es dem Anwender, unabhängig vom Stromnetz völlig frei von Kabeln zu schweißen. Auch die äußere Form wurde komplett darauf ausgerichtet, den rauen Anforderungen im Schweißalltag gerecht zu werden. Die intelligente Kombination von Metall und Kunststoff verbindet Mobilität mit Robustheit.

Begründung der Jury
AccuPocket ist ein außerordentlich handliches Gerät, dessen Kompaktheit durch die rote Einfassung der Vorder- und Rückseite eindrucksvoll unterstrichen wird.

Leica Rugby 840
Construction Laser
Baulaser

Manufacturer
Leica Geosystems AG,
Heerbrugg, Switzerland
Design
BUDDE BURKANDT DESIGN
(Janine Budde, Marco Burkandt),
Munich, Germany
Web
www.leica-geosystems.com
www.buddeburkandt.de

The Leica Rugby 840 is a self-levelling construction laser for professional exterior and interior applications. Extremely rugged, shock-resistant and waterproof, it is designed to provide perfect measurement results with reliable precision even under the harshest conditions. The well-conceived, partially rubberised housing provides, next to shock-resistance, a safe grip in any position. All functions can be simply and directly controlled from the central interface at the front of the device. A mobile solar panel flexibly charges the built-in battery on the go.

Statement by the jury
The stable construction of the Leica Rugby 840 displays favourably compact dimensions. A wireless construction makes this laser suitable for a broad range of applications.

Der Leica Rugby 840 ist ein selbstnivellierender Baulaser für den professionellen Einsatz im Außen- und Innenbereich. Besonders robust, stoßfest und wasserdicht, ist er dafür konzipiert, dem Nutzer mit verlässlicher Präzision auch unter rauester Bedingungen perfekte Ergebnisse zu liefern. Das durchdachte, partiell gummierte Gehäuse gewährleistet neben der Stoßfestigkeit auch den sicheren Halt in allen Gebrauchspositionen. Die Funktionen werden einfach und direkt über das zentrale Bedienfeld an der Vorderfront gesteuert. Ein mobiles Solarpanel ermöglicht es, den eingebauten Akku flexibel unterwegs zu laden.

Begründung der Jury
Die stabile Bauweise des Leica Rugby 840 wird durch eine kompakte Größe abgerundet. Die kabellose Konstruktion macht den Baulaser vielseitig einsetzbar.

Weller Consumer Iron
Soldering Iron
Lötkolben

Manufacturer
Weller Tools GmbH,
Besigheim, Germany
In-house design
Gert Mittmann,
Uwe Loch
Design
prodesign GmbH
(Alfred Fordon),
Limburg/Lahn, Germany
Web
www.weller-tools.com

This soldering iron is characterised by its ergonomically shaped, two-component handle, which rests comfortably in the user's hand. The front end has a triangular shape for easier positioning of the soldering iron tip. The three integrated LEDs enable the soldering surface to be lit in an optimal way. At the same time, the integrated lighting frees up the user's hands for the soldering work.

Statement by the jury
The LED technology integrated into the soldering iron is cleverly thought-out. Red and black colour contrasts communicate self-confidence and safety.

Der Lötkolben zeichnet sich durch einen ergonomisch geformten 2-Komponenten-Griff aus, der gut in der Hand liegt. Im vorderen Bereich ist er dreieckig, dies erleichtert die optimale Ausrichtung der Lötspitze auf die zu lötende Stelle. Die drei integrierten LEDs ermöglichen ein optimales Ausleuchten der Lötstelle. So kann sich der Nutzer ganz auf die Arbeit konzentrieren, und es ist keine weitere Lichtquelle nötig.

Begründung der Jury
Die integrierte LED-Technologie im Lötkolben ist hervorragend durchdacht. Die rot-schwarzen Farbkontraste strahlen Selbstbewusstsein und Sicherheit aus.

IGNITER BM4/BR4
Ignition Blower
Zündgebläse

Manufacturer
Leister Technologies AG,
Kägiswil, Switzerland
In-house design
Andreas Fürling, Hans Arnold
Web
www.leister.com

These ignition blowers were specifically developed for installation in pellet and woodchip heating systems. They achieve output ranging from 600 watts to 3.4 kW. The interfaces have been selected in such a way that the ignition blowers can be effortlessly installed into any heating boiler. The integrated installation recesses on the side of the housing, the freely configurable coupler plug and the air hose connection adapter with internal thread enable optimal installation in all categories of heating boilers.

Statement by the jury
The functional construction and the precise manufacturing of the Igniter BM4/BR4 ignition blowers allow for very simple and fast assembly and disassembly.

Die Zündgebläse wurden speziell für den Einbau in Pellets- und Hackschnitzelheizungen entwickelt. Sie erreichen Leistungen von 600 Watt bis 3,4 kW. Die Schnittstellen wurden so gewählt, dass sich die Zündgebläse ohne Probleme in jeden Heizkessel einbauen lassen. Die integrierten Montageaufnahmen seitlich am Gehäuse, der frei konfigurierbare Gerätestecker und der Luftschlauch-Anschlussadapter mit Innengewinde sorgen für einen optimalen Einbau in Heizkessel aller Kategorien.

Begründung der Jury
Die funktionale Bauweise und präzise Verarbeitung der Zündgebläse Igniter BM4/BR4 ermöglichen einen überaus einfachen und schnellen Ein- und Ausbau.

HL Stick
Compact Hot Air Gun
Kompakt-Heißluftgebläse

Manufacturer
Steinel Vertrieb GmbH,
Herzebrock-Clarholz, Germany
Design
Eckstein Design
(Eckstein Design Crew),
Munich, Germany
Web
www.steinel.de
www.eckstein-design.com

Thanks to its small size and low weight, the HL Stick compact hot air gun enables the completion of complex jobs with the highest precision. The safe base moreover allows for hands-free heating. The compact design makes it possible for the HL Stick to easily reach even difficult-to-access places, such as the inside of models and other cavities. An integrated LED work lamp provides support for highly precise heating in poorly lit areas.

Statement by the jury
The different colours and surface characteristics of its housing components give the HL stick a dynamic, high-performance appearance.

Mit dem Kompakt-Heißluftgebläse HL Stick lassen sich dank der handlichen Größe und des geringen Gewichts komplexeste Arbeiten mit höchster Präzision ausführen. Die sichere Standfläche erlaubt außerdem freihändiges Erwärmen. Durch die kompakte Bauweise kommt der HL Stick überall hin, selbst an schwer zugängliche Stellen, wie z. B. das Innere von Modellen und andere Hohlräume. An schlecht beleuchteten Stellen bietet das integrierte LED-Arbeitslicht Hilfestellung für hochpräzises Erwärmen.

Begründung der Jury
Die unterschiedlichen Farb- und Oberflächeneigenschaften der Gehäusekomponenten verleihen dem HL Stick ein dynamisch-leistungsstarkes Aussehen.

Hilti TE-CD / TE-YD
Hollow Drill Bits
Hohlbohrer

Manufacturer
Hilti Corporation,
Schaan, Liechtenstein

In-house design
Hilti Corporation

Design
Proform Design,
Winnenden, Germany

Web
www.hilti.com
www.proform-design.de

reddot design award
best of the best 2013

A clean solution
Hollow drilling, especially in concrete, produces
a lot of fine dust. The work environment gets dirty
and the drill hole needs to be cleaned afterwards,
so that dowels will be held firmly in place. The
Hilti TE-CD / TE-YD drill system allows dust-free
conditions in this field of application. It is designed
to achieve the greatest possible efficiency and
effectiveness, especially in repetitive drilling
applications. This innovative concept combines a
hollow drill with a vacuum cleaner in one functional
and user-friendly tool: drilling dust is extracted
by the vacuum system while drilling is still in
progress. The result of this combination, besides a
virtually dust-free environment, is a faster drilling
and work process. For the user, this means healthier
work, without dust exposure. Anchors can be
inserted into the hole immediately after the drilling
and will hold reliably and firmly in place. The Hilti
TE-CD / TE-YD drill system is ergonomically well
thought out and ensures fatigue-free working.
Its clear design vocabulary with the brand-specific
colouring lends it an elegant appeal, while its
operation is almost self-explanatory for users.
Through design, work conditions in the construction
sector have been considerably improved.

Saubere Lösung
Bei dem Vorgang des Hohlbohrens, insbesondere in
Beton, fällt viel Feinstaub an. Dieser verschmutzt die
Umgebung und das Bohrloch muss nachträglich gesäu-
bert werden, damit die Dübel später gut halten können.
Das Bohrsystem Hilti TE-CD / TE-YD ermöglicht in
diesem Bereich ein staubfreies Arbeiten. Es ist ausgelegt
für die größtmögliche Effizienz und Effektivität speziell
bei Serienanwendungen. Dieses innovative Konzept
verbindet einen Hohlbohrer mit einem Staubsauger in
einer funktionalen und anwenderfreundlichen Einheit:
Das Bohrmehl wird bereits während der Bohrphase in
das Staubabsaugsystem gesaugt. Das Resultat dieser
Verbindung ist, neben der völligen Staubfreiheit, ein
schnelleres Bohren und Arbeiten. Für den Bohrenden
bedeutet es ein gesünderes Arbeiten ohne Staubbelas-
tung. In die sauberen Bohrlöcher können anschließend
sofort Anker eingesetzt werden, die zuverlässig und
sicher halten. Das Bohrsystem Hilti TE-CD / TE-YD ist
ergonomisch durchdacht und erlaubt ein ermüdungsfreies Arbeiten. Seine klare Formensprache mit der
markentypischen Farbgebung verleiht ihm Eleganz und
seine Funktionsweise erklärt sich dem Anwender gut
selbst. Durch Design werden hier Arbeitsbedingungen
im Bausektor erheblich verbessert.

Statement by the jury
The absolutely convincing feature of the Hilti
TE-CD / TE-YD hollow drill bits is the possibility of
directly extracting drilling dust. At the same time,
the function of the drill head was optimised to
prevent it from blocking inside the drill hole. This
product is functional as well as ergonomically
sophisticated, so that it makes drilling easier. It can
be used together with any existing system.

Begründung der Jury
Absolut überzeugend ist bei den Hohlbohrern Hilti
TE-CD / TE-YD die Möglichkeit der direkten Absaugung
des Baustaubs. Gleichzeitig wurde die Funktion des
Bohrkopfes so optimiert, dass eine Blockierung innerhalb des Bohrlochs vermieden wird. Dieses Produkt ist
funktional wie ergonomisch gut durchdacht, sodass es
das Bohren vereinfacht. Jeder Anwender kann es mit
seinem vorhandenen System nutzen.

Wiha BitBuddy
Bit-Box

Manufacturer
Wiha Werkzeuge GmbH,
Schonach, Germany
In-house design
Gerd Heizmann
Web
www.wiha.com

The BitBuddy impresses with its simple and effective handling of frequently changing screwing applications. The bits are removed directly from the box by being inserted into the bit holder and are replaced in the same way. In addition, the bit is automatically returned to the correct position in the box when bits are changed. This approach serves to ensure order so that fewer parts are lost. The box is made from fibreglass-reinforced polyamide and is complemented with a practical belt clip.

Statement by the jury
Besides its excellent functionality, this bit box impresses with its precise finish and hard-wearing materials.

Der BitBuddy besticht durch ein einfaches und effektives Handling für häufig wechselnde Schraubanwendungen. Die Bits werden direkt aus der Box in den Bithalter entnommen und wieder zurückgelegt. Zudem wird der Bit beim Wechsel automatisch wieder an die richtige Stelle in der Box abgelegt. Dadurch sorgt das Aufbewahrungssystem für mehr Ordnung und weniger verlorene Teile. Die Box besteht aus schlagzähem, glasfaserverstärktem Polyamid und wird durch einen praktischen Gürtelclip ergänzt.

Begründung der Jury
Die Bit-Box überzeugt nicht nur durch ihre ausgezeichnete Funktionalität sondern auch durch ihre besonders präzise Verarbeitung und die unempfindlichen Materialien.

Enduro Trijet
Hammer Drill Bit
Hammerbohrer

Manufacturer
ITW Heller GmbH,
Dinklage, Germany
In-house design
Rainer Lampe
Web
www.hellertools.com

The Enduro Trijet hammer drill bit follows an innovative design concept, which makes it an efficient problem solver for any kind of manual work. Thanks to its three-winged carbide tip, the hammer drill bit achieves sound performance and results. Its concrete drilling life is very high, as is the bit's durability should reinforcement bars be inadvertently hit.

Statement by the jury
The sophisticated coil shape of this hammer drill bit enables fast and straight drilling, as well as satisfactory removal of debris from the drilling hole.

Der Hammerbohrer Enduro Trijet folgt einem innovativen Gestaltungskonzept, das ihn zum effizienten Problemlöser für jegliche handwerkliche Arbeit macht. Denn durch seine drei geschwungenen Hartmetallschneiden erzielt das Werkzeug außergewöhnliche Leistungen und überzeugende Ergebnisse bei der Anwendung. Die Lebensdauer bei Betonbohrungen ist extrem hoch ebenso wie die Stabilität bei Armierungstreffern.

Begründung der Jury
Die ausgefeilte Wendelform des Hammerbohrers ermöglicht ein sehr schnelles und gerades Bohren sowie einen guten Bohrmehltransport.

Wiha iTorque
Torque Screwdriver
Drehmomentschraubendreher

Manufacturer
Wiha Werkzeuge GmbH,
Schonach, Germany
In-house design
Matthias Schmidt
Design
Henssler und Schultheiss
Fullservice Productdesign GmbH
(Martin Schultheiss, Heinrich Henssler),
Schwäbisch Gmünd, Germany
Web
www.wiha.com
www.henssler-schultheiss.de

Wiha iTorque is one of the first intelligent mechatronic torque screwdrivers in the world. The integrated counter records each application and displays its actual number. In conjunction with the alarm function, which can be specially adjusted, the monitoring of testing equipment is particularly easy. A visually appealing, clear digital display helps to avoid reading errors. Moreover, users can take advantage of the permanent laser marking of their own company logo.

Statement by the jury
The mechatronic technology integrated into the Wiha iTorque greatly supports process reliability. The laser marking feature allows for individualisation of the tool.

Wiha iTorque ist einer der ersten intelligenten mechatronischen Drehmomentschraubendreher weltweit. Der integrierte Zähler registriert jede Anwendung und zeigt deren tatsächliche Anzahl an. Zusammen mit der Alarmfunktion, die speziell festgelegt werden kann, wird so die Prüfmittelüberwachung besonders leicht. Das digitale Display besticht durch seine klare Anzeige, durch die Ablesefehler vermieden werden. Für den Anwender ist eine individuelle Laserbeschriftung z. B. mit dem Firmenlogo möglich.

Begründung der Jury
Die im Wiha iTorque integrierte Mechatronik ist von hohem Nutzen für die Prozesssicherheit. Die Laserbeschriftung ermöglicht eine individuelle Kennzeichnung.

stripper-screwdriver
Screwdriver
Schraubendreher

Manufacturer
Yih Cheng Factory Co., Ltd.,
Nantou, Taiwan
In-house design
Jack Lin
Web
www.lancertools.com.tw
Honourable Mention

This tool is a combination of screwdriver, wire bender and wire stripper in one. With its 3-in-1 principle, overhead work is simplified considerably. To remove insulation, the wire is inserted into the handle and fixed by the stripper button at its end. The measurement scale indicates the length of wire to be cut off. By rotating and simultaneously pressing the button, the insulation is cleanly removed.

Statement by the jury
The design of this screwdriver is very practice-oriented and functional, qualities that are supported by the powerful colour scheme.

Dieses Werkzeug ist Schraubendreher, Drahtbieger und Drahtabisolierer in einem. Sein 3-in-1-Prinzip bedeutet eine große Erleichterung bei Überkopfarbeiten. Zum Abisolieren wird der Draht in den Griff geschoben und durch den Knopf am Griffende fixiert. Die Messskala zeigt an, wie viel von der Länge des Drahtes abgeschnitten wird. Durch Drehen und gleichzeitiges Drücken des Griffes wird das Isolierteil anschließend sauber entfernt.

Begründung der Jury
Die Aufmachung des Schraubendrehers ist sehr praxisorientiert und funktional, was auch die kräftigen Farben widerspiegeln.

NANOVIB
Vibration-Damping Hammer
Vibrationsdämpfender Hammer

Manufacturer
Fiskars France,
Arvillard, France
In-house design
Emmanuel Rado,
Chris Thelisson
Web
www.leborgne.fr

With its patented technology, Nanovib reduces the vibrations that occur when using a hammer and that often lead to injury of the musculoskeletal system. Among the advantageous product features is a magnet to hold the nail for single-handed hammer use. Furthermore, the lateral striking surfaces allow nailing in confined spaces. The sharp upper edges are designed for scraping off concrete residues, while the handle's end, made of hard polymer, makes it possible to work on materials without damaging them.

Statement by the jury
Nanovib's different materials, which improve its grip, are distinctively reflected in the dynamic lines and colour contrasts.

Durch seine patentierte Technologie verringert Nanovib die Vibrationen, die beim Gebrauch eines Hammers häufig zu Problemen des Bewegungsapparates führen. Ein weiterer Vorteil ist ein Magnet, der den Nagel hält, sodass der Nutzer die Hände frei hat und gezielter arbeiten kann. Außerdem ermöglichen es die seitlichen Schlagflächen, Nägel auf begrenztem Raum einzuschlagen. Mit den oberen scharfen Kanten lassen sich Betonreste abkratzen, und mit dem Griffende aus hartem Polymer können Materialien bearbeitet werden, ohne sie zu beschädigen.

Begründung der Jury
Die Materialunterschiede von Nanovib sorgen für eine verbesserte Griffigkeit und spiegeln sich markant in den spannungsvollen Linien und Farbkontrasten wider.

GARANT Electronic Torque Wrench
GARANT Elektronischer Drehmomentschlüssel

Manufacturer
Hoffmann GmbH Qualitätswerkzeuge,
Munich, Germany
Design
Böhler CID Corporate Industrial Design
(Thomas Breun),
Fürth, Germany
Web
www.hoffmann-group.com
www.boehler-design.com

The Garant electronic torque wrench ensures optimum functional reliability in the fields of aviation, medical technology and mechanical engineering. The control wheel enables intuitive single-handed operation and setting of the required fastening torque, as well as easy scrolling through the menu. The traffic light function of the LED display gives the user precise feedback on the status of the desired torque. Data can be read and analysed on a PC using the USB port.

Statement by the jury
This electronic torque wrench features cutting-edge technology, which is reflected by its stylistically consistent design.

Der elektronische Drehmomentschlüssel von Garant gewährleistet optimale Funktionssicherheit bei Verschraubungen in den Bereichen Luftfahrt, Medizintechnik oder Maschinenbau. Das Encoderrad ermöglicht eine intuitive Einhandbedienung, das Einstellen der geforderten Anzugmomente sowie ein einfaches Scrollen durch das Menü. Die LED-Ampelanzeige gibt visuelle Rückmeldung darüber, ob der gewünschte Drehmoment erreicht ist. Über die USB-Schnittstelle lassen sich Daten am PC einlesen und auswerten.

Begründung der Jury
Der elektronische Drehmomentschlüssel ist technisch auf dem neuesten Stand, was sich auch in seiner stilsicheren Gestaltung zeigt.

SMART wrench
Ratchet Wrench
Schraubenschlüssel

Manufacturer
Ocean Allied Co., Ltd.,
Taichung, Taiwan
In-house design
Jacob Lin, Allen Lin
Web
www.oaprecision.com
Honourable Mention

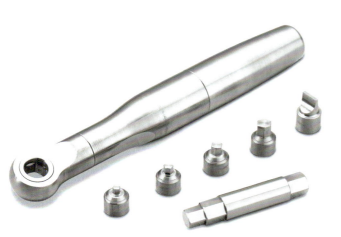

The Smart wrench is made from a special titanium alloy, which provides an excellent combination of corrosion resistance and strength. Furthermore, the material is particularly lightweight and biocompatible. The tool features ergonomic grooves, making it pleasant and easy to use. The housing can be opened and serves as a storage space for various hex socket bits.

Statement by the jury
The intelligent solution of storing the hex socket bits inside this extremely high-quality wrench successfully addresses the concern of losing individual parts.

Der Schraubenschlüssel Smart wrench besteht aus einer speziellen Titanlegierung, die eine ausgezeichnete Kombination aus Korrosionsbeständigkeit und Widerstandskraft aufweist. Zudem ist das Material ausgesprochen leicht und biokompatibel. Das Werkzeug verfügt über ergonomische Rillen, die die Handhabung angenehm leicht machen. Das Gehäuse kann außerdem geöffnet werden und dient als Aufbewahrungsort für verschiedene Inbuseinsätze.

Begründung der Jury
Durch die intelligente Aufbewahrung der Inbuseinsätze im Inneren des überaus hochwertigen Schraubenschlüssels geht kein Einzelteil mehr verloren.

Manufacturer
Atlas Copco Tools AB,
Nacka, Sweden
In-house design
Ola Stray
Web
www.atlascopco.com

The patented ETT-STR nut runner is an offset screwdriver and pistol tool in one. With the two triggers, users can first pre-assemble components with the upper pistol grip and then fully mount them with the lower handle. Both handles have been developed to ergonomically support different hand positions. A careful choice of materials lends the nut runner the character of a precise measuring tool, in combination with a robust industrial appearance. Several gyro sensors and torque transducers improve precision and ergonomics. Functional operating elements are emphasised by a yellow hue and accentuated with a stainless-steel finish. Rounded lines and recesses, as well as finely sectioned areas, lend the tool strong expression and underscore its performance and precision.

Der patentierte Tensorschrauber ETT-STR ist Winkelschrauber und Pistolenwerkzeug in einem. Durch die beiden Starttaster können Bauteile zunächst mit dem oberen Pistolenhandgriff vormontiert und anschließend mit dem unteren Handgriff fertig montiert werden. Die beiden Handgriffe wurden dazu entwickelt, unterschiedliche Handgriffpositionen ergonomisch zu unterstützen. Die sorgfältige Materialauswahl verleiht dem Schrauber den Charakter eines präzisen Messinstruments, kombiniert mit einem robusten industriellen Erscheinungsbild. Mehrere Gyroskope und Messsensoren verbessern die Genauigkeit und die Ergonomie. Funktionelle Bedienelemente wurden gelb betont oder mit einem Edelstahl-Finish akzentuiert. Abgerundete Linien und Vertiefungen sowie fein abgegrenzte Bereiche verleihen dem Werkzeug einen starken Ausdruck und unterstreichen seine Leistung und Präzision.

Statement by the jury
The ETT-STR nut runner is a particularly versatile tool with a design that stands out through functional simplicity and strong ergonomics.

Begründung der Jury
Der Tensorschrauber ETT-STR ist ein besonders wandlungsfähiges Werkzeug, das gestalterisch durch funktionelle Simplizität und hohe Ergonomie besticht.

D Handle
D-Griff
Handsaw
Handsäge

Manufacturer
Albrecht Drees GmbH & Co. KG,
Remscheid, Germany
In-house design
Udo Zirden
Design
evolution industrialdesign,
block seibel gbr (Martin Block),
Remscheid, Germany
Web
www.aldre.de
www.evolution-industrialdesign.com

The D Handle handsaw offers the possibility of inserting different saw blades with the same shank into one handle. Thanks to the specially developed spring lock system inside the handle, blades can be changed in a matter of seconds by pressing a button. The hard component is made from impact-resistant ABS plastic, giving the handle high rigidity and protecting fingers from injury. The soft red plastic serves as a comfortable, non-slip grip zone.

Statement by the jury
This hand saw's design meets high ergonomic and safety standards. Also, the distinctive D-shape offers strong recognition value.

Die Handsäge D-Griff bietet die Möglichkeit, unterschiedlich gezahnte Sägeblätter mit gleicher Schaftaufnahme in einen Griff einzusetzen. Aufgrund eines speziell entwickelten Federschließsystems im Griffinneren kann ein Sägeblatt per Knopfdruck innerhalb von Sekunden gewechselt werden. Die Hartkomponente aus steifestem ABS-Kunststoff verleiht der Säge eine hohe Steifigkeit und schützt die Finger vor Verletzungen. Der rote Weichkunststoffbereich bietet einen komfortablen und rutschsicheren Griff.

Begründung der Jury
Bei der Handsäge wurden ein hohes Anspruchsdenken an die Ergonomie und Sicherheit realisiert. Die D-Form hat einen besonderen Wiedererkennungswert.

X-PRO EVO
Cable Scissors
Kabelschere

Manufacturer
BM S.p.A.,
Rozzano, Italy
In-house design
R&D Department
Web
www.bm-group.com
Honourable Mention

The X-Pro Evo cable scissors for electricians are engineered to keep manual effort to a minimum during use. This is achieved via two intuitive mechanisms: the security lock and the central spring. The lower handle is shaped as an open ring and ensures a comfortable and secure grip. The handle shaft is covered with a two-component plastic, featuring a soft-touch material on one side and a non-slip surface on the other.

Statement by the jury
These cable scissors facilitate comfortable and easy cutting. The ergonomics are underscored visually by a dynamic colour contrast.

Die Kabelschere X-Pro Evo ist darauf ausgelegt, der Kraftaufwand beim Benutzen der Schere möglichst gering zu halten. Hierzu tragen zwei Mechanismen bei: der Sicherheitsverschluss und die Mittelfeder. Der untere Griff ist wie ein offener Ring geformt und bietet hohen Haltekomfort. Der gesamte Griff besteht aus einem 2-Komponenten-Kunststoff, auf der einen Seite aus einem Softtouch-Material und auf der anderen Seite aus einem Antirutschbelag.

Begründung der Jury
Die Kabelschere ermöglicht ein angenehmes, weiches Schneiden. Die Ergonomie wird durch den schwungvollen Farbkontrast visuell unterstrichen.

CC Long Tapes
Tape Measure Series
Bandmaß-Serie

Manufacturer
Hultafors Group AB,
Bollebygd, Sweden
Design
Veryday
(Pelle Reinius, Peter Ejvinsson,
Hans Himbert),
Bromma, Sweden
Web
www.hultafors.com
www.veryday.com

The tape measure series CC Long Tapes comprises eight models in four different casings. The units provide optimal handling regardless of hand size and working position. The handle, which is made of polymer and rubber, can be conveniently gripped with either the right or left hand. The case is held in one hand while the measuring tape is unravelled to conduct measurements or spooled back in using a specially shaped crank. The design also features a hole-hanger for safe work at heights.

Statement by the jury
The ergonomics of this tape measure series are very well thought out. The bold colour design imparts professionalism and high value.

Die Bandmaß-Serie CC Long Tapes umfasst acht Modelle in vier verschiedenen Gehäuseausführungen. Sie bieten eine optimale Nutzung unabhängig von Handgröße und Arbeitsposition. Der Griff aus Polymer und Gummi lässt sich mit der rechten wie mit der linken Hand gut greifen. So kann das Gehäuse festgehalten und das Band herausgezogen werden, um die Messung vorzunehmen, oder das Band wird mithilfe der speziell geformten Kurbel aufgewickelt. Zudem verfügt es über einen Aufhänger für sicheres Arbeiten in der Höhe.

Begründung der Jury
Die Ergonomie dieser Bandmaß-Serie ist vortrefflich durchdacht. Die selbstbewusste Farbgestaltung drückt Professionalität und Wertigkeit aus.

Air Hose Reel
Luftschlauchtrommel

Manufacturer
Yongkang Haili Industrial Co., Ltd.,
Yongkang, China
In-house design
Zheng Shi
Web
www.hailichina.com

This air hose reel is available in lengths ranging from 5 to 20 metres and in a variety of configurations. In the version with automatic position locking, the hose can be pulled out to the length needed and locked in that position. Pulling on the hose again automatically retracts it into the reel. An alternative model with ratchet lock is also available. A metal bracket makes it easy to install the reel on a wall, ceiling or floor.

Statement by the jury
The air hose reel embodies functionality and dynamics. The colour and material contrasts in particular give rise to this energetic appearance.

Die Luftschlauchtrommel ist in Längen von 5 bis 20 Metern sowie in verschiedenen Ausführungen erhältlich. Bei der Variante mit automatischer Positionsverriegelung kann der Schlauch bis zur gewünschten Länge ausgezogen und festgestellt werden. Zieht man ein weiteres Mal am Schlauch, wird er automatisch aufgerollt. Alternativ steht ein Modell mit einer Einrast-Arretierung mittels Zahnsperre zur Verfügung. Mithilfe des Metallhalters lässt sich die Trommel an der Wand befestigen.

Begründung der Jury
Die Luftschlauchtrommel verkörpert Funktionalität und Dynamik. Insbesondere die Farb- und Materialkontraste machen den Anblick so schwungvoll.

clinggo spot
Fastening Hold
Befestigungsaufnahme

Manufacturer
Meese GmbH,
Bergisch Gladbach, Germany
In-house design
Felix Meese,
Ludwig Meese
Web
www.meese-ideas.de

The fastening hold clinggo spot is integrated into tunnel lining segments shortly prior to casting. The formwork plugs are made of tempered steel and are virtually wear-free. First they are screwed into the tunnel lining segments. Then the clinggo spot is manually attached to the formwork elements and the concrete is poured. Due to the signal colour of the membrane remaining visible, the fastening holds are readily discernible, even in the artificial light of the construction site.

Statement by the jury
The two individual components of the clinggo spot are precisely matched. The striking colouring contrasts well with the concrete.

Die Befestigungsaufnahme clinggo spot wird bei Tunnelausbauten in die Schalung der Ausbauelemente integriert. Die nahezu verschleißfreien Schalungsknöpfe aus vergütetem Stahl werden zunächst in die Tunnelverschalungen geschraubt. Dann wird der clinggo spot manuell auf die Schalungsknöpfe aufgesteckt und der Betonguss aufgebracht. Durch ihre Signalfarbe bleiben die Befestigungsaufnahmen auch im Kunstlicht der Baustelle weithin gut zu erkennen.

Begründung der Jury
Die beiden Einzelkomponenten des clinggo spot sind präzise aufeinander abgestimmt. Die auffällige Farbgebung hebt sich hervorragend vom Beton ab.

Professional Extreme
Welding Table
Schweißtisch

Manufacturer
Bernd Siegmund GmbH,
Großaitingen, Germany
In-house design
Evelyn Bergmann,
Stephanie Laritz
Design
Selic Industriedesign
(Mario Selic),
Augsburg, Germany
Web
www.siegmund-group.com

The Professional Extreme is designed for optimum flexibility and precision in accurately positioning and fixing construction components down to the millimetre. It is used as a welding, clamping and measuring table in metalworking shops and industrial firms. Made from pre-hardened steel in combination with a plasma-nitrided surface finish, the table is highly resistant against scratching, rust and weld-spatter adhesion. Thanks to round edges, the fixing bolts easily glide into the bore holes without tilting. This moreover prevents the edges from being damaged when coming into tough contact with tools, while also providing a comfortable feel. Furthermore, the welding table offers significantly more clamping possibilities due to the double bore hole lines running along the edges.

Der Professional Extreme ist auf optimale Flexibilität und Präzision beim millimetergenauen Positionieren und Fixieren von Bauteilen ausgelegt und wird als Schweiß-, Spann- oder Messtisch in Schlossereien und Industriebetrieben eingesetzt. Er besteht aus einem vorgehärteten Stahl in Verbindung mit einer durch Plasmanitrieren veredelten Oberfläche. Dadurch ist der Tisch extrem widerstandsfähig gegen Kratzer, Rost und Schweißspritzer. Dank der abgerundeten Kanten rutschen Fixierbolzen problemlos in die Bohrungen, ohne zu verkanten. Sie verhindern zudem ein Verschlagen der Kanten beim rustikalen Ansetzen von Werkzeugen und sorgen für ein angenehmes haptisches Gefühl. Darüber hinaus bietet der Schweißtisch durch eine umlaufende Doppellochreihe an den Seitenwangen erheblich mehr Spannmöglichkeiten.

GUIDE 10, 12, 14 & 16
Working Gloves
Schutzhandschuhe

Manufacturer
Skydda Protecting People Europe AB,
Ulricehamn, Sweden
In-house design
Åsa Lindberg-Svensson
Web
www.guide.eu

The working gloves of the Guide collection are softer, thinner and more durable than classic leather glove models. In the production process, the manufacturer resorted to the experience and know-how of a South Korean company which, since the 1990s, has produced basketballs from the material Serino for the world market. The gloves are made of a soft version of the Serino material and thus ensure a secure grip. Bright colour accents and contours lend dynamism to the gloves.

Die Schutzhandschuhe aus der Guide-Kollektion sind weicher, dünner und strapazierfähiger als klassische Ledermodelle. Bei ihrer Herstellung wird auf die langjährige Erfahrung und das Know-how eines südkoreanischen Werks zurückgegriffen, das seit den 1990ern aus dem Material Serino Basketbälle für den Weltmarkt fertigt. Die Handschuhe bestehen aus einer weichen Variante von Serino und sorgen damit für einen sicheren Griff. Durch die bunten Farbakzente und -konturen wirken sie freundlich und dynamisch.

UVEX One
Safety Shoes
Sicherheitsschuhe

Manufacturer
UVEX Arbeitsschutz GmbH,
Fürth, Germany
In-house design
UVEX Safety Group Footwear
Design
scherfdesign,
Cologne, Germany
Web
www.uvex-safety.com
www.scherfdesign.com

Uvex One is a comfortable, multi-functional safety shoe that combines the sporty-technical design language of a running shoe with maximum safety. The reduced shaft, with a low proportion of sewn parts, and the directly soled outsole enable highly efficient, automated and sustainable production. The combination of a breathable mesh material and specifically perforated microfibre ensures a comfortable foot climate. The adhesive-free and lightweight outsole construction, made from directly moulded polyurethane with TPU components, supports the natural rolling movement of the foot, providing excellent cushioning and slip-resistance. Moreover, the bungee lacing with no loose ends reduces the risk of accidents while ensuring an excellent fit.

Statement by the jury
The appearance of the Uvex One shoe is stylishly posited between work and casual shoe. It is well manufactured and provides outstanding wearing comfort.

Uvex One ist ein multifunktionaler, komfortabler Sicherheitsschuh, der die sportlich-technische Formensprache eines Laufschuhs mit maximaler Sicherheit kombiniert. Der reduzierte Schaft mit geringem Nähanteil und eine direktbesohlte Laufsohle ermöglichen eine hocheffiziente, automatisierte und nachhaltige Produktion. Die Kombination aus atmungsaktivem Mesh-Material mit gezielt perforierter Mikrofaser sorgt für ein angenehmes Fußklima. Die klebstofffreie Laufsohlenkonstruktion in Leichtbauweise aus direkt angeschäumtem Polyurethan mit TPU-Komponenten unterstützt die natürliche Abrollbewegung, darüber hinaus besitzt sie gute Dämpfungseigenschaften und ist rutschsicher. Eine Gummizugschnürung ohne frei hängende Enden reduziert das Unfallrisiko bei hervorragender Passform.

Begründung der Jury
Die Anmutung des Uvex One bewegt sich stilsicher zwischen Arbeits- und Freizeitschuh. Er ist sehr gut verarbeitet und bietet einen exzellenten Tragekomfort.

Security clothes

3M Peltor X
Hearing Protection
Gehörschutz

Manufacturer
3M,
Värnamo, Sweden
Design
Veryday
(Oskar Juhlin, Martin Birath,
Hans Nyström, August Michael,
Stefan Strandberg, Fredrik Ericsson),
Bromma, Sweden
Web
www.3m.com
www.veryday.com

The 3M Peltor X hearing protection features cups and a headband that are adjustable, guaranteeing a perfect fit at all times. The especially small and light cups are dual moulded for improved protection and comfort. The headband is fashioned from a rust-free, stainless-steel wire, covered in a breathable material, with a spring load that can withstand even long-term strain. The tension in the headband is carefully balanced without exerting too much pressure.

Der Gehörschutz 3M Peltor X verfügt über verstellbare Schalen und Kopfbügel, die stets für eine ideale Passform sorgen. Die besonders kleinen und leichten Cups sind für einen besseren Schutz und Komfort zweifach geformt. Der Kopfbügel besteht aus rostfreiem Stahldraht, der mit einem atmungsaktiven Material umhüllt ist und dessen Federkraft auch anhaltende Beanspruchung aushält. Die Spannung im Bügel ist sorgfältig ausbalanciert, ohne zu viel Druck auszuüben.

Statement by the jury
Safety and comfort share equal weight in the design of the 3M Peltor X. The colourful edge gives this hearing protection solution a distinctive visual highlight.

Begründung der Jury
Sicherheit und Komfort wurden bei der Ausstattung des 3M Peltor X gleichermaßen berücksichtigt. Die farbige Kante gibt dem Gehörschutz den visuellen Clou.

Dräger HPS 7000
Firefighter Helmet
Feuerwehrhelm

Manufacturer
Dräger Safety AG & Co. KGaA,
Lübeck, Germany
In-house design
Lennart Wenzel,
Matthias Willner
Design
Formherr Industriedesign
(Jens Bingenheimer),
Braunschweig, Germany
Web
www.draeger.com
www.formherr.de

The Dräger HPS 7000 is one of the lightest firefighter helmets in its class. The ergonomic design evenly distributes weight on the head and relieves neck muscles. The padded four-point harness safely and easily accommodates any head shape. Moreover, head size can be adjusted quickly via an easily accessible sizing wheel. Reflective strips improve the visibility under unfavourable conditions and allow for individual marking.

Statement by the jury
The notably light Dräger HPS 7000 firefighter helmet can be customised to comfortably fit any head size and guarantees enhanced safety.

Der Dräger HPS 7000 gehört zu den leichtesten Feuerwehrhelmen seiner Klasse. Das ergonomische Design verteilt das Gewicht gleichmäßig auf dem Kopf und entlastet die Nackenmuskulatur. Durch die gepolsterte 4-Punkt-Bänderung kann der Helm sicher und einfach an jede Kopfform angepasst werden. Über ein leicht erreichbares Verstellrad ist die Kopfweite schnell einzustellen. Reflexstreifen erhöhen die Sichtbarkeit bei schlechten Sichtverhältnissen und ermöglichen eine persönliche Kennzeichnung.

Begründung der Jury
Der ausgesprochen leichte Feuerwehrhelm Dräger HPS 7000 lässt sich der Kopfform bequem anpassen und bietet gleichzeitig sicheren Halt.

Wolflite XT Handlamp
Rechargeable Handlamp
Akkuhandleuchte

Manufacturer
Wolf Safety Lamp Company,
Sheffield, GB
Design
Renfrew Group International,
Rocket Studios
(Bruce Renfrew, Shaun Philips),
Leicester, GB
Web
www.wolf-safety.co.uk
www.renfrewgroup.com

The Wolflite XT rechargeable handlamp is ideally suited for use in hazardous industrial fields, satisfying the highest requirements of explosion protection. It is equipped with high-power LEDs and can be operated in both normal and power-save mode. A battery level indicator integrated into the lamp housing informs the user of the current charging status. The lamp is made of very rugged, anti-static thermoplastic material and is fully dust- and watertight.

Statement by the jury
The design of the handy Wolflite XT rechargeable handlamp distinguishes itself through the use of high-quality materials and state-of-the-art technology.

Die Akkuhandleuchte Wolflite XT eignet sich hervorragend für den Einsatz in Industriebereichen mit hohem Anspruch an den Explosionsschutz. Sie verfügt über High-Power-Leuchtdioden und ist wahlweise im Normalbetrieb oder Sparmodus zu verwenden. Eine im Lampengehäuse integrierte Anzeige informiert über den aktuellen Ladezustand. Die Leuchte besteht aus sehr robustem antistatischem Thermoplast und ist gegen das Eindringen von Staub und Strahlwasser geschützt.

Begründung der Jury
Die Gestaltung der handlichen Akkuhandleuchte Wolflite XT zeichnet sich durch die Verwendung hochwertiger Materialien und State-of-the-Art-Technologien aus.

Neon
Work Chair
Arbeitsstuhl

Manufacturer
Interstuhl GmbH & Co. KG,
Meßstetten-Tieringen, Germany
Design
Phoenix Design GmbH & Co. KG,
Stuttgart, Germany
Web
www.interstuhl.de
www.phoenixdesign.com

Modern factory workstations take on more of the qualities characteristic of an office workplace. Neon is a seating solution that is especially geared towards shifting work conditions in industry and laboratories. With intuitive controls in the form of white icons and the possibility of replacing the seat and the backrest easily, this work chair is designed for changing users. A distinct circumferential band made of elastic synthetic gives the chair the robustness required for tough use. Moreover, it is available in different signal colours.

Moderne Werkarbeitsplätze weisen immer stärker Büroqualitäten auf. Bei Neon handelt es sich um eine Sitzlösung, die speziell auf diese sich verändernden Arbeitsbedingungen im Industrie- und Laborumfeld eingeht. Durch intuitiv verständliche Bedienelemente in Form von weißen Icons sowie die Möglichkeit, Sitz- und Rückenpolster mit wenigen Handgriffen auszutauschen, ist der Arbeitsstuhl auch für wechselnde Nutzer geeignet. Das markante umlaufende Band aus elastischem Kunststoff gibt dem Arbeitsstuhl die nötige Robustheit für den täglichen Einsatz und ist in verschiedenen Signalfarben erhältlich.

GARANT Industrial Work Chair Series
GARANT Industrie-Arbeitsstuhl-Serie

Manufacturer
Hoffmann GmbH Qualitätswerkzeuge,
Munich, Germany
Design
Böhler CID Corporate Industrial Design
(Simon Müller, Christoffer Sens),
Fürth, Germany
Web
www.hoffmann-group.com
www.boehler-design.com
Honourable Mention

The Garant industrial work chair series was developed against the backdrop of increased demands on mobility. It also addresses demands for the enhanced ergonomic support of various postures and anatomical proportions. As a design motif, the spine is visually heightened by an aluminium cover. The selection of different surface materials provides high flexibility in terms of user application areas. In addition to the work chair, the series features a work stool and leaning aid.

Die Industrie-Arbeitsstuhl-Serie entstand mit dem Wissen um die erhöhten Anforderungen an die Bewegungsfreiheit sowie an die ergonomische Unterstützung vielfältiger Sitzhaltungen und anatomischer Proportionen. Das Rückgrat wird durch die Aluminiumabdeckung optisch betont und ist so auch Gestaltungsmotiv. Die Auswahl an verschiedenen Oberflächenmaterialien eröffnet eine große Flexibilität bezüglich der Einsatzbereiche. Neben dem Arbeitsstuhl komplettieren Arbeitshocker und Stehhilfe die Serie.

elneos
Device and Workstation System
Geräte- und Arbeitsplatzsystem

Manufacturer
erfi Ernst Fischer GmbH + Co. KG,
Freudenstadt, Germany
In-house design
David Köhler,
Prof. Gerd Flohr
Web
www.erfi.de

elneos consists of the elneos five device system and the elneos connect workstation system. The 7" multi-touch display of the former is operated via gestures. The connection sockets are embedded flush with the large glass surface and therefore well protected. The intelligent connection panel indicates the current operational status, with the ring lighting up in different colours according to activated function. The sophisticated profile system and the elneos connect connector enable the integration and uninterrupted routing of cables and the regulation of working height using a stable hydraulic height adjuster. It also interconnects to form a bridge across the tables. The workplace lighting and the indicator light use modern RGB LED technology and sensors.

elneos besteht aus dem Gerätesystem elneos five und dem Arbeitsplatzsystem elneos connect. Das 7" große Multi-Touch-Display von elneos five wird durch Gesten bedient. Die Anschlussbuchsen sind flächenbündig in die durchgehende Glasfront eingelassen und dadurch gut geschützt. Das intelligente Anschlussfeld zeigt den Aktiv- oder Inaktivzustand an, und die Ringbeleuchtung der Buchsen leuchtet je nach Funktion in verschiedenen Farben. Das ausgefeilte Profilsystem und der Connector von elneos connect ermöglicht die Aufnahme und unterbrechungsfreie Führung von Kabelwerk, die Justierung der Arbeitshöhe über eine stabile Hydraulikhöhenverstellung sowie das Verbinden zu einer tischüberführenden Brücke. Die Arbeitsplatzbeleuchtung und ein Indikationslicht werden durch moderne RGB-LED-Technologie und Sensorik umgesetzt.

Statement by the jury
Neon green colour accents give elneos, which otherwise impresses with its puristic and airy design language, an independent character.

Begründung der Jury
Durch die neongrünen Farbakzente bekommt elneos, das ansonsten durch eine puristische und luftige Formensprache überzeugt, einen eigenständigen Charakter.

Working place systems

VC 40-U (M) &
VC 20-U (M)
Hybrid Vacuum Cleaners
Hybridsauger

Manufacturer
Hilti Corporation,
Schaan, Liechtenstein
In-house Design
Design
Proform Design,
Winnenden, Germany
Web
www.hilti.com
www.proform-design.de

The VC 40-U (M) and the VC 20-U (M) are vacuum cleaners for wet and dry applications, especially designed to meet the needs of the construction industry. In addition to conventional power-cord operation, these appliances also allow for mobile, cordless use with standard batteries from the Hilti range and a built-in battery charger. Even in battery mode, they conveniently fulfil the extremely high demands placed on suction performance when removing dust or sludge. Thanks to a comprehensive range of accessories, these vacuum cleaners are suitable for a wide range of applications in various trades. Both hybrid vacuum cleaners help to ensure a clean and virtually dustless working environment at all times and thus set new standards in terms of health and safety.

VC 40-U (M) und VC 20-U (M) sind für die Anforderungen der Bauindustrie entwickelte Nass-/Trockensauger. Die Geräte erlauben neben dem klassischen Netzbetrieb auch den mobilen, kabellosen Einsatz mit regulären Akkus aus dem Hilti-Sortiment inklusive integriertem Ladegerät. Selbst im Akkubetrieb werden die extrem hohen Leistungsanforderungen beim Aufsaugen von Staub oder Schlamm problemlos erfüllt. Durch das umfangreiche Zubehör können die Sauger an die speziellen Bedürfnisse der unterschiedlichen Branchen angepasst werden. Beide Hybridsauger sorgen stets für eine praktisch staubfreie und saubere Arbeitsumgebung und setzen damit Maßstäbe in Bezug auf Gesundheit und Sicherheit.

Statement by the jury
The compact dimensions and rounded contours of these hybrid vacuum cleaners are both practical and aesthetically appealing.

Begründung der Jury
Mit ihren kompakten Abmessungen und den abgerundeten Konturen sind die Hybridsauger ästhetisch ansprechend und praktisch zugleich.

S 20 ECO Power
Vacuum Cleaner
Kesselsauger

Manufacturer
Fakir Hausgeräte GmbH,
Vaihingen/Enz, Germany
nilco Reinigungsmaschinen GmbH,
Vaihingen/Enz, Germany
Design
Brandis Industrial Design,
Nuremberg, Germany
Web
www.fakir.de
www.nilco.de
www.brandis-design.com

The S 20 Eco Power vacuum cleaner is a high-performance, low-noise professional device for dry vacuuming. The circular rubber ring, rubber rollers and shockproof plastic housing protect both cleaner and furniture. Together with the suction tube made of brushed metal, they contribute to the longevity of the device. An energy-saving compressor motor with two power levels and a wide operating range guarantee excellent cleaning results

Statement by the jury
The aesthetics of this vacuum cleaner are clear and unassuming. Mobility and efficiency are underscored by the compact dimensions.

Der Kesselsauger S 20 Eco Power ist ein leistungsstarkes und sehr geräuscharmes Profigerät zur trockenen Bodenpflege. Der umlaufende Gummiring, gummierte Rollen sowie der stoßfeste Kunststoff schonen Gerät und Möbel gleichermaßen. Zusammen mit dem Saugrohr aus gebürstetem Metall tragen sie zudem zur Langlebigkeit des Gerätes bei. Der energiesparenden Kompressormotor mit zwei Leistungsstufen und der große Aktionsradius bringen ausgezeichnete Reinigungsergebnisse.

Begründung der Jury
Die Ästhetik des Kesselsaugers ist klar und unaufdringlich. Seine Wendigkeit und Effizienz wird durch die kompakte Größe unterstrichen.

Poseidon 5-6-7
Professional High-Pressure Cleaner
Profi-Hochdruckreiniger

Manufacturer
Nilfisk-Advance A/S,
Hadsund, Denmark
Design
Weinberg & Ruf Produktgestaltung
(Andreas Weinberg, Martin Ruf),
Filderstadt, Germany
Web
www.nilfisk-alto.com
www.weinberg-ruf.com

The Poseidon 5-6-7 range comprises innovative, professional high-pressure cleaners for the agricultural sector as well as for the construction and automotive industries. A 30-mm steel frame with a protective bumper provides the required stability and robustness. A foldable handle and integrated holes for lifting enable optimal storage and transportation. Large wheels and the swivel castor at the front ensure high manoeuvrability. The front flap can be quickly and easily removed to obtain direct access to the motor and the pump.

Statement by the jury
Functional details like the front bumper and compartments for accessories lend the high-pressure cleaners a pragmatic touch.

Die Baureihe Poseidon 5-6-7 umfasst innovative Profi-Hochdruckreiniger für die Landwirtschaft, Bau- und Automobilindustrie. Durch den 30-mm-Stahlrahmen mit Front-Prallschutz besitzen sie die nötige Stabilität und Robustheit. Ein klappbarer Fahrbügel und integrierte Griffmulden ermöglichen optimale Lagerung und einfachen Transport. Große Räder und die Lenkrolle vorne sorgen für hohe Mobilität. Die Fronthaube kann schnell und einfach abgenommen werden, um direkten Zugang zu Motor und Pumpe zu erhalten.

Begründung der Jury
Funktionale Details wie der Front-Prallschutz und die Utensilienfächer verleihen den Hochdruckreinigern einen pragmatischen Anstrich.

Mxr
Floor Scrubber-Dryer
Scheuersaugmaschine

Manufacturer
Fimap SpA,
Santa Maria di Zevio, Italy
In-house design
Giorgia Mel
Web
www.fimap.com

Mxr is a ride-on floor scrubber-dryer for professional floor cleaning. Thanks to its compact size, it can replace a walk-behind model. Instead of pushing the machine, the user can assume a comfortable sitting position. Moreover, the unit is particularly environmentally friendly: it is mostly made of recyclable materials and the number of components has been reduced. Harmonious proportions and a special steering-wheel shape give the machine an individual look and provide the user with a comfortable driving experience. All controls are positioned on the steering wheel, making it easier to operate. The scrubber-dryer is exceptionally quiet, which makes it suitable for versatile use in a wide variety of settings.

Mxr ist eine Scheuersaugmaschine zum Aufsitzen, die bei der professionellen Bodenreinigung zum Einsatz kommt. Dank ihrer kompakten Größe kann sie ein Nachläufermodell ersetzen. Anstatt die Maschine schieben zu müssen, kann der Bediener also eine bequeme Sitzposition einnehmen. Das Arbeitsgerät ist besonders umweltfreundlich: Es wurden vorzugsweise wiederverwertbare Materialien verwendet und die Anzahl der Bauteile wurde verringert. Die harmonischen Proportionen und die besondere Lenkradform verleihen der Maschine Individualität und dem Bediener ein angenehmes Fahrerlebnis. Alle Steuerungen sind am Lenkrad angebracht, was die Bedienung übersichtlich macht. Die Maschine ist ausgesprochen leise und kann so jederzeit und überall eingesetzt werden.

GD
Industrial Tumble Dryer
Industrieller Wäschetrockner

Manufacturer
grandimpianti I.L.E. Ali S.p.A.,
Belluno, Italy
In-house design
Patrizia Terribile
Web
www.grandimpianti.com

The GD industrial tumble dryer clearly stands out from conventional industrial laundry appliances thanks to its elegant and softly curved form. Operation of the machine is simple and intuitive, and the dryer is moreover quiet, highly efficient and environmentally friendly. Internal air flow, thermal insulation and the completely automated drying process are all designed to achieve optimal drying results.

Statement by the jury
GD appeals with its simple design, which is highlighted by a monochrome colour scheme and large window.

Der industrielle Wäschetrockner GD hebt sich durch seine elegante und sanft geschwungene Form deutlich von den üblichen industriellen Wäschetrocknern und Waschautomaten ab. Die Bedienung ist einfach und intuitiv, der Trockner ist im Betrieb leise, leistungsstark und umweltschonend. Der innere Luftstrom, die Wärmeisolierung und der durch die elektronische Steuerung vollkommen automatisierte Trocknungsprozess führen zu besten Trocknungsergebnissen.

Begründung der Jury
GD gefällt durch seine formale Schlichtheit, die durch die monochrome Farbgebung und das großdimensionierte Fenster unterstrichen wird.

MAG-System
Floor Cleaning System
Bodenreinigungssystem

Manufacturer
Vermop Salmon GmbH,
Gilching, Germany
In-house design
Michael Egger
Web
www.vermop.com

This modular floor cleaning system allows users to connect four types of mops to one handle, thus covering a variety of cleaning applications. The use of magnets in the connector simplifies the process of attaching and removing the mop head. Its use is back-friendly for cleaning staff, and it prevents them from coming into contact with the soiled mop. The converging angled edges visually merge the connector with the stick.

Statement by the jury
The Mag-System displays an exceptionally ergonomic design. All functional parts are clearly identified by colour accents.

Mit dem modularen Bodenreinigungssystem kann der Anwender mit nur einem Stiel und Halter vier Mopptypen aufspannen und so diverse Reinigungsvarianten abdecken. Die Verwendung von Magneten im Halter erleichtern das Auf- und Abspannen der Mopps. Die Reinigungskraft arbeitet rückenschonend und kommt nicht mit dem schmutzigen Mopp in Kontakt. Durch die aufeinander zulaufenden abgeschrägten Kanten scheint der Halter mit dem Stiel zu verschmelzen.

Begründung der Jury
Die Gestaltung des Mag-Systems ist in herausragender Weise auf Ergonomie ausgelegt. Dank der Farbakzente sind die Funktionsteile eindeutig zuzuordnen.

PV Premium
Photovoltaic In-Roof System
Photovoltaik Indach-System

Manufacturer
Monier Braas GmbH,
Oberursel, Germany

In-house design
Alexander Flebbe

Design
Stafier Holland bv
(Ruben Beijer),
Zevenaar, Netherlands

Web
www.braas.de
www.stafier.com

reddot design award
best of the best 2013

Sunny prospects

The energy from the sun holds capacities that go far beyond the total energy needed all over the world. The PV Premium photovoltaic in-roof system is an innovative system which offers new possibilities for architecture in the field of solar energy. Instead of installing solar panels on top of the roof, the modules of this system replace the roof tiling and are integrated directly inside the roof surface. They seamlessly integrate with the rest of the tiling, lending the roof a harmonious and straightforward appearance. This innovative anthracite-coloured in-roof system is suitable for both refurbishing old buildings and installation in new buildings. Mounting is easy and self-explanatory, and the system is available for Tegalit roofs as well as the roof tile systems Frankfurter tile, Double-S and Taunus tile. One module replaces six tiles in width and one tile in height. The PV Premium photovoltaic in-roof system was tested and shown to be rainproof inside a wind tunnel and protects the roof even under extreme weather conditions. Its aesthetic appearance blends in with functional details such as a good back ventilation, which increases the efficiency of the system. Thus, it offers a wide variety of possibilities for naturally integrating solar technology into roof design – and yet another argument for using solar energy.

Sonnige Aussichten

Die Kraft der Sonne birgt Kapazitäten, die weit über das hinausreichen, was weltweit an Energie benötigt wird. Das Photovoltaik Indach-System PV Premium ist ein innovatives System, welches der Architektur neue Möglichkeiten im Bereich der Solarenergie bietet. Statt Solarmodule auf ein Dach zu setzen, werden die Module dieses Systems anstelle der Dacheindeckung in die Dachfläche integriert. Da sie sich von dieser nicht abheben, entsteht insgesamt ein harmonisches und geradliniges Deckbild. Das innovative anthrazitfarbene Indach-System kann bei der Altbausanierung wie auch bei Neubauten eingeplant werden. Es lässt sich einfach und selbsterklärend montieren und ist für die Dachstein-Modelle Tegalit, Frankfurter Pfanne, Doppel-S und Taunus Pfanne erhältlich. Ein Modul ersetzt hierbei sechs Dachsteine in der Breite und einen Dachstein in der Höhe. Das Photovoltaik Indach-System PV Premium wurde im Windkanal auf Regensicherheit getestet und erfüllt seine Schutzfunktion auch unter Extrembedingungen. Seine ästhetische Anmutung verbindet sich mit funktionalen Details wie einer guten Hinterlüftung, die seine Effektivität steigern. Auf diese Weise bieten sich vielfältige Möglichkeiten, die Solartechnologie in die Dachgestaltung zu integrieren – ein weiteres Argument für die Nutzung der Sonnenenergie.

Statement by the jury

The PV Premium photovoltaic in-roof system offers architects and planners versatile new possibilities. This PV system does not stand out from the roof and thus allows for a continuous and even roof surface. It results in impressive new aesthetics, which is combined with a high level of performance in energy production.

Begründung der Jury

Planern und Architekten eröffnet das Photovoltaik Indach-System PV Premium vielseitige neue Möglichkeiten. Dieses PV-System hebt sich nicht vom Dach ab und erlaubt somit eine durchgehend plane Gesamtfläche. Es entsteht eine beeindruckend neue Ästhetik, die sich mit einem hohen Maß an Leistungsfähigkeit bei der Energiegewinnung verbindet.

Photovoltaics

**Plug Connector
for Photovoltaic Plants**
Steckverbinder
für Photovoltaikanlagen

Manufacturer
Weidmüller Interface GmbH & Co. KG,
Detmold, Germany
Design
s3 designkonzepte
(Klaus Schmidt)
Web
www.weidmueller.com

Plug in, rotate, done – this is the simple principle behind the pre-assembled, single-component PV-Stick plug connector. Developed especially for direct cabling in the field, the connector enables rapid and cost-saving assembly without the need for additional special tools. High-quality workmanship and reliable performance combined with intuitive operation guarantee dependable connections for inverters and photovoltaic modules in the field.

Stecken, drehen, fertig. Auf diesem einfachen Prinzip beruht der vormontierte, einteilige Steckverbinder PV-Stick. Er wurde speziell für die Direktverkabelung im Feld entwickelt und ermöglicht eine schnelle und kostensparende Montage ohne zusätzliches Spezialwerkzeug. Hochwertige Verarbeitung und garantierte Funktionsfähigkeit kombiniert mit intuitiver Bedienbarkeit versprechen zuverlässige Verbindungen von Wechselrichtern und Photovoltaikmodulen im Feld.

Statement by the jury
The relief-like, delicate surface details of the PV-Stick successfully highlight its functional, high-quality character.

Begründung der Jury
Die reliefartigen, sauber ausgearbeiteten Oberflächendetails des PV-Sticks unterstreichen seinen funktionellen und hochwertigen Charakter.

REFUsol 333K
Photovoltaic Central Inverter
Photovoltaik-Zentralwechselrichter

Manufacturer
REFUsol GmbH,
Metzingen, Germany
In-house design
Marcel Pfeiffer
Web
www.refusol.com

The REFUsol 333K is a highly efficient central inverter for outdoor usage in large-scale photovoltaic power plants. Due to its power density and outdoor functionality, it provides an alternative to the standard brick-and-mortar solutions currently used for central inverters. The concept of increased voltage reduces energy losses about 33 per cent. The usage of many inverters in parallel reduces grid connection costs.

Der REFUsol 333K ist ein hocheffizienter Zentralwechselrichter für den Außeneinsatz in Photovoltaik-Großanlagen. Aufgrund seiner Leistungsdichte und Witterungsbeständigkeit stellt er eine Alternative zu bislang gängigen, umbauten Lösungen bei Zentralwechselrichtern dar. Das Konzept der erhöhten Spannungen reduziert die Leitungsverluste um ein Drittel. Der mögliche Parallelbetrieb reduziert die Netzanschlusskosten.

Statement by the jury
The REFUsol 333K for outdoor usage translates its technical innovative spirit into a compact housing and distinctive colour design.

Begründung der Jury
Der outdoorfähige REFUsol 333K überführt seinen technischen Innovationsgeist in ein kompaktes Gehäuse und eine prägnante Farbgestaltung.

Vitocal 161-A
Warm Water Heat Pump
Warmwasser-Wärmepumpe

Manufacturer
Viessmann Werke GmbH & Co. KG,
Allendorf/ Eder, Germany
Design
Phoenix Design GmbH & Co. KG
(Andreas Haug),
Stuttgart, Germany
Web
www.viessmann.com
www.phoenixdesign.com

The warm water heat pump Vitocal 161-A is a sustainable and environmentally friendly unit that derives heat for warm water from the ambient room temperature. The resulting cooler air is used to ventilate and cool living quarters. The cylinder's powerful design is contrasted by the calm, geometric form of the vertical panel that hides the technical elements. This results in operating controls that are easy to understand and have been reduced to the essentials. The compact, space-saving housing contains not only the heat pump but also a 300-litre boiler. This integrative design concept contributes to a reduction of complexity and provides maximum benefits and convenience for the user.

Bei der Warmwasser-Wärmepumpe Vitocal 161-A handelt es sich um ein nachhaltig und umweltfreundlich arbeitendes Gerät, das aus der vorhandenen Raumwärme Warmwasser aufbereitet. Die dabei entstehende abgekühlte Luft wird zur Belüftung und Kühlung von Wohnräumen verwendet. Die kraftvolle Formgebung des Zylinders kontrastiert mit der ruhigen, geometrischen Form des vertikalen Paneels, in dem sich technische Elemente verbergen. So zeigt sich dem Nutzer ein aufgeräumtes Gerät mit einem leicht verständlichen, auf das Notwendige reduzierte Bedienelement. Neben der Wärmepumpe enthält das platzsparende, kompakte Gehäuse einen 300-Liter-Warmwasserspeicher. Dieses integrative Gestaltungskonzept trägt dazu bei, die Komplexität zu reduzieren und dem Anwender größtmöglichen Nutzen und Komfort zu bieten.

Statement by the jury
With its high-quality and cohesive design, the heat pump Vitocal 161-A conveys a style that is both puristic and timeless.

Begründung der Jury
Die Wärmepumpe Vitocal 161-A wirkt durch ihre hochwertige und einheitliche Gestaltung zeitlos und puristisch.

Klix-Kühlrohr-180
Ceiling Radiator
Deckenradiator

Manufacturer
Klix Deckenradiatoren GmbH,
Rottweil, Germany
In-house design
Uwe Klix
Web
www.klix-deckenradiatoren.com

The Klix-Kühlrohr-180 ceiling radiator is conceived to provide passive heating and cooling in halls and offices. The extruded aluminium profile comes with an anodised or lacquered surface. The design's main feature is the groove for a non-drip condensate drain off at water temperatures down to 1 degree Celsius. This allows the available cooling temperatures from geothermal and ice storage systems to be used more efficiently.

Das Klix-Kühlrohr-180 sorgt für den passiven Heiz- und Kühlbetrieb in Hallen und Büros. Das Strangpressprofil besteht aus Aluminium mit einer eloxierten oder lackierten Oberfläche. Gestalterisches Hauptmerkmal ist die Rinne für eine tropfenfreie Kondensatführung bei Wassertemperaturen bis 1 Grad Celsius. Dadurch können die vorhandenen Kühltemperaturen aus geothermischen Anlagen und Eisspeichern besser genutzt werden.

Statement by the jury
With its small absorption area, the design of the Klix-Kühlrohr-180 provides reliable cooling. It is thus particularly economical and also ecologically friendly.

Begründung der Jury
Die Bauweise des Klix-Kühlrohrs-180 bietet mit einer geringen Absorbfläche eine zuverlässige Kühlleistung. Es ist dadurch besonders ökonomisch und ökologisch.

PrimusCenter
Domestic Water Center
Hauswasser-Center

Manufacturer
Honeywell GmbH,
Mosbach, Germany
In-house design
Ralf Hilbers
Design
AFRISO-EURO-INDEX GmbH
(Nadine Grob, Thomas Heinz),
Güglingen, Germany
Web
www.honeywell.com
www.afriso.com

PrimusCenter unites all the functions of a conventional drinking water point-of-entry in a very small space. The pressure-reducing valve lowers the pre-pressure to a steady level in order to protect the system and to achieve an economical use of water. During drinking water distribution, the water filter prevents the entry of dirt particles into the system, thus protecting valves, machines, continuous-flow water heaters, etc., from malfunctions caused by dirt. Furthermore, the domestic water center can be modularly extended and, thanks to its adjustable mounting system, it can be installed quickly with only three screws. With its clear form and discreet colour scheme, the unit also blends in perfectly with the design of modern technical, cellar or utility spaces.

PrimusCenter vereinigt sämtliche Funktionen einer herkömmlichen Hauswasser-einführung auf kleinstem Raum. Der Druckminderer reduziert den Vordruck zum Schutz der Installation und zum wirtschaftlichen Wasserverbrauch auf einen gleichmäßigen anlagenspezifischen Druck. Der Wasserfilter verhindert bei der Trinkwasserverteilung das Einspülen von Schmutzpartikeln in die Hausinstallation und schützt somit Ventile, Maschinen, Durchlauferhitzer usw. vor schmutzbedingten Funktionsstörungen. Darüber hinaus kann das Hauswasser-Center modular erweitert werden und lässt sich dank seines verstellbaren Befestigungssystems mit nur drei Schrauben schnell montieren. Durch die klare Form und dezente Farbgebung fügt es sich zudem perfekt in die Gestaltung moderner Technik-, Keller- oder Hauswirtschaftsräume ein.

Automatic Flat and Angle Bracket Sets
Automatik-Laschen und -Winkelsätze
Aluminium Profile Connection
Aluminium-Profil-Verbindung

Manufacturer
item Industrietechnik GmbH,
Solingen, Germany
In-house design
item Design/Entwicklung
Web
www.item24.com

Thanks to patent-pending functional geometry, these preassembled sets of automatic flat and angle brackets are easy to fit into any necessary place using just one hand. High-strength industrial construction can therefore be erected and disassembled in a particularly fast way. During installation the screws fit into place automatically; when tightened they form a strong connection between the aluminium profiles.

Dank der zum Patent angemeldeten Funktionsgeometrie lassen sich die vormontierten Automatik-Laschen und -Winkelsätze mit einem Handgriff an jeder Stelle positionieren. Hoch belastbare industrielle Konstruktionen können außerordentlich schnell aufgebaut oder demontiert werden. Die Schrauben platzieren sich automatisch richtig. Die Aluminiumprofile sind nach dem Anziehen der Schrauben fest verbunden.

Wärtsilä Propulsion Control System
Wärtsilä Antriebssteuerungssystem

Manufacturer
Wärtsilä Finland Oy,
Espoo, Finland

In-house design
Wärtsilä Industrial Design Team

Web
www.wartsila.com

Everything under control

On a ship, all actions between man and machine need to be absolutely clearly defined. This is one of the reasons for the often strict hierarchy on board. The Wärtsilä propulsion control system is an innovative user interface, which improves interactions and safety aboard ships of all sizes. It unites levers and touch-screen interfaces in a clear and user-friendly design and adapts to all the possible propulsion configurations a modern ship can have. Its clearly arranged user interface allows for intuitive control in all situations and conditions. It provides users with a comprehensive view of all instruments and gives important information exactly when needed. Thus, the Wärtsilä propulsion control system guarantees a high degree of safety and quick control both at sea and in port. In addition, its sophisticated modularity makes installing, commissioning, configuration and maintenance simple and efficient, thus saving on time and costs. Thanks to a fundamental reinterpretation in terms of design, the Wärtsilä propulsion control system combines a high degree of effectiveness and elegance.

Gut im Blick

Auf einem Schiff müssen alle Aktionen zwischen Mensch und Maschine unmissverständlich geregelt sein. Dies ist einer der Gründe für die meist strenge Hierarchie an Bord. Das Wärtsilä Antriebssteuerungssystem ist eine innovative Bedienerschnittstelle, die die Interaktionen und die Sicherheit an Bord von Schiffen aller Größen verbessert. Es vereint Hebel- und Touchscreen-Schnittstellen in einer klaren und nutzerfreundlichen Gestaltung und kann sich allen Antriebskonfigurationen eines modernen Schiffs anpassen. Seine übersichtlich gestaltete Bedienoberfläche ermöglicht dem Nutzer eine intuitive Kontrolle der Situationen und Bedingungen. Er erhält eine weitreichende Übersicht und es stehen ihm wichtige Informationen im jeweils passenden Moment zur Verfügung. Auf diese Weise gewährleistet das Wärtsilä Antriebssteuerungssystem ein hohes Maß an Sicherheit und Schnelligkeit auf See oder im Hafen. Eine durchdachte Modularität ermöglicht zudem eine einfache und effiziente Installation, Inbetriebnahme, Konfiguration und Wartung dieses Systems. Das führt dazu, dass Kosten und Zeit eingespart werden können. Durch eine grundlegende gestalterische Neuinterpretation entstand hier ein Antriebssteuerungssystem, welches ein hohes Maß an Effektivität mit Eleganz verbindet.

Statement by the jury

The design of this propulsion control system achieves the reduction of complexity in an impressive manner. Levers and touch-screen interfaces are ergonomically well thought out and all information is clearly arranged. The clear design vocabulary as well as the quality of the user interfaces of this propulsion control system create a comfortable environment.

Begründung der Jury

Der Gestaltung des Wärtsilä Antriebssteuerungssystems gelingt auf beeindruckende Weise die Reduktion von Komplexität. Die Hebel und Touchscreen-Schnittstellen sind ergonomisch durchdacht und alle Informationen übersichtlich angeordnet. Die klare Formensprache wie auch die Qualität der Nutzeroberflächen dieses Antriebssteuerungssystems schaffen ein angenehmes Umfeld.

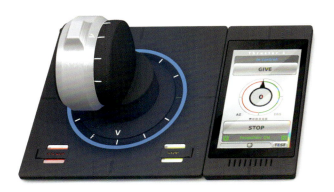

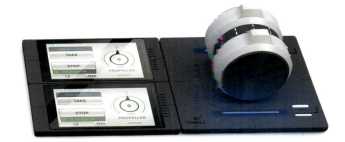

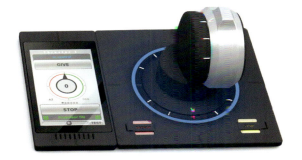

CenFlex
Superfinish Machine

Manufacturer
Supfina Grieshaber GmbH,
Wolfach, Germany
Design
Design Tech
(Jürgen R. Schmid),
Ammerbuch, Germany
Web
www.supfina.com
www.designtech.eu

The CenFlex superfinish machine is designed for the surface treatment of workpieces and motorsports components. Its curved form gives the machine a highly distinct appearance. The flush sliding doors seamlessly fit into the front, which is fashioned from dark real glass. Outstanding accessibility of the machining area keeps all working areas in reach for technicians at all times.

Die Superfinish-Maschine CenFlex kann für die Oberflächenbearbeitung von Werkstücken wie auch von Motorsport-Komponenten eingesetzt werden. Ihre geschwungene Form verleiht der Maschine einen hohen Wiedererkennungswert. Die versenkbaren Schiebetüren fügen sich nahtlos in die Front aus dunklem Echtglas ein. Der Bearbeitungsraum ist sehr gut zugänglich und sämtliche Arbeitsbereiche sind vom Bediener unmittelbar zu erreichen.

Statement by the jury
A seamless construction consisting of dark glass, light lateral surfaces and red colour accents lends CenFlex a particularly elegant and high-quality appearance.

Begründung der Jury
Die fugenlose Konstruktion aus dunklem Glas, hellen Seitenflächen und roten Farbakzenten verleiht CenFlex ein besonders edles und hochwertiges Äußeres.

Manufacturer
XYZTEC,
Panningen, Netherlands
Design
GBO DESIGN,
Helmond, Netherlands
Web
www.xyztec.com
www.gbo.eu

The Condor Sigma bond tester for materials testing is characterised by a clear and simple overall shape with smooth surfaces and rounded edges, which is further highlighted by a distinctive colour scheme. Thanks to its open architecture, the machine can handle a wide spectrum of sample sizes. The accompanying software enables intuitive operation. The revolving measurement unit (RMU) at the front of the machine facilitates a quick and easy exchange of tools.

Der Bondtester Condor Sigma für Materialtests besitzt eine einfache Hauptform mit glatten Oberflächen und abgerundeten Ecken, deren Klarheit durch ein markantes Farbschema noch weiter unterstrichen wird. Durch die offene Architektur der Maschine kann ein breites Spektrum an Mustergrößen gehandhabt werden. Die Software ermöglicht eine intuitive Bedienung. Dank der sogenannten Revolving Measurement Unit (RMU) auf der Vorderseite der Maschine ist ein schneller und einfacher Werkzeugwechsel möglich.

The clear, geometric structure and strong colour accents make Condor Sigma a bond tester that features not only functional but also aesthetic qualities.

Die klare geometrische Gliederung und kräftige Farbakzente machen Condor Sigma zu einem Bondtester, der nicht nur über funktionale, sondern auch über ästhetische Qualitäten verfügt.

Hartl Crusher
Bucket Crusher
Schaufelbrecher

Manufacturer
Hartl Engineering & Marketing GmbH,
Mauthausen, Austria
Design
RDD Design Network
(Rainer Atzlinger),
Ried, Traunkreis, Austria
Web
www.hartl-crusher.com
www.rdd.at

The Hartl Crusher is a fully fledged, robust bucket crusher. The technically robust design of this crushing unit, which is integrated into an excavator bucket, guarantees high performance and reliability for crushing natural stone or recycling building rubble. The compact crusher unit is mounted and connected to an excavator by means of a quick-coupling system. It is driven via the hydraulics of the excavator. The bucket crusher convinces with maximum throughput performance at minimal wear expenditure, as well as with a high-grade cubic and constant final grain. The side panels feature a lightweight construction, are optimised for material strength, and can be dismounted quickly and easily by hand.

Statement by the jury
The Hartl Crusher convinces with its high-quality components and a stream-lined design that facilitates the picking up of stones and other materials.

Der Schaufelbrecher Hartl Crusher ist ein vollwertiger, robuster Backenbrecher. Die technisch robuste Konstruktion dieser in einer Baggerschaufel integrierten Brechereinheit garantiert hohe Leistung und Zuverlässigkeit bei der Zerkleinerung von Naturgestein sowie beim Recycling von Baurestmassen. Die kompakte Brechereinheit wird mithilfe eines Schnellwechselsystems an einem Bagger montiert und angeschlossen. Der Antrieb wird über die Hydraulik des Baggers gewährleistet. Der Schaufelbrecher überzeugt durch hohe Durchsatzleistungen bei minimalen Verschleißkosten sowie durch ein hochwertiges kubisches und konstantes Endkorn. Die seitlichen Verkleidungsteile sind im Sinne des Leichtbaus gefertigt, auf ihre Materialstärke optimiert und mit wenigen Handgriffen zu demontieren.

Begründung der Jury
Hartl Crusher überzeugt durch seine qualitativ hochwertigen Komponenten und die stromlinienförmige Gestaltung, die die Gesteins- und Materialaufnahme erleichtert.

Electric Chain Hoist Series
Elektrokettenzugfamilie

Manufacturer
Konecranes Plc,
Hyvinkää, Finland
In-house design
Johannes Tarkiainen
Design
PM Design
(Olivier Hong),
Roissy-en-France, France
Web
www.konecranes.com
www.pmdesign-france.com

This multi-branded electric chain hoist series is suited to handling loads from 63 to 2,500 kg and features a modular construction. Each chain hoist is protected by a robust aluminium housing, which can be easily opened for maintenance work. The ergonomic pendant control is rubber-coated to ensure a firm grip. When the pendant control is not in use, it can be fixed via an integrated magnet to any metal structure. The hook block possesses a specially formed recessed handle, while the upper part of the block also works as a trigger for the electrical limit switch.

Statement by the jury
The ergonomic details of this electric chain hoist series offer exceptional ease of use. The aluminium housing is robust and functional.

Die Multi-Marken-Elektrokettenzugfamilie ist für das Handling von Lasten von 63 bis 2.500 kg geeignet und modular aufgebaut. Der Kettenzug ist jeweils durch ein robustes Aluminiumgehäuse geschützt, das sich für Wartungsarbeiten leicht öffnen lässt. Die Handsteuerung ist gummiert, um einen festen, ergonomischen Griff zu gewährleisten. Wenn das Handsteuergerät nicht benötigt wird, kann es durch den integrierten Magneten überall an Metallstrukturen befestigt werden. Das Hakengeschirr ist mit einer speziell geformten Griffmulde versehen. Das obere Ende des Geschirrs dient auch als Auslöser für die elektrische Hakenendabschaltung.

Begründung der Jury
Die ergonomischen Details der Elektrokettenzugfamilie bieten außergewöhnlichen Bedienkomfort. Das Aluminiumgehäuse ist robust und funktional.

HighLight
Distributed Control System
Prozessleitsystem

Manufacturer
Hangzhou Winmation Automation Co., Ltd.,
Hangzhou, China
In-house design
Axel Lohbeck
Design
LKK Design Co., Ltd.
(Jia Wenlong, Lian Zhen,
Xu Xianbin, Zhang Ke, Li Bin),
Shanghai, China
Web
www.winmation.com
www.lkkdesign.com

The design of the HighLight distributed control system incorporates the triangle of the company logo to create a cohesive design language and high brand recognition. The system components are based on several patented technologies which improve performance and stability of the entire system. Special attention has likewise been paid to ergonomic aspects. The plugs, for instance, can be handled with one hand; they are arranged in a line, making the slots easy to access and convenient for plug in or removal. Data transfer with dual tone multi-frequency signalling optimises work processes, thus enhancing efficiency.

In der Gestaltung des Prozessleitsystems HighLight wurde das Dreieck des Firmenlogos aufgegriffen und in die Architektur integriert, wodurch eine einheitliche Formensprache und hohe Unternehmensidentifikation entsteht. Die Systemkomponenten basieren auf mehreren patentierten Technologien, was die Leistungsfähigkeit und Stabilität des gesamten Systems verbessert. Auch Aspekte der Ergonomie wurden in besonderem Maße berücksichtigt. So lassen sich die Anschlussstecker beispielsweise einhändig bedienen und sind hintereinander aufgereiht, damit die Steckplätze leicht zu erreichen und die Stecker bequem anzuschließen oder herauszunehmen sind. Die Datenübermittlung mit dem Mehrfrequenzwahlverfahren optimiert die Arbeitsabläufe und steigert dadurch die Effizienz.

Statement by the jury
This distributed control system evinces a compact, robust and functional design. At the same time, its dark housing and blue lights give it an elegant and modern touch.

Begründung der Jury
Das Prozessleitsystem erscheint kompakt, robust und funktional. Gleichzeitig wirkt es durch das dunkle Gehäuse und die blauen Leuchten elegant und modern.

FE35
Tablet Press
Tablettenpresse

Manufacturer
Fette Compacting GmbH,
Schwarzenbek, Germany
Design
Dominic Schindler Creations GmbH
(Dominic Schindler),
Lauterach, Austria
Web
www.fette-compacting.com
www.dominicschindler.com

The FE35 offers one of the shortest product changeover times for machines of its class and is thus an ideal platform for increasingly flexible tablet production. The single rotary press can be fitted with up to 51 compression stations, thus enabling the production of 370,000 tablets per hour. Thanks to its optimised viewing area, the structure provides very good access to its components. All working steps required to change the rotor have been fully automated and can be carried out in absence of tools. The optimised frame structure makes this a low-vibration tablet press and also reduces noise emissions. The housing is made of a certified high-performance synthetic material.

Die FE35 bietet eine der kürzesten Produkt-wechselzeiten der Anlagen ihrer Leistungs-klasse und ist damit die ideale Plattform für eine zunehmend flexible Tablettenpro-duktion. Als Einfachrundläufer kann die Maschine mit bis zu 51 Stempelstationen ausgerüstet werden und ermöglicht damit die Produktion von bis zu 370.000 Tablet-ten pro Stunde. Die Maschine besitzt einen optimalen Einsichtbereich und gewährt einen sehr guten Zugang zu den Baugrup-pen. Beim Rotorwechsel sind sämtliche Arbeitsschritte automatisiert und werk-zeugfrei. Die optimierte Rahmenstruktur macht die Tablettenpresse vibrationsarm und verringert die Geräuschemissionen. Das Gehäuse besteht aus einem zertifizier-ten Hochleistungskunststoff.

MegaPack4
Cylinder Bundle
Gasflaschenbündel

Manufacturer
Messer Group GmbH,
Bad Soden, Germany
Design
octopus.design
(Reinhard Zetsche),
Munich, Germany
Web
www.messergroup.com
www.octopus-design.de
Honourable Mention

Cylinder bundles like MegaPack4 enable the transport of technical gases such as oxygen or nitrogen to manufacturing plants. It is characterised by innovations in safety, ergonomics and environmental protection, which have been incorporated into an easily understandable design. With its central control panel, ergonomic content indicator and tamper-evident seal as well as secure impact protection, the unit is geared specifically to the user. The cylinder bundles are stackable and also feature colour coding to identify types of gas by means of a variety of distinctive colours. The unique construction, with a roll bar functioning as a crumple zone, facilitates a very lightweight concept, making the transport more ecologically friendly and economical.

Gasflaschenbündel wie MegaPack4 erlauben den Transport von technischen Gasen wie Sauerstoff oder Stickstoff an Fertigungsbetriebe. Es zeichnet sich durch Innovationen in Sicherheit, Ergonomie und Umweltschutz aus, die in ein nachvollziehbares Design überführt wurden. Mit dem zentralen Bedienpanel, seiner ergonomischen Inhaltsanzeige und Originalitätsverschluss sowie dem sicheren Stoßschutz ist es in besonderer Weise auf den Anwender zugeschnitten. Das Flaschenbündel ist stapelbar und mit einer Farbcodierung versehen, welche mit verschiedenen Kennfarben die Gasarten definiert. Durch die einzigartige Konstruktion mit einem Überrollbügel als Knautschzone konnte das Gewicht reduziert werden, was den Transport ökologischer und ökonomischer macht.

INGENIA
PVD Coating System
PVD-Beschichtungsanlage

Manufacturer
OC Oerlikon Balzers AG,
Balzers, Liechtenstein
Design
Design Form Technik
(J. Peter Klien),
Triesenberg, Liechtenstein
Web
www.oerlikon.com
www.design-form-technik.com

The Ingenia PVD coating system unites maximum performance and superb engineering in the smallest possible space. It is equipped with the latest developments in PVD technology. The system impresses with its clear, compact design and high-grade materials. The combination of anodised aluminium and dark coated surfaces gives the system a contemporary presence in high-tech environments. Furthermore, the generously arranged handle and control elements facilitate the fluent operation of the system.

Statement by the jury
With its immaculately smooth surfaces in monochrome colours, the Ingenia PVD coating system conveys distinguished elegance and premium quality.

Die PVD-Beschichtungsanlage Ingenia vereint maximale Leistung und hervorragende Ingenieursarbeit auf kleinstem Raum. Sie ist mit den neusten Errungenschaften der PVD-Technologie ausgestattet. Die Anlage besticht durch eine klare, kompakte Formgebung und hochwertige Materialien. Die Kombination von eloxiertem Aluminium mit den dunkel beschichteten Oberflächen verleiht dem System eine zeitgemäße Präsenz im Hightech-Bereich. Darüber hinaus erlauben die großzügig angeordneten Griff- und Bedienelemente ein einfaches Handling der Anlage.

Begründung der Jury
Durch die makellos glatten Oberflächen in monochromen Farben strahlt die PVD-Beschichtungsanlage eine repräsentative Eleganz und Wertigkeit aus.

Comexi F2
Flexographic Printing Press
Flexodruckmaschine

Manufacturer
Comexi,
Girona, Spain
Design
Loop
(Guido Charosky, Álvaro Quintanilla),
Sant Cugat del Vallès, Spain
Web
www.comexigroup.com
www.loop-cn.com

The Comexi central impression flexographic printing press F2 enhances perfect printing even with water based inks for medium and long print runs in the flexible packaging industry. Its eco-sensitive design improves energetic efficiency. It features a user friendly work environment with tool free maintenance system, equipped with the latest software and electronic technology. The modular platform guarantees flexibility and versatility configuration, enabling future technological upgrades at any stage.

Die Zentralzylinder-Flexodruckmaschine Comexi F2 ermöglicht einen perfekten Druck mit wasserbasierten Tinten für mittlere und lange Auflagenserien in der flexiblen Verpackungsindustrie. Ihr umweltfreundliches Design verbessert die Energieeffizienz. Durch die werkzeugfreie Wartung schafft das System ein benutzerfreundliches Arbeitsumfeld, das mit modernster Software und Elektrotechnologie ausgestattet ist. Die modulare Plattform gewährleistet Flexibilität und Vielseitigkeit, sodass technische Nachrüstungen jederzeit realisierbar sind.

Statement by the jury
The rectangular geometry and functional simplicity of this flexographic printing press highlights its functionality and performance in a superior way.

Begründung der Jury
Die rechtwinklige Geometrie und sachliche Schlichtheit der Flexodruckmaschine unterstreicht auf hervorragende Weise ihre Funktionalität und Leistungsfähigkeit.

DATRON M8Cube
CNC Milling Machine
CNC-Fräsmaschine

Manufacturer
DATRON AG,
Mühltal, Germany
In-House design
Frank Wesp
Web
www.datron.de

The Datron M8Cube is a high-performance CNC milling machine. Its minimalist form and contrasting colour scheme deliberately direct attention to the operational process, that is, to the milling area. Machine activity is clearly visible through three large viewing windows. A slim user terminal with an integrated touchpad enables the user to easily execute the milling programmes. The integration of clearly positioned signal lights in the user terminal and on the machine cabinet provides clear information about the machine's operational status. A space-saving flap concept and the large chip drawer integrated into the machine's housing ensure ergonomic and fatigue-free working conditions.

Die Datron M8Cube ist eine Hochleistungs-CNC-Fräsmaschine. Durch die reduzierte Form und die kontrastreiche Farbwahl wurde der Fokus bewusst auf den Bearbeitungsprozess, sprich auf den Fräs-Bereich, gelenkt. Dieser ist durch die Zuhilfenahme von großen Sichtscheiben von drei Seiten sehr gut einsehbar. Ein schlankes Bedienterminal mit integriertem Touchpad ermöglicht eine unkomplizierte Ausführung der Fräsprogramme. Die Integration von klar positionierten Signalleuchten in Bedienterminal und Portalgehäuse geben im Produktionsbetrieb eindeutigen Aufschluss über den Maschinenstatus. Durch das platzsparende Türkonzept und den im Maschinengehäuse integrierten großzügigen Spänewagen ist ein ergonomisches und ermüdungsfreies Arbeiten gewährleistet.

KUKA flexibleCUBE
Welding Cell
Schweißzelle

Manufacturer
KUKA Systems GmbH,
Augsburg, Germany
In-house design
Sebastian Mocker,
Martin Greppmeier,
Yong-Hak Cho
Design
Grewer Industriedesign
(Michael Grewer),
Augsburg, Germany
Web
www.kuka-systems.com
www.grewer-industriedesign.de

As a manufacturing extension for existing automated systems, the Kuka flexibleCube welding cell represents a first step towards automation. As such, it can be easily and seamlessly integrated into the manufacturing process. The cell is characterised by optimally adjusted components, the use of the latest robot control and welding technologies, as well as tried and tested standards from the area of inert-gas welding. The operating and visualising software, which is fully integrated into the robot control, makes the welding cell easy to use. With its modular design and wide variety of automation options, the system enables users to dynamically react to changes in the manufacturing environment.

Die Schweißzelle Kuka flexibleCube lässt sich als Einstieg in die Automatisierung oder als Fertigungserweiterung zu bestehenden Automatisierungsanlagen einfach und nahtlos in den Fertigungsprozess integrieren. Sie zeichnet sich durch optimal aufeinander abgestimmte Komponenten aus, durch die Verwendung modernster Robotersteuerungs- und Schweißtechnologien sowie bewährte Standards aus dem Schutzgasschweißen. Eine vollständig in die Robotersteuerung integrierte Bedien- und Visualisierungssoftware erleichtert den Umgang mit der Schweißzelle. Das System ist modular aufgebaut und bietet zahlreiche Optionen zur Automatisierung. Damit versetzt es den Anwender in die Lage, dynamisch auf Veränderungen im Fertigungsumfeld zu reagieren.

W+D TIMOS
Totally Integrated
Mail Output Solution
Vollintegrierte
Produktionsstraße für
Mailings

Manufacturer
W+D Winkler+Dünnebier GmbH,
W+D Direct Marketing Solutions GmbH,
Neuwied, Germany
Design
Braake Design,
Stuttgart, Germany
Web
www.w-d.de
www.braake.com

The W+D Timos is a totally integrated
mail output solution designed for the
personalised production and printing
of envelopes; it also features automated
inserting of the accompanying letters.
Its modular housing reflects the
individual workflow steps. Large glass
hoods and partial glass fronts provide
an excellent view of the production
process running in opposite directions
on two levels. Furthermore, the
operating terminals enable intuitive
process control in every area of the
machine.

Die W+D Timos ermöglicht die persona-
lisierte Herstellung, Bedruckung sowie
Kuvertierung von Briefumschlägen inklusive Anschreiben. Das modular aufgebaute
Gehäuse spiegelt dabei den Verlauf der
einzelnen Produktionsschritte nach außen
wider. Große Glashauben und partielle
Glasfronten bieten hervorragenden Einblick in den gegenläufigen Prozessfluss auf
zwei Ebenen. Zusätzlich gewährleisten die
Bedienterminals eine intuitive Prozesskont-
rolle an jedem Bereich der Maschine.

EOL
Transmission Test Rig
Getriebeprüfstand

Manufacturer
ZF Friedrichshafen AG,
Passau, Germany
Design
Braake Design
Stuttgart, Germany
Web
www.zf.com
www.braake.com

This transmission test rig is designed to carry out the oil filling of automatic transmissions for transversely mounted engines and to perform final tests. The newly developed noise protection cladding significantly reduces background noise and protects the user during the testing process. The large safety glass doors provide optimal access and excellent visibility for checking the test components. The monitor arm is rotatable, moveable in a linear path and has very good ergonomic characteristics with regard to service and operation.

Statement by the jury
The transmission test rig's minimalist and calm design language impressively communicates its reliability and efficiency.

Der Getriebeprüfstand übernimmt die Befüllung und Endkontrolle von Automatikgetrieben für quer eingebaute Motoren. Die neu entwickelte Schallschutzverkleidung dient der akustischen Entkoppelung und dem Schutz des Bedieners während des Prüfvorgangs. Die großen Sicherheitsglastüren ermöglichen einen optimalen Zugang und hervorragenden Einblick für die visuelle Kontrolle der Testkomponenten. Der Monitorarm ist schwenkbar, linear verfahrbar und bietet eine sehr gute Ergonomie für Service und Bedienung.

Begründung der Jury
Die reduzierte und ruhige Formensprache des Getriebeprüfstands kommuniziert eindrücklich seine Zuverlässigkeit und Effizienz.

COLORMAN e:line
Offset Printing Press for Newspapers
Rollenoffsetdruckmaschine für Zeitungen

Manufacturer
manroland web systems GmbH,
Augsburg, Germany
Design
The Kaikai Company
(Tim R. Wichmann, Christian Jaeger),
Munich, Germany
Web
www.manroland-web.com
www.thekaikaico.com

The Colorman e:line embodies low energy consumption, ergonomics and efficiency. Its highlights moreover include good accessibility and maintenance-friendliness. Operation of the entire press line takes place at the control console level, which underscores its enhanced user-friendliness. A resource-conserving design using smart materials represents a highlight of the Colorman e:line concept. The new manroland operating concept makes the user a pilot who has all production processes under complete control.

Statement by the jury
The Colorman e:line captivates with its dynamic appearance, which results from distinctly contoured seams featuring soft strips of light.

Die Colorman e:line ist energiesparend, ergonomisch und effizient. Sie zeichnet sich zudem durch ihre gute Zugänglichkeit und Wartungsfreundlichkeit aus. Die gesamten Anlage wird von der Leitstandsebene aus gesteuert, was die hohe Anwenderfreundlichkeit unterstreicht. Die ressourcenschonende Gestaltung mit Smart-Materials stellt ein Highlight im Konzept der Colorman e:line dar. Das neue manroland-Bedienkonzept macht den Anwender zum Piloten, der alle Produktionsprozesse souverän steuert.

Begründung der Jury
Die Colorman e:line fasziniert durch ihr dynamisches Erscheinungsbild, das durch die markanten Fugen mit dem sanften Lichtband erzeugt wird.

Powerball Lightweight Arm LWA 4P

Six-axis Lightweight Robot
Sechs-Achs-Leichtbauroboter

Manufacturer
SCHUNK GmbH & Co. KG,
Lauffen/Neckar, Germany
In-house design
Roxo Tschakarow
Design
Busse Design + Engineering
(Michael Tinius),
Elchingen, Germany
Web
www.schunk.com
www.busse-design.com

The Powerball Lightweight Arm LWA 4P is one of the most powerful lightweight arms in the world. It is designed for stationary and mobile use. Despite a dead weight of only 12 kg, it can dynamically handle a payload of up to 6 kg, and its wrist moves deftly in tight spaces. Due to the use of the latest generation of torque motors, the average energy consumption of this agile helper amounts on average to an economical 80 watts of power, thus saving resources.

Der Powerball Lightweight Arm LWA 4P ist einer der leistungsdichtesten Leichtbauarme der Welt und für den stationären wie auch mobilen Einsatz geeignet. Trotz eines Eigengewichts von nur 12 kg kann er Lasten bis 6 kg dynamisch handhaben. Sein Handgelenk bewegt sich auch auf engstem Raum geschickt. Dank Torquemotoren der neuesten Generation verbraucht der wendige Helfer durchschnittlich sparsame 80 Watt und arbeitet so mit Ressourcen schonendem Energieeinsatz.

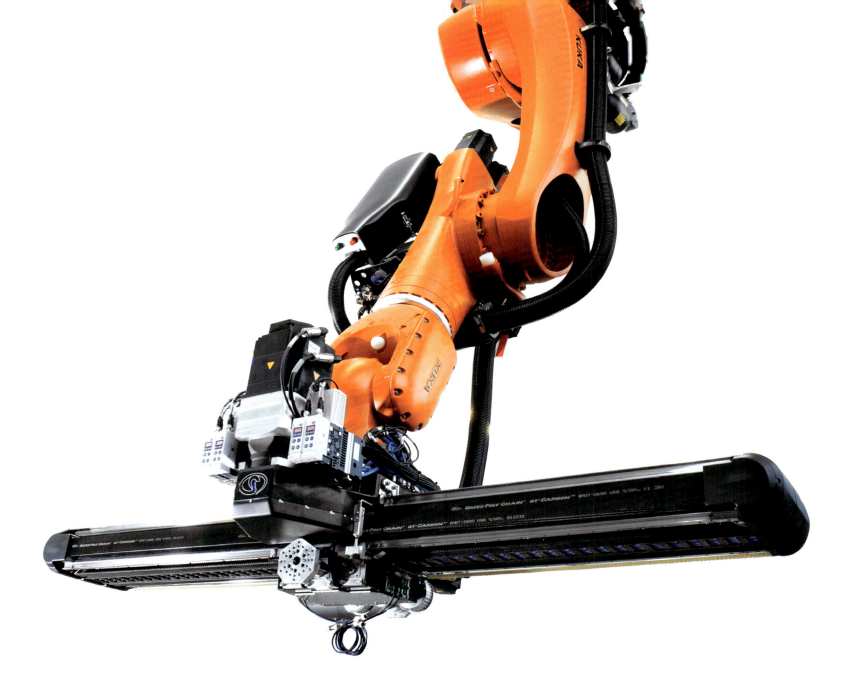

KUKA Cobra
Robotic Arm
Roboterarm

Manufacturer
KUKA Systems GmbH,
Augsburg, Germany
In-house design
Michael Büchler,
Sebastian Mocker,
Yong-Hak Cho
Web
www.kuka-systems.com

The Kuka Cobra is a very fast system for linking presses. An additional linear and pivot axis has been mounted to the robotic arm, which loads and unloads the presses. This increases output, reduces the distance between presses and combines the dynamic qualities of a linear axis with the flexibility of a robot. Moreover, the Kuka Cobra is made of carbon, which turns the system into a particularly advanced and lightweight construction.

Die Kuka Cobra ist ein sehr schnelles System zur Pressenverkettung. Am Roboterarm ist eine zusätzliche Linear- und Schwenkachse montiert, die die Pressen be- und entlädt. Dadurch wird die Ausbringungsleistung erhöht, der Pressenabstand verringert und die Dynamik einer Linearachse mit der Flexibilität eines Roboters kombiniert. Die Kuka Cobra besteht darüber hinaus aus Carbon, was das System zu einer besonders zukunftsweisenden Leichtbaukonstruktion macht.

Statement by the jury
Rounded lines and a clear design characterise the look of the Kuka Cobra and successfully highlight its efficiency and agility.

Begründung der Jury
Abgerundete Linien und eine konstruktive Klarheit bestimmen die Erscheinung der Kuka Cobra und betonen gekonnt ihre Effizienz und Wendigkeit.

Servus ARC3
Autonomous Robotic Carrier
Autonomer Transportroboter

Manufacturer
Servus Intralogistics GmbH,
Dornbirn, Austria
Design
Wolfgang Brändle,
Hohenems, Austria
Web
www.servus.info

Servus Arc3 is an autonomous and intelligent robotic carrier for automated assembly and both in-house and warehouse logistics. It is the key component of a modular, decentralised system consisting of tracks, junctions, lifts, shelving, transfer stations etc. Its components are designed so that a maximum variety of sizes can be processed with a minimum of variable elements. Every Servus Arc3 consists of two drive modules and a basic module, with two loading equipment sections integrated between them. A load can be received or deposited at any point along the route. In addition, the robotic carrier features traction control, an anti-lock braking system and energy recirculation.

Statement by the jury
The autonomous robotic carrier impresses with its modular structure and its functional, robust design language, which underscores its technical character.

Servus Arc3 ist ein autonomer und intelligenter Transportroboter für Montageautomation, Inhouse- und Lagerlogistik. Er ist zentraler Teil eines modularen dezentralen Systems, bestehend aus Fahrstrecken, Weichen, Hebern, Regalen, Übergabestationen etc. Seine Bauteile sind so konzipiert, dass mit einem Minimum an variablen Elementen ein Maximum an verschiedenen Größen realisiert werden kann. Jeder Servus Arc3 besteht aus zwei Antriebsmodulen und einem Basismodul, dazwischen sind zwei Lademittel-Stränge integriert. Eine Ladung kann an jeder beliebigen Stelle entlang der Strecke aufgenommen oder abgegeben werden. Zudem verfügt der Roboter über Antischlupfregelung, Antiblockiersystem und Energierückführung.

Begründung der Jury
Der Transportroboter überzeugt durch seinen modularen Aufbau und die funktional-robuste Formensprache, die seinen technischen Charakter unterstreicht.

Granulate Conveyor
Granulatsicheres Transportband

Manufacturer
Mettler Toledo Garvens,
Giesen, Germany
In-house design
Rüdiger Kliefoth,
Sascha Goly
Web
www.mt.com

This conveyor was specifically designed for granulate products such as sugar, flour or washing powder. The extra-wide conveyor belt protects all working parts, such as the driving and idle rollers, from penetration by the granules. The conveyor body has been milled from one solid block to reduce unnecessary crevices. The toothed belt is situated in a completely sealed housing to provide further protection from particles, thus promoting safety during assembly and operation.

Statement by the jury
The well-planned construction design and choice of materials for the conveyor belt indicate clearly how demands for handling granulate materials are met in a comprehensive way.

Das Transportband wurde speziell für granulate Produkte wie Zucker, Mehl oder Waschpulver entwickelt. Der extra breite Transportgurt schützt bewegliche Teile wie die Antriebs- und Laufrollen vor dem Eindringen des Granulats. Der Bandkörper ist aus einem soliden Block gefräst und weist keine unnötigen Zwischenräume auf. Der Zahnriemen befindet sich zum Schutz vor Partikeln in einem vollständig geschlossenen Gehäuse, was zudem Sicherheit bei der Montage und Bedienung gewährleistet.

Begründung der Jury
In der planvollen Bauart und Materialwahl des Transportbandes zeigt sich deutlich, wie umfassend die Anforderungen an den Umgang mit Granulaten bedacht wurden.

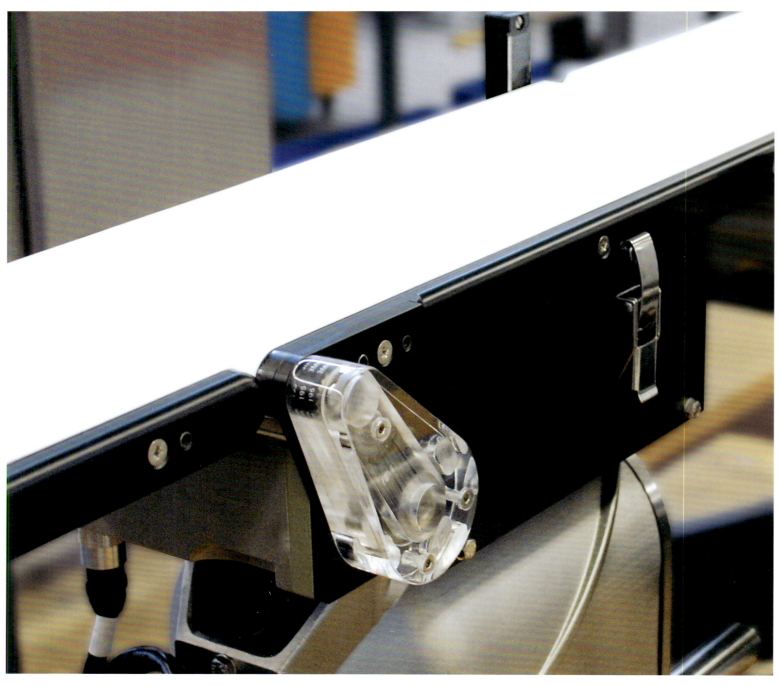

Scan&Fly
Self Bag Drop
Gepäckautomat

Manufacturer
Type22,
Delft, Netherlands
In-house design
Reinout vander Meûlen
Design
VanBerlo
(Bas Bruining),
Delft, Netherlands
Web
www.type22.aero
www.vanberlo.nl

With Scan&Fly, passengers can effortlessly check-in their own baggage. Thanks to its compact design and stainless-steel support frame, the unit fits seamlessly into any airport interior. A sensor recognises when a passenger is near and immediately starts the self-service check-in process. The touch-screen and the intuitive user interface ensure a smooth procedure. A built-in camera captures and links images of passengers and their luggage items, making it much easier to match lost baggage with its owner.

Statement by the jury
A design reduced to the essentials and a clear arrangement make the operation of this self-service bag drop easy to understand and follow.

Mit Scan&Fly können Fluggäste ihr Gepäck problemlos selbst einchecken. Dank seiner kompakten Bauweise und der Edelstahl-Tragrahmen lässt sich der Automat in jedes Flughafenumfeld integrieren. Ein Sensor erkennt Passagiere und startet daraufhin den Check-in-Vorgang. Der Touchscreen und die intuitive Benutzeroberfläche sorgen für einen flüssigen Ablauf des Vorgangs. Eine eingebaute Kamera macht Bilder und verknüpft Personen mit ihrem Gepäckstück, sodass verloren gegangenes Frachtgut schnell wieder zuzuordnen ist.

Begründung der Jury
Durch die auf das Wesentliche reduzierte Gestaltung und den sauber gegliederten Aufbau von Scan&Fly ist seine Benutzung leicht nachvollziehbar.

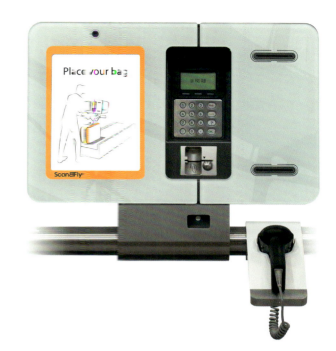

Toshiba Currency Sorter
IBS-1000
Banknote Sorter
Banknoten-Sortiergerät

Manufacturer
Toshiba Corporation,
Social Infrastructure Systems Company,
Tokyo, Japan
In-house design
Tadashi Kurokawa,
Chieko Yasui,
Tomiaki Ishihara,
Ryoichi Ishiura
Web
www.toshiba.co.jp

The IBS-1000 banknote sorter is characterised by compact measurements and a high sorting speed that is among the world's fastest at 1,000 notes per minute. It consists of sorting, extension, and sealing modules, all of which can be flexibly combined. An ergonomic design increases operating comfort while the user is seated and also the visibility of the banknote discharge slot, while simultaneously improving the ease of banknote removal.

Statement by the jury
With the sorter module's dynamically sloping contour and the continuous blue band along its base, this banknote sorter sets fascinating accents.

Das Banknoten-Sortiergerät IBS-1000 zeichnet sich durch kompakte Abmessungen und eine der höchsten Sortiergeschwindigkeiten der Welt von 1.000 Geldscheinen pro Minute aus. Es besteht aus einem Sortier-, einem Erweiterungs- und einem Versiegelungsmodul, die flexibel kombiniert werden können. Die ergonomische Gestaltung erhöht den Komfort bei der Bedienung im Sitzen sowie die Überschaubarkeit des Geldschein-Ausgabeschlitzes und erleichtert das Entnehmen der Banknoten.

Begründung der Jury
Mit der dynamisch geneigten Linie des Sortiermoduls und der durchgängig blauen Abschlussleiste setzt das Banknoten-Sortiergerät spannende Akzente.

RCS 400
Retail Cash System
Münzsortiermaschine
für den Einzelhandel

Manufacturer
SCAN COIN AB,
Malmö, Sweden
Design
Knightec AB
(Jörgen Westin, Mattias Widerstedt),
Stockholm, Sweden
Web
www.scancoin.com
www.knightec.se

The RCS 400 coin-processing machine was conceived as a back-office solution for retail businesses. Emphasis has been placed on colours so as to make utilisation of the machine as self-explanatory as possible. Points of interaction have been made bright, while the passive areas are kept in a dark grey hue. The simple plug-and-play system requires only two cables: one for connection to the power supply and one for a computer.

Statement by the jury
The RCS 400 automated coin-processing machine convinces with its easy-to-understand form and friendly rounded edges that underline its compactness.

Die Münzsortiermaschine RCS 400 ist für den Backoffice-Bereich im Einzelhandel konzipiert. Für eine selbsterklärende Benutzerführung wurde ein Schwerpunkt auf die Farbgebung gelegt. Hellere Teile weisen auf die Interaktionspunkte hin, die passiven Bereiche sind in einem dunkleren Grauton gehalten. Es ist ein Plug-und-Play-System, daher werden lediglich zwei Kabel benötigt: eins für die Stromversorgung und eins für den Anschluss an einen PC.

Begründung der Jury
Die Münzsortiermaschine besticht durch ihre benutzerfreundliche Gestaltung und die abgerundeten Kanten, die ihre Kompaktheit unterstreichen.

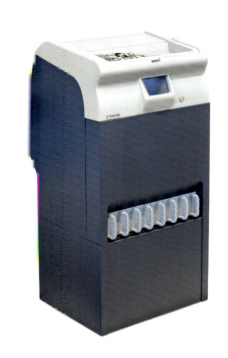

IKA LR 1000
Laboratory Reactor
Laborreaktor

Manufacturer
IKA-Werke Staufen GmbH & Co. KG,
Staufen, Germany
In-house design
Web
www.ika.com

The IKA LR 1000 laboratory reactor,
featuring modular construction, is
capable of optimising chemical reaction
processes on a laboratory scale. The
reactor is equipped with an one litre
container and a stirrer. A heating unit is
integrated under the reactor container,
while a temperature sensor controls the
temperature of the medium. Rotational
speed and temperature are shown
on a clearly arranged digital display.
All processes can be monitored and
documented from the PC via a USB
port. The reactor is dishwasher safe
and thus easy to clean.

Der Laborreaktor IKA LR 1000 in Modul-
bauweise eignet sich zur Optimierung
chemischer Reaktionsvorgänge im
Labormaßstab. Er ist mit einem 1-Liter-
Reaktorbehälter und einem Rührer
ausgestattet. Eine Heizung ist im System
unter dem Reaktorbehälter integriert. Ein
Temperaturfühler reguliert die Temperatur
des Mediums. Drehzahl und Temperatur
werden auf einem übersichtlichen Digitaldisplay angezeigt. Alle Prozesse lassen sich
über eine USB-Schnittstelle vom PC aus
steuern und dokumentieren. Der Reaktor
ist spülmaschinengeeignet und dadurch
leicht zu reinigen.

Statement by the jury
The IKA LR 1000 is appealing due to
its unified, cohesive and consistently
reduced look with a digital display that
is seamlessly integrated into the rear
panel.

Begründung der Jury
IKA LR 1000 gefällt durch sein in sich
geschlossenes und konsequent reduziertes
Erscheinungsbild mit einem nahtlos in der
Rückwand integrierten Digitaldisplay.

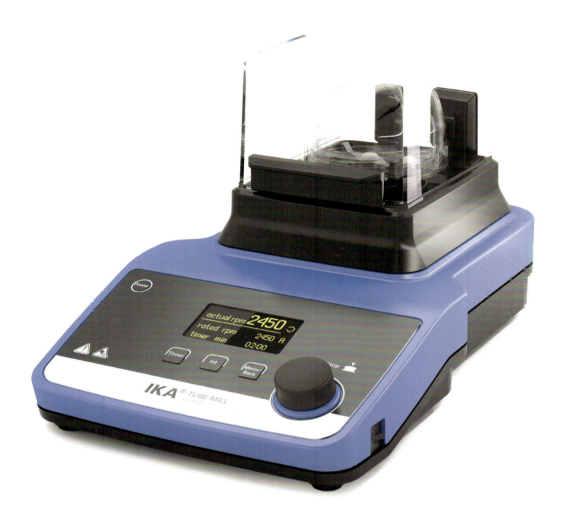

IKA Tube Mill
Labormühle

Manufacturer
IKA-Werke GmbH & Co. KG,
Staufen, Germany
In-house design
Web
www.ika.com

The IKA Tube Mill is designed for milling toxic, infectious or other critical substances. With its single-use grinding tubes, the once necessary need to thoroughly clean the mill after use becomes obsolete. The milled samples can be labelled, transported, stored and cooled for further processing without transfer to another container. Both the grinding tubes and the covering hood are made of transparent material so that the milling process can be monitored. A clearly arranged OLED display and a timer also simplify handling. The reinforced tube mill hood enhances safety and reduces noise.

Die IKA Tube Mill ist eine Labormühle zum Vermahlen von toxischen, infektiösen und anderen kritischen Stoffen. Dank der Einweg-Mahlbecher entfällt die aufwendige Reinigung des Geräts nach dem Gebrauch. Die Mahlproben können zur Weiterbearbeitung ohne Umfüllen beschriftet, transportiert, gelagert und gekühlt werden. Sowohl die Mahlbecher als auch die Abdeckhaube sind aus transparentem Material gefertigt, sodass der Mahlprozess überwacht werden kann. Ein übersichtliches OLED-Display und ein Timer vereinfachen die Bedienung zusätzlich. Die verstärkte Haube der Labormühle sorgt für Sicherheit und Schallreduzierung

Statement by the jury
The construction of the IKA Tube Mill allows for very good visibility during use. The novel utilisation of single-use grinding tubes significantly contributes to operational safety.

Begründung der Jury
Durch ihren Aufbau ist die Labormühle IKA Tube Mill sehr gut einsehbar. Die fortschrittliche Verwendung von Einweg-Mahlbechern trägt entscheidend zur Sicherheit bei.

NEBV
Connecting Cables
Verbindungsleitungen

Manufacturer
Festo AG & Co. KG,
Esslingen, Germany

In-house design
Jörg Peschel

Web
www.festo.com

reddot design award
best of the best 2013

Clear Connections

Complex pneumatic automation units generally require several interlinked control elements. The NEBV connecting cables provide the electrical link between the control unit and the valve terminals. There are two options, each of which respond to different needs: a black cable conceived in particular for standard applications and a grey cable for applications that have higher safety requirements. In its design and choice of materials, this grey cable complies with the "Clean-Design" demands of particular manufacturing environments such as food processing. Both cables have a 45 degree outlet, which makes them suitable for use in a wide variety of installation situations. A convincing detail in the design is the thickening at the end of the cable, which clearly indicates the grip area of the cable. With their consistent design, the NEBV connecting cables perfectly integrate into their work environment – they combine their functionality with a high degree of formal clarity.

Klare Verbindung

Komplexe pneumatische Automatisierungseinheiten benötigen meist mehrere miteinander verknüpfte Steuerelemente. Die Verbindungsleitung NEBV stellt dabei die elektrische Verbindung zwischen Steuerungseinheit und Ventilinseln dar. Es gibt zwei Ausführungen welche den unterschiedlichen Anforderungen Rechnung tragen: Eine schwarze Leitung ist speziell für Standardanwendungen konzipiert; für höhere Schutzgradanforderungen steht eine graue Leitung zur Verfügung. Diese entspricht durch ihre Gestaltung und die Wahl der Materialien den „Clean-Design" Anforderungen für spezielle Fertigungsumgebungen, wie etwa im Nahrungsmittelbereich. Beide Leitungen sind mit einem 45-Grad-Abgang gestaltet, weshalb sie so eingesetzt werden können, dass sie verschiedene Einbausituationen abdecken können. Ein schlüssiges Detail ihrer Gestaltung ist eine Verdickung des Leitungsendes, welches gut erkennbar den eigentlichen Griffbereich definiert. Dieser ist auch ergonomisch so durchdacht, dass der Nutzer ihn gut greifen kann. Mit ihrer stimmigen Gestaltung integrieren sich die Verbindungsleitungen NEBV perfekt in ihr Arbeitsumfeld – sie verbinden ihre Funktionalität dabei mit einem hohen Maß an formaler Klarheit.

Statement by the jury

The design concept of the NEBV connecting cables meets the complexity of their requirements with a convincing functionality. The thickened end of the cables is an impressive design solution; it is self-explanatory and lies well in the hand. The clear design vocabulary of these connecting cables shows conciseness and reflects the corporate design of the Festo company.

Begründung der Jury

Das Gestaltungskonzept der Verbindungsleitungen NEBV begegnet der Komplexität der gestellten Anforderungen mit einer schlüssigen Funktionalität. Ihr verdicktes Leitungsende ist gestalterisch beeindruckend gelöst; es ist selbsterklärend und liegt gut in der Hand. Die klare Formensprache dieser Verbindungsleitungen zeigt Prägnanz und spiegelt das Corporate Design des Unternehmens Festo wider.

EMCA
Motor for Positioning Tasks
Positionierantrieb

Manufacturer
Festo AG & Co. KG,
Esslingen, Germany
In-house design
Simone Mangold
Web
www.festo.com

The Emca unites a variety of functions in a compact housing concept: wear-free EC motor, power electronics with position controller, optional holding brake and fieldbus connection. Its modular design facilitates the attachment of various gear units. Sophisticated connection technology and strong protection ratings against water and dirt make it ideal for industrial use. The design emphasises the motor's compactness and provides a framework for the whole unit. A clearly defined connection area supports intuitive operation by the user.

Der Emca vereint in einem kompakten Gehäusekonzept einen verschleißfreien EC-Motor, Leistungselektronik mit Positioniercontroller, optionale Haltebremse und Feldbusanschluss. Der Anbau verschiedener Getriebe ist in einem modularen Baukasten vorgesehen. Eine ausgeklügelte Anschlusstechnik sowie hohe Schutzklassen gegen Wasser und Schmutz prädestinieren ihn für den industriellen Einsatz. Das Design unterstreicht die Kompaktheit und bildet den Rahmen für die gesamte Einheit. Ein klar definierter Anschlussbereich unterstützt die intuitive Bedienung durch den Nutzer.

HE DB Mini
Manual On-off Valve
Handeinschaltventil

Manufacturer
Festo AG & Co. KG,
Esslingen, Germany
In-house design
Jörg Peschel
Web
www.festo.com

This manual on-off valve is an addition to the series DB Mini providing a simple way to regulate the pressurisation and venting of pneumatically operated machines. It is designed for use in production systems where the focus is on simplicity and minimising costs. Intuitive operation is achieved with the blue-coloured slider as frame, which is an integral part of the housing while encompassing it at the same time. When the valve is closed, the frame also offers the option of locking it.

Das Handeinschaltventil ist eine Ergänzung für die DB-Mini-Baureihe. Mit dem Ventil lässt sich auf einfache Weise das Be- und Entlüften von mit Druckluft betriebenen Maschinen regeln. Die Zielanwendung des Ventils sind Produktionsanlagen, bei denen Einfachheit und Kosten im Fokus stehen. Eine intuitive Bedienung wird durch die blaue Farbigkeit und die Form des Schiebers erreicht. Dieser bildet einen Rahmen, welcher das Gehäuse umgreift und gleichzeitig einen Teil dessen bildet. Bei geschlossenem Zustand bietet der Rahmen zudem die Möglichkeit, das Ventil mit einem Schloss zu sichern.

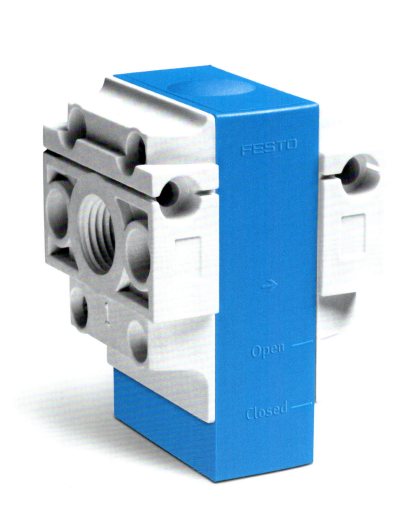

DRRD

Pneumatic Rotary Drive
Pneumatischer Drehantrieb

Manufacturer
Festo AG & Co. KG,
Esslingen, Germany
In-house design
Simone Mangold
Web
www.festo.com

Pneumatic rotary drives with twin pistons operate based on the rack-and-pinion principle, converting linear motion into rotary motion. The new generation DRRD pneumatic rotary drive features an innovative bearing system that, due to its large diameter with integrated pinion and direct integration into the housing, provides high robustness and accuracy. Available in a variety of sizes with torques ranging from 0.2 to 110 Nm, the DRRD provides a wide range of applications in the fields of handling and assembly technology. The minimalist design visually emphasises the gear racks' direction of motion.

Pneumatische Drehantriebe mit Doppelkolben arbeiten nach dem Zahnstangen-Ritzel-Prinzip. Diese wandeln die Linearbewegung in eine Drehbewegung um. Die neue Generation DRRD bietet eine innovative Lagerung, die durch den großen Lagerdurchmesser mit integriertem Ritzel und dem direkten Einbau ins Gehäuse eine hohe Robustheit und Genauigkeit ermöglicht. Durch die Vielzahl der Baugrößen von 0,2 bis 110 Nm Drehmoment, bietet der DRRD ein breites Einsatzspektrum in den Bereichen Handhabung und Montagetechnik. Das reduzierte Design unterstützt visuell die Bewegungsrichtung der Zahnstangen.

PATROL
Sound Generator
Schallgeber

Manufacturer
Pfannenberg GmbH,
Hamburg, Germany
In-house design
Philipp Müller
Web
www.pfannenberg.com

The Patrol sound generator covers a range from 85 to 120 dB. Thanks to a special signal spectrum, which includes many low frequencies, its acoustic resonance is so strong that fewer devices are required to produce the same effect. Installation of the device can be quickly carried out by just one person. It is practically impossible to incorrectly assemble and install the unit. The combination devices, which include flashing lights, feature perfectly aligned visual and acoustic signals.

Statement by the jury
Due to its cubic shape, Patrol requires very little space, while a simplified installation process makes the sound generator particularly user-friendly.

Der Schallgeber Patrol deckt den Bereich 85 bis 120 dB ab. Durch sein besonderes Signalspektrum mit hohem Tieftonanteil ist sein akustisches Signal so weitreichend, dass weniger Geräte für die gleiche Alarmwirkung nötig sind. Für die schnelle Installation ist nur eine Person nötig. Eine fehlerhafte Montage ist praktisch unmöglich. Beim Kombigerät mit Blitzleuchte sind die optischen und akustischen Signale ideal aufeinander abgestimmt.

Begründung der Jury
Durch seine Kubusform braucht Patrol sehr wenig Platz. Die vereinfachte Montagetechnik macht den Schallgeber besonders benutzerfreundlich.

XM 2
Honing Tool
Honwerkzeug

Manufacturer
Kadia Produktion GmbH + Co.,
Nürtingen, Germany
Design
Design Tech
(Jürgen R. Schmid),
Ammerbuch, Germany
Web
www.kadia.de
www.designtech.eu

The Honing Tool XM 2 provides high stability and true running accuracy for drilling applications. The monobloc core is made of high-strength steel. In order to make the slide-on sleeves more easily distinguishable from one another, each displays the tool designation in its special depression, which at the same time prevents it from rolling away. The sleeve's matte surface creates a deliberate contrast to the tool's polished shaft.

Statement by the jury
The components of this honing tool are precisely correlated and simultaneously retain their own character thanks to the contrasting of matte and glossy surfaces.

Das Honwerkzeug XM 2 bietet hohe Stabilität und exakte Rundlaufgenauigkeit beim Bohren. Der Werkzeugkern in Monoblock-Bauweise besteht aus hochfestem Stahl. Um die separat aufschiebbaren Hülsen besser unterscheiden zu können, ist jede von ihnen auf dafür vorgesehenen Vertiefungen beschriftet. Die Vertiefungen verhindern zugleich ein Wegrollen des abgelegten Werkzeugs. Die matte Oberfläche der Hülse erzeugt einen bewussten Kontrast zum polierten Schaft des Werkzeugs.

Begründung der Jury
Bei diesem Honwerkzeug sind die Komponenten präzise aufeinander abgestimmt und bewahren durch den Matt-/Glänzend-Kontrast dennoch einen eigenen Charakter.

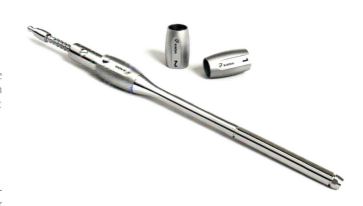

TriVAX
Electrohydraulic Valve Actuator
Elektrohydraulischer Armaturenantrieb

Manufacturer
HOERBIGER GmbH,
Altenstadt, Germany
Design
Design Tech
(Jürgen R. Schmid),
Ammerbuch, Germany
Web
www.hoerbiger.com
www.designtech.eu

The valve actuator TriVax combines technology from the fields of hydraulics, mechanics and electronics in a single, compact casing, thus reducing the complexity of earlier solutions. The flanges for explosion protection have been integrated completely into the casing for the first time, and the actuator is extendable on a modular basis. Innovative details like the switch, which can be locked in any position, increase the safety of operation. All controls found in the user interface are coloured blue.

Statement by the jury
The components of this valve actuator are integrated almost invisibly into the casing. As such, the TriVax sets a new aesthetic design standard in valve actuator technology.

Der Armaturenantrieb TriVax vereint Hydraulik, Mechanik und Elektronik in einem einzigen kompakten Gehäuse und reduziert so die Komplexität früherer Anlagen. Die Flansche für den Explosionsschutz wurde erstmals komplett ins Gehäuse integriert. Der Antrieb ist jederzeit modular erweiterbar. Innovative Details wie der in jeder Position abschließbare Schalter erhöhen zudem die Betriebssicherheit. Alle Bedienelemente des User-Interfaces sind blau eingefärbt.

Begründung der Jury
Die Komponenten des Armaturenantriebs sind nahezu unsichtbar in den Korpus integriert. Dadurch setzt TriVax ein neues ästhetisches Zeichen in der Antriebstechnik.

Power Mini
Primary Switched-Mode Power Supply
Primär getaktetes Schaltnetzteil

Manufacturer
BLOCK Transformatoren-Elektronik GmbH,
Verden, Germany
In-house design
Alexander Kesmann
Design
Bernd Huth,
Hannover, Germany
Web
www.block-trafo.de
www.berndhuth.de

The Power Mini device series ensconces efficient switched-mode power supply in slim plastic cases, covering the low to medium power range from 25 to 100 watts. The grey colour was developed under a variety of lighting conditions in a light laboratory and derives from the well-established blue colour of the Block brand. The clear structure and arrangement of the connections and controls provide an enhanced degree of functionality and ease of use, in addition to fostering a high-quality impression.

Statement by the jury
Thanks to its reduced, premium design and a modern colour concept, the Power Mini device series conveys a stylish and elegant appearance.

Die Geräteserie Power Mini umfasst effiziente Schaltnetzteile im schlanken Kunststoffgehäuse. Sie decken den unteren und mittleren Leistungsbedarf von 25 bis 100 Watt ab. Der Grauton wurde unter verschiedensten Beleuchtungssituationen im Lichtlabor aus dem etablierten Blau der Marke Block entwickelt. Die klare Gliederung und Anordnung der Anschlüsse und Bedienelemente sorgen für hohe Funktionalität und Bedienkomfort und vermitteln eine hochwertige Anmutung.

Begründung der Jury
Dank der reduzierten, aber hochwertigen Gestaltung und dem modernen Farbkonzept ist die Geräteserie Power Mini eine stilvoll-elegante Erscheinung.

Sunny Multigate-US
Communication Gateway
Kommunikationsschnittstelle

Manufacturer
SMA Solar Technology AG,
Niestetal, Germany
Design
industrialpartners GmbH
(Jens Arend),
Frankfurt/Main, Germany
Web
www.sma.de
www.industrialpartners.de

Sunny Multigate-US is a smart termination point designed to connect photovoltaic systems to the power distribution grid. The photovoltaic plant is thus connected directly to the Sunny Portal monitoring system via Ethernet communication, which enables customers to have quick access to the most important plant data with real-time monitoring. This concept makes Sunny Multigate-US a favourable solution for intelligent energy monitoring.

Statement by the jury
The reduced and well-arranged design appearance of the Sunny Multigate-US ensures clear intelligibility in terms of both functionality and usage.

Sunny Multigate-US ist der intelligente Anschlusspunkt einer Photovoltaikanlage an das öffentliche Stromnetz. Die Solaranlage ist somit via Ethernet-Kommunikation direkt mit dem Überwachungssystem Sunny Portal verbunden. Dies erlaubt über Echtzeit-Monitoring auf Moduleebene den bequemen Zugriff auf die wichtigsten Anlagendaten. Damit ist Sunny Multigate-US eine optimale Lösung für die intelligente Energieüberwachung.

Begründung der Jury
Das reduzierte und aufgeräumte Erscheinungsbild von Sunny Multigate-US sorgt für klare Verständlichkeit in Funktion und Anwendung.

SCHUKOultra II
Plug and Connector
Steckvorrichtung

Manufacturer
ABL SURSUM,
Bayerische Elektrozubehör GmbH & Co. KG,
Lauf/Pegnitz, Germany
Design
Büro für Konstruktion & Gestaltung
(Prof. Hagen Kluge),
Buch am Erlbach, Germany
Web
www.abl-sursum.com
www.hagenkluge.de

SCHUKOultra II is a connector for use under the harshest conditions in trade and industry. Two high-performance plastic materials were specially developed to complement each other, thus ensuring that this plug and connector set is highly resistant to both high and low temperatures. The two-component exterior design of the grips exhibits a strong interplay of hard and soft plastics. The diamond pattern found in the soft components gives the connector a special non-slip grip.

Statement by the jury
The design of this plug and connector set places strong emphasis on ergonomics. All components are seamlessly interconnected

SCHUKOultra II ist eine Steckvorrichtung für den Einsatz unter erschwerten Bedingungen im Handwerk und in der Industrie. Zwei aufeinander abgestimmte und eigens dafür entwickelte Hochleistungskunststoffe machen Stecker und Kupplung sehr hitze- und kältebeständig. Die 2-Komponenten-Technologie der Griffe sorgt für ein gutes Zusammenspiel von hartem und weichem Kunststoff. Durch das Rautenmuster ist die Weichkomponente besonders griffig.

Begründung der Jury
Bei der Gestaltung der Steckvorrichtung wurde sehr viel Wert auf die Ergonomie gelegt. Alle Komponenten sind nahtlos miteinander verbunden.

SRB1021
Storage Rack Battery Module
Storage Rack Batterie-Modul

Manufacturer
ads-tec GmbH,
Leinfelden-Echterdingen, Germany
In-house design
Roman Molchanov,
Matthias Bohner
Web
www.ads-tec.de
Honourable Mention

The SRB1021 is a real-time power storage module based on lithium-ion technology, which assures the temporary storage of renewable energy. The 19" battery features three independent safety circuits and a permanent temperature monitor. In addition, its special construction prevents the risk of loose screws causing a short circuit. The design of the battery module is defined by a high-quality front, which is dominated by the specially developed safety plugs. Plus and minus symbols indicate how to install the system and highlight the purpose of the product.

Statement by the jury
The concept of this battery module comprises a large number of safety measures packed into a clear language of form with high-quality appeal.

Die SRB1021 ist ein echtzeitfähiges elektrisches Speichermodul auf Lithium-Ionen-Basis, mit dem sich regenerative Energien zwischenspeichern lassen. Die Batterie im 19"-Format verfügt über drei unabhängige Sicherheitskreise und eine permanente Temperaturüberwachung. Die spezielle Konstruktion verhindert zudem, dass lose Schrauben einen Kurzschluss auslösen können. Das Erscheinungsbild des Batterie-Moduls wird von der hochwertigen Front bestimmt, in der die eigens entwickelten Sicherheitsstecker dominieren. Plus und Minus zeigen an, wie montiert wird und worum es sich bei dem Gerät handelt.

Begründung der Jury
Das Batterie-Modul bietet etliche sicherheitstechnische Maßnahmen, die sich in einer hochwertig anmutenden und klaren Gestaltung präsentieren.

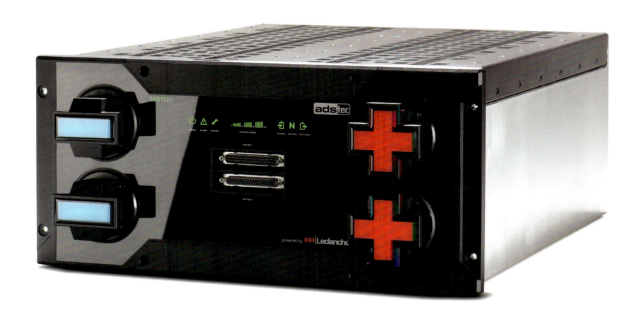

Sarix IL Series Box Cameras and Micro Domes
IP Camera
Netzwerkkamera

Manufacturer
Pelco by Schneider Electric,
California, USA
In-house Design
Vincent Wong, Junwei Geng
Design
Yang Design
(Jamy Yang, Liu Qian, Sven Grumbrecht),
Shanghai, China
Web
www.pelco.com
www.yang-design.com

This series of high-definition IP cameras is available with two different caps and can be mounted either to a wall or a ceiling. Its form is inspired by triangular traffic warning signs and is a metaphor for surveillance in urban space. In addition, the triangular prism is a space-saving solution at a minimal volume. When installed in the ceiling, the camera with the dome cap is positioned inside the ceiling void, with the cap generating an elegant ceiling finish.

Diese Serie von hochauflösenden Netzwerkkameras ist mit zwei verschiedenen Aufsätzen erhältlich und kann an der Wand oder an der Decke verwendet werden. Die Form ist von einem dreieckigen Verkehrswarnschild inspiriert und steht als Sinnbild für die Überwachung im urbanen Raum. Das dreieckige Prisma ist zudem eine platzsparende Lösung bei einem minimalen Volumen. Wenn die Kamera mit dem Domaufsatz in der Decke installiert wird, befindet sie sich im Deckenhohlraum und der Aufsatz bildet einen eleganten Deckenanschluss.

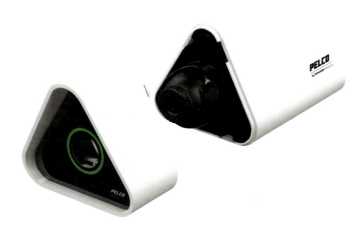

SeeSV-S205
Sound Camera
Tonkamera

Manufacturer
SM Instruments Co., Ltd.,
Daejeon, South Korea
Design
I²DEA, Department of Industrial Design,
KAIST (Prof. Seok-Hyung Bae),
Daejeon, South Korea
Web
www.smins.co.kr
http://i2dea.kaist.ac.kr

The SeeSV-S205 handheld sound camera records sound impressions in real time. A high-resolution optical camera is then used to display the sound waves as colour contours on a computer screen. The device is used, for instance, in the development or repair of vehicles to spot the source of various kinds of noise. The pentagonal form of the body harmonises with the five spiral arrays of high-sensitivity digital microphones. The camera weighs only 1.78 kg, is 39 cm wide and 38 cm high. Its central handle at the back facilitates one-handed operation, while the side handles provide a stable stand.

Die tragbare Tonkamera SeeSV-S205 nimmt Klangeindrücke in Echtzeit auf. Die Schallwellen werden dann über eine hochauflösende optische Kamera als Farbkonturen auf einem Computer dargestellt. Das Gerät wird z. B. bei der Reparatur oder Entwicklung von Fahrzeugen eingesetzt, um die Quelle von Stör- oder Nebengeräuschen aufzuspüren. Die fünfeckige Form des Gehäuses harmoniert mit den fünf spiralförmig angeordneten hochempfindlichen Digitalmikrofonen. Die Kamera wiegt nur 1,78 kg, ist 39 cm breit und 38 cm hoch. Ihr Mittelgriff auf der Rückseite erlaubt den Einhandbetrieb, während die seitlichen Halter für einen stabilen Stand sorgen.

Statement by the jury
The exciting aesthetics of this sound camera result from its pentagonal shape and the pattern of concentric circles visualising the sound waves.

Begründung der Jury
Die aufregende Ästhetik der Tonkamera ergibt sich aus der fünfeckigen Form und dem Muster aus konzentrischen Kreisen, das die Schallwellen visualisiert.

CP3919
Control Panel
Bedienpanel

Manufacturer
Beckhoff Automation GmbH,
Verl, Germany
Design
Design AG (Frank Greiser),
Rheda-Wiedenbrück, Germany
Web
www.beckhoff.com
www.design-ag.de

The CP3919 is a freely suspended control panel for operating machines and systems. Designed without screws or projections, it can also be used in the fields of food production and pharmaceutical packaging. In addition to its modern elegance, the panel possesses the necessary robustness, which is for instance evident in its protected glass edges. The waterproof plug and mechanical system are housed in the support arm in a space-saving way.

Statement by the jury
The reduced and flush design of the CP3919 is both practical and elegant. It also skilfully highlights the quality of the control panel.

Das CP3919 ist ein frei hängendes Bedien-panel zur Steuerung von Maschinen und Anlagen. Es besitzt weder Schrauben noch Vorsprünge und kann daher auch in der Lebensmittelfertigung und Medi-kamentenverpackung eingesetzt werden. Neben einer zeitgemäß eleganten Gestal-tung bietet das Panel auch die notwendi-ge Robustheit, z. B. durch die geschütz-ten Glaskanten. Der wasserdichte Stecker und die Mechanik sind platzsparend im Tragarm untergebracht.

Begründung der Jury
Die reduzierte und flächenbündige Er-scheinung des CP3919 ist sowohl zweck-mäßig als auch edel. Sie betont gekonnt die Wertigkeit des Bedienpanels.

SANGO
**Electronic
Point of Sale Terminal**
Verkaufsstellensystem

Manufacturer
AURES Technologies,
Lisses, France
In-house design
Patrick Cathala
Design
ID'S (Bertrand Médas),
Lyon, France
Web
www.aures.com
www.id-s.fr

Sango is an electronic point of sale (EPOS) terminal that frees up the space beneath the touchscreen. The monitor is suspended from a support arm that also houses the CPU. This has been achieved by using a monitor stand fashioned only from die-cast aluminium based on the principle of an exoskeleton. Partly concealed, the stand is clad with an interchangeable coloured polycarbonate panel available in seven colours.

Statement by the jury
The exposed stand contributes to the visual effect of the monitor hovering in mid-air, while also turning this point of sale terminal into a sculptural object.

Bei dem Verkaufsstellensystem Sango ist der Raum unter dem Touchscreen zur freien Nutzung verfügbar. Der Monitor wird von einer Tragvorrichtung mit Ausleger gehalten, in der auch die Zentraleinheit untergebracht ist. Dies ist möglich, weil ein nach dem Prinzip des Exoskeletts ganz aus Aluminiumspritzguss bestehendes Monitorgestell verwendet wurde. Das Gestell liegt zum Teil frei und ist mit einem in sieben Farben verfüg-baren Wechselrahmen aus Polycarbonat verkleidet.

Begründung der Jury
Das freigelegte Gestell trägt zum visu-ellen Schwebezustand des Monitors bei und macht das Verkaufsstellensystem zum skulpturalen Objekt.

VMT8000 Series
Terminal PC

Manufacturer
acs-tec GmbH,
Leinfelden-Echterdingen, Germany
In-house design
Axel Fett,
Matthias Bohner
Web
www.ads-tec.de

The VMT8000 series is used as a terminal on forklift trucks and vehicles, as well as on production lines and machines. These terminal PCs are intuitively operated and withstand the rough conditions in industrial workplaces. This is guaranteed by the touchscreen made of hardened glass, the rigid aluminium housing, which functions as a passive cooler, and the solid, glass-reinforced polyamide front. The series is also shock-protected and waterproof. With only two special screw connections, which can withstand even strong shocks and vibrations such as when installed on forklifts, the terminal can be pivoted and adjusted to any viewing angle using various mountings.

Die VMT8000-Serie kommt als Staplerterminal und Fahrzeugterminal zum Einsatz sowie an Anlagen und Maschinen. Die Terminal-PCs sind intuitiv zu bedienen und halten den rauen Umgebungsbedingungen an industriellen Arbeitsplätzen stand. Dafür sorgen der Touchscreen aus gehärtetem Glas, das unverwüstliche Aluminiumgehäuse, das als passiver Kühler dient, und die massive glasverstärkte PA-Front. Die Serie ist zudem schlagfest und wasserdicht. Mit nur zwei Spezialverschraubungen, die auch starke Erschütterungen etwa auf einem Stapler aushalten, kann das Terminal auf diversen Halterungen geneigt und im richtigen Blickwinkel eingestellt werden.

Statement by the jury
The fascinating aspect of this terminal PC series is the transfer of modern technology into a highly rigid and robust construction.

Begründung der Jury
Das Faszinierende an den Terminal-PCs ist die Überführung moderner Technologie in eine enorm stabile und widerstandsfähige Baukonstruktion.

OPC8000 Series
Panel PC

Manufacturer
ads-tec GmbH,
Leinfelden-Echterdingen, Germany
In-house design
Axel Fett,
Matthias Bohner
Web
www.ads-tec.de

The OPC8000 series is a family of panel PCs for stationary installation in factories, on machines or in switching cabinets. In order to withstand extremely tough operating conditions in industrial environments, this series combines a service-friendly and robust construction with a hardened multi-touch glass front that can be operated by finger, while wearing gloves or with a stylus. Thanks to the quick-snap mounting system, the PC is simply inserted into the mounting slot before being bolted down to ensure that the front is splash-proof.

Die OPC8000-Serie ist eine Panel-PC-Familie zur stationären Montage in Werkhallen, an Maschinen und in Schaltschränken. Um unter extrem harten Einsatzbedingungen in der Industrie zu bestehen, verbindet die Serie einen servicefreundlichen und robusten Geräteaufbau mit einer gehärteten Multitouch-Glasfront, die mit Fingern, Handschuhen oder Spezialstiften bedient werden kann. Durch die Quick-Snap-Montage muss der PC nur in den Einbauausschnitt gedrückt werden.

Statement by the jury
These panel PCs surprise with a congenial mix of practice-oriented robustness and modern touchscreen aesthetics.

Begründung der Jury
Die Panel-PCs überraschen durch eine kongeniale Mischung aus praxisorientierter Robustheit und zeitgemäßer Touchscreen-Ästhetik.

Microtector III & Polytector III
Portable Gas Detector
Tragbares Gaswarngerät

Manufacturer
GfG Gesellschaft für Gerätebau,
Dortmund, Germany
Design
Lengyel Design
(Stefan Lengyel),
Essen, Germany
Web
www.gasmessung.de
www.lengyel.de

This portable gas detector comes in two versions: Microtector III without a pump and Polytector III with an integrated suction pump. Importance has been placed on reducing the size and weight of the unit to provide a lighter overall experience for the user when carrying protective equipment. The display is slightly inclined forward and can be rotated using a simple key combination to provide comfortable readability from any holding or carrying position. The unit warns of gas hazards with a 103 dB horn. In addition, the display provides clear information about the current hazard situation using the traffic light colours red, yellow and green.

Das tragbare Gaswarngerät gibt es in den Ausführungen Microtector III ohne Pumpe und Polytector III mit integrierter Ansaugpumpe. Besonderer Wert wurde darauf gelegt, die Gerätegröße und das Gewicht zu reduzieren, um die Schutzausrüstung für den Träger nicht unnötig schwer zu machen. Das Display ist leicht nach vorn geneigt und kann mit einer einfachen Tastenkombination gedreht werden, um ein bequemes Ablesen der Anzeige aus jeder Halte- oder Trageposition zu ermöglichen. Bei Gasgefahren warnt das Gerät mit einer 103 dB lauten Sirene. Zusätzlich gibt das Display durch die Ampelfarben Rot, Gelb und Grün Aufschluss über die aktuelle Gefährdungslage.

Leica DISTO™ X310
Laser Distance Meter
Laser-Distanzmessgerät

Manufacturer
Leica Geosystems AG,
Heerbrugg, Switzerland
Design
BUDDE BURKANDT DESIGN
(Janine Budde, Marco Burkandt),
Munich, Germany
Web
www.leica-geosystems.com
www.buddeburkandt.de

Thanks to a reinforced housing and sturdy interior, the Leica Disto X310 laser distance meter can withstand falls from a height of up to two metres. The device is water-jet-protected and dust-tight. Next to the familiar measurement functions, its integrated tilt sensor also offers the option of taking indirect distance and height measurements. The measurements can be taken reliably from all of its six sides.

Statement by the jury
The laser distance meter fascinates with its robustness and functionality. Red lines clearly structure the device and lend it a less austere presence.

Das Laser-Distanzmessgerät Leica Disto X310 hält dank seiner widerstandsfähigen Außenhülle und des stabilen Innenlebens Stürzen aus bis zu zwei Metern Höhe stand. Das Gerät ist strahlwassergeschützt und staubdicht. Neben den gewohnten Messfunktionen bietet es durch den integrierten Neigungssensor zusätzlich die Möglichkeit für indirekte Distanz- und Höhenmessungen. Die Messungen lassen sich von allen sechs Seiten sicher durchführen.

Begründung der Jury
Das Laser-Distanzmessgerät besticht durch Robustheit und Funktionalität. Die roten Linien sorgen für eine klare Unterteilung und lockern das Gesamtbild auf.

Leica DISTO™ D210
Laser Distance Meter
Laser-Distanzmessgerät

Manufacturer
Leica Geosystems AG,
Heerbrugg, Switzerland
Design
BUDDE BURKANDT DESIGN
(Janine Budde, Marco Burkandt),
Munich, Germany
Web
www.leica-geosystems.com
www.buddeburkandt.de

The Leica Disto D210 laser distance meter is a handy entrance-level model for the professional field. It offers the user many reliable, optimised functions for obtaining fast and exact measurements. A compact design combined with the tool's precise performance demonstrates portable dependability. The structured body and emphasis on functionally important keys enable simple, virtually self-explanatory operation.

Statement by the jury
The clear design language of the laser distance meter reflects its high technical precision and outstanding user-friendliness.

Das Laser-Distanzmessgerät Leica Disto D210 ist ein handliches Einstiegsmodell für den Profibereich. Es hilft mit vielen optimierten Funktionen zuverlässig bei der schnellen und exakten Vermessung. Die kompakte Form kombiniert mit der präzisen Leistung bildet die Zuverlässigkeit des Geräts ab. Der strukturierte Aufbau und die Hervorhebung funktional wichtiger Tasten ermöglichen eine einfache und nahezu selbsterklärende Bedienung.

Begründung der Jury
In der klaren Formensprache des Laser-Distanzmessgeräts spiegeln sich seine hohe technische Präzision und die ausgezeichnete Bedienungsfreundlichkeit wider.

Prisma
Ultrasonic Flaw Detector
Ultraschall-Fehlerprüfgerät

Manufacturer
Sonatest Ltd,
Milton Keynes, GB
In-house design
Stewart Lamont,
Yvan Gosselin
Design
LA Design
(Pete Holdcroft, Matthew Brown),
Hertfordshire, GB
Web
www.sonatest.com
www.la-design.co.uk

Prisma is an ultrasonic flaw detector incorporating real-time imaging and 3D scanning representation. It covers all testing applications for material inspection in a single device, features integral data recording and uses a rigid, shock-mounted internal chassis surrounded by an impact-absorbing enclosure. The external yellow band increases visibility and contains the major connectors, D-rings and a retractable handle. The device is easy to set up and use, even in extreme environments, since it features a multi-position stand and hook, as well as a tripod mount and harness attachments.

Statement by the jury
The Prisma ultrasonic flaw detector merges state-of-the art technology with a robust construction and modern colour aesthetics.

Prisma ist ein Ultraschall-Fehlerprüfgerät mit integrierter Echtzeit-Bildgebung und 3D-Scan-Darstellung. Es deckt alle Testanwendungen für die Materialprüfung in einem Gerät ab und zeichnet die Daten automatisch auf. Das interne Chassis ist gefedert und von einem stoßdämpfenden Gehäuse umgeben. Die gelbe Umrandung verbessert die Sichtbarkeit und enthält die wichtigsten Anschlüsse, D-Ringe und einen ausziehbaren Griff. Das Gerät kann auch unter extremen Bedingungen überall aufgestellt bzw. eingesetzt werden, dafür stehen der in mehrere Positionen ausrichtbare Fuß, eine Stativhalterung und Gurtbefestigungen zur Verfügung.

Begründung der Jury
Das Ultraschall-Fehlerprüfgerät Prisma vereint State-of-the-Art-Technologien mit einer robusten Bauweise und einer modernen Farbästhetik.

Arcode
Controller, Driver and Evacuation System for Traction Lifts
Steuereinheit, Antriebselektronik und Evakuierungssystem für Lastenaufzüge

Manufacturer
Arkel Elektrik Elektronik Tic. Ltd. Sti.,
Istanbul, Turkey
Design
Arman Design and Development,
Istanbul, Turkey
Web
www.arkel.com.tr
www.armantasarim.com

Arcode unites a controller, a driver and an evacuation system for traction lifts in one device. It thus simplifies complex structures and is also convenient to install. Although the system comprises a large number of electronic and mechanical components, it exhibits a notably compact design. Moreover, the system is equipped with clearly laid out, user-friendly connector interfaces whose functions are indicated by LEDs.

Statement by the jury
Arcode convinces with its very compact dimensions and appealing orange-coloured, high-gloss front design with elegant contours and curves.

Arcode vereint Steuereinheit, Antriebselektronik und Evakuierungssystem für Lastenaufzüge in sich. Es vereinfacht dadurch komplexe Strukturen und ist zudem besonders leicht und bequem zu montieren. Obwohl das Gerät aus zahlreichen elektronischen und mechanischen Bauteilen besteht, weist es eine sehr kompakte Form auf. Es ist mit sehr übersichtlichen und benutzerfreundlichen Schnittstellen ausgestattet, deren Funktionen durch LEDs kenntlich gemacht werden.

Begründung der Jury
Arcode überzeugt durch seine sehr kompakten Abmessungen und eine ansprechend gestaltete orangeglänzende Bedienfront mit Bögen und Abrundungen.

Summit pointer SMT-15R
Precision Gauge Block
Präzisionsendmaß

Manufacturer
Nishimura JIG Co., Ltd.,
Kanazawa City,
Ishikawa Prefecture, Japan
In-house design
Akira Nishimura
Design
Scala Design und technische Produktentwicklung (Peter Theiss),
Böblingen, Germany
Web
www.nishimura-jig.jp
www.scala-design.de

Summit pointer SMT-15R is a precision gauge block for measuring workpieces with bevelled or broken edges. The tool is positioned on a right-angled edge and, with its constant, precise radius, enables the user to accurately approach theoretical edges with measuring instruments, drill bits and milling tools. The contact surfaces are clearly indicated by their bevelled edges and, with their integrated neodymium magnets, provide a secure hold on metallic workpieces.

Statement by the jury
The smooth, seamless and carefully finished design of this precision gauge block emphasises its high accuracy.

Summit pointer SMT-15R ist ein Präzisionsendmaß zum Vermessen von Werkstücken mit abgefasten oder gebrochenen Kanten. Das Werkzeug wird an einer 90-Grad-Kante angesetzt und bietet durch den konstant präzisen Radius die Möglichkeit, imaginäre Kanten exakt mit Mess-, Bohr- und Fräsköpfen anzufahren und einzumessen. Die Aufsetzflächen sind durch abgeschrägte Kanten klar ablesbar und sorgen mit integrierten Neodym-Magneten für sicheren Halt an metallischen Werkstücken.

Begründung der Jury
Die glatte, fugenlose und sehr sorgfältig verarbeitete Gestaltung des Präzisionsendmaßes unterstreicht seine hohe Genauigkeit.

Teranex 2D Processor
Video Standards Converter
Video-Normwandler

Manufacturer
Blackmagic Design Pty Ltd,
Melbourne, Australia
In-house design
Blackmagic Industrial Design Team
Web
www.blackmagicdesign.com

The Teranex 2D Processor is a video standards converter that allows a wide variety of formats to be converted without a loss in quality. This device is also certified to create content for the Apple Store. What is more, it is half the size of existing video standards converters and particularly quiet in operation. The left-to-right workflow is complemented by high-contrast backlit buttons that are colour-coded to facilitate fast, intuitive operation in live, high-pressure, outdoor broadcast environments. The aluminium fascia is engineered to provide strength and rigidity for rack mounting, whilst also reducing light bleed between buttons for improved clarity and legibility. The anodised finish provides a tactile contrast to the high-grip silicone buttons and thus helps to minimise operator errors.

Der Teranex 2D Processor ist ein Video-norm-Wandler, der die Konvertierung zwischen einer breiten Auswahl von Formaten ohne Qualitätsverlust ermöglicht. Das Gerät ist sogar für die Erstellung von Inhalten für den Apple Store zertifiziert. Es ist zudem nur halb so groß wie die meisten anderen Normwandler und dazu besonders geräuscharm. Die Bedienab-folge von links nach rechts ist zusammen mit den beleuchteten Tasten und der Farbkodierung ideal auf eine intuitive, schnelle Benutzung während zeitkriti-scher Liveübertragungen abgestimmt. Die Aluminiumfront sorgt sowohl für Stabi-lität bei der Schaltschrank-Montage als auch für eine gut lesbare Anzeige durch reduzierte Lichtstreuung zwischen den einzelnen Tasten. Die eloxierte Oberfläche ist haptisch gut von den griffigen Silikon-Tasten zu unterscheiden, was Bedienungs-fehler verringert.

Statement by the jury
The Teranex 2D processor shows an elegant, slim and clearly arranged appearance. Thanks to its compact size, it is easy to transport.

Begründung der Jury
Der Teranex 2D Processor wirkt elegant, schlank und übersichtlich. Dank seiner kompakten Größe lässt er sich außerdem gut transportieren.

Battery Converter
Video Format Converter
Videoformat-Konverter

Manufacturer
Blackmagic Design Pty Ltd,
Melbourne, Australia
In-house design
Blackmagic Industrial Design Team
Web
www.blackmagicdesign.com

Thanks to its built-in battery, this battery converter provides uninterrupted signal conversion for live outdoor productions without the need for external power. It features time-saving automatic detection of SDI or HDMI video formats and utilises the latest high-frequency technology to transmit flawless HD or SD video across long cable lengths. The compact aluminium casing is palm-sized and thus fits into any camera bag, yet the converter is also strong enough to be run over by a truck and keep working. The laser-etched graphics are scratch-resistant, whilst the silicone feet provide a high-grip base. Integrated LEDs indicate the built-in battery's charge level.

Der Battery Converter ermöglicht dank seines eingebauten Akkus den unterbrechungsfreien Einsatz an Drehorten ohne externe Stromversorgung. Er verfügt über eine zeitsparende automatische Erkennung von SDI- oder HDMI-Video-Formaten und nutzt modernste Technologien um HD- und SD-Videos störungsfrei über lange Kabel zu übertragen. Das kompakte Aluminiumgehäuse passt auf eine Handfläche und somit in jede Kameratasche. Dabei ist der Converter robust genug um es unbeschadet zu überstehen, wenn er von einem LKW überrollt werden würde. Die Lasergravuren sind kratzfest, und die Gummifüße sorgen für Standfestigkeit. Der Ladezustand der Akkus wird durch LEDs angezeigt.

Statement by the jury
This battery converter is practically indestructible: due to its robust construction, it can endure exceptionally heavy wear and tear.

Begründung der Jury
Der Battery Converter ist praktisch unverwüstlich. Dank seiner robusten Ausstattung hält er außergewöhnliche starke Belastungen aus.

Intelligent Banknote Recycler
Intelligenter Banknoten-Recycler

Manufacturer
GLORY LTD.,
Himeji, Japan
In-house design
Hirofumi Tougo,
Takayuki Kuroda
Web
www.glory-global.com

The RBG-200 intelligent banknote recycler is a compact multi-purpose device for use at cashier workspaces in financial institutions. It can recognise and handle notes of various sizes from around the world and even accepts notes of poor quality. The recycler includes a function for recording the serial numbers of banknotes and can record up to 128 denominations, so that multi-currency operations are possible simultaneously. Thanks to the auto verification feature, the device can recount the amount of cash without human intervention, which drastically reduces the amount of work for bank staff. Furthermore, a visual presentation has been added to the operator monitor to further enhance ease of use.

Der intelligente Banknoten-Recycler RBG-200 ist ein kompaktes Mehrzweckgerät für die Verwendung im Kassenbereich von Geldinstituten. Er kann mit internationalen Banknoten unterschiedlicher Größe umgehen und akzeptiert sogar Geldscheine schlechter Qualität. Er besitzt eine Funktion zur Aufzeichnung der Seriennummern der Banknoten und kann bis zu 128 Stückelungsarten verarbeiten, sodass mehrere Währungen gleichzeitig durchlaufen können. Dank der automatischen Überprüfungsfunktion kann das Gerät den Bestand ohne menschliches Eingreifen nachzählen, was für die Bankmitarbeiter den Arbeitsaufwand erheblich reduziert. Außerdem wird der Bedienmonitor für eine noch bessere Verständlichkeit durch grafische Darstellungen ergänzt.

Statement by the jury
The fascia of this banknote recycler is marked by a distinctively contoured blue frame that adds a dynamic aspect to everyday work in the banking sector.

Begründung der Jury
Die Front des Banknoten-Recyclers wird durch einen markant geschwungenen blauen Rahmen eingefasst und setzt damit einen dynamischen Akzent im Bankalltag.

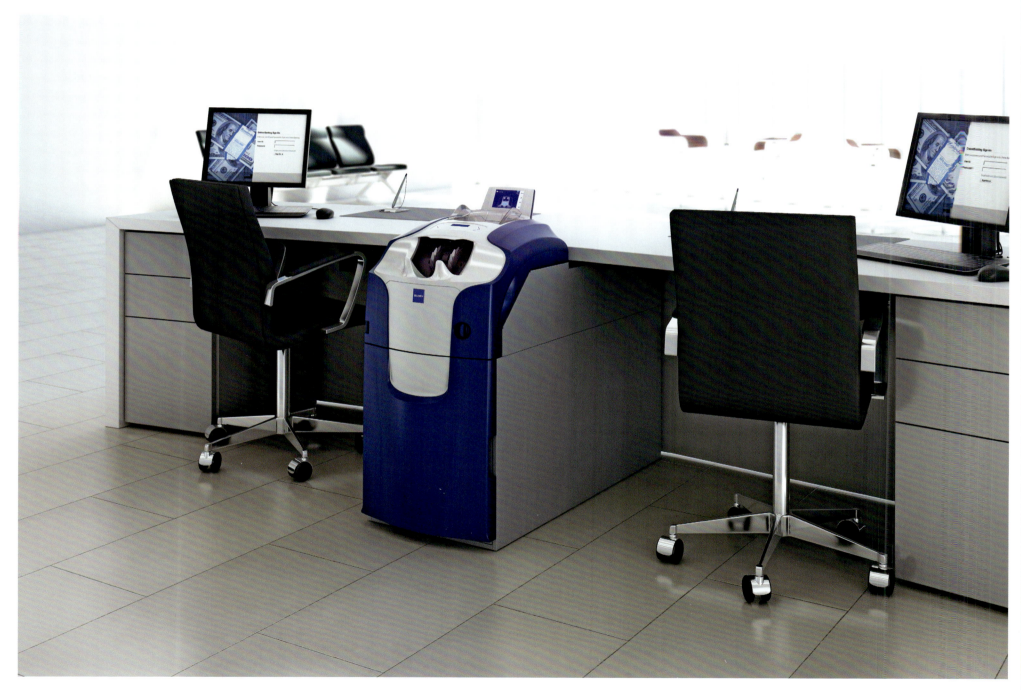

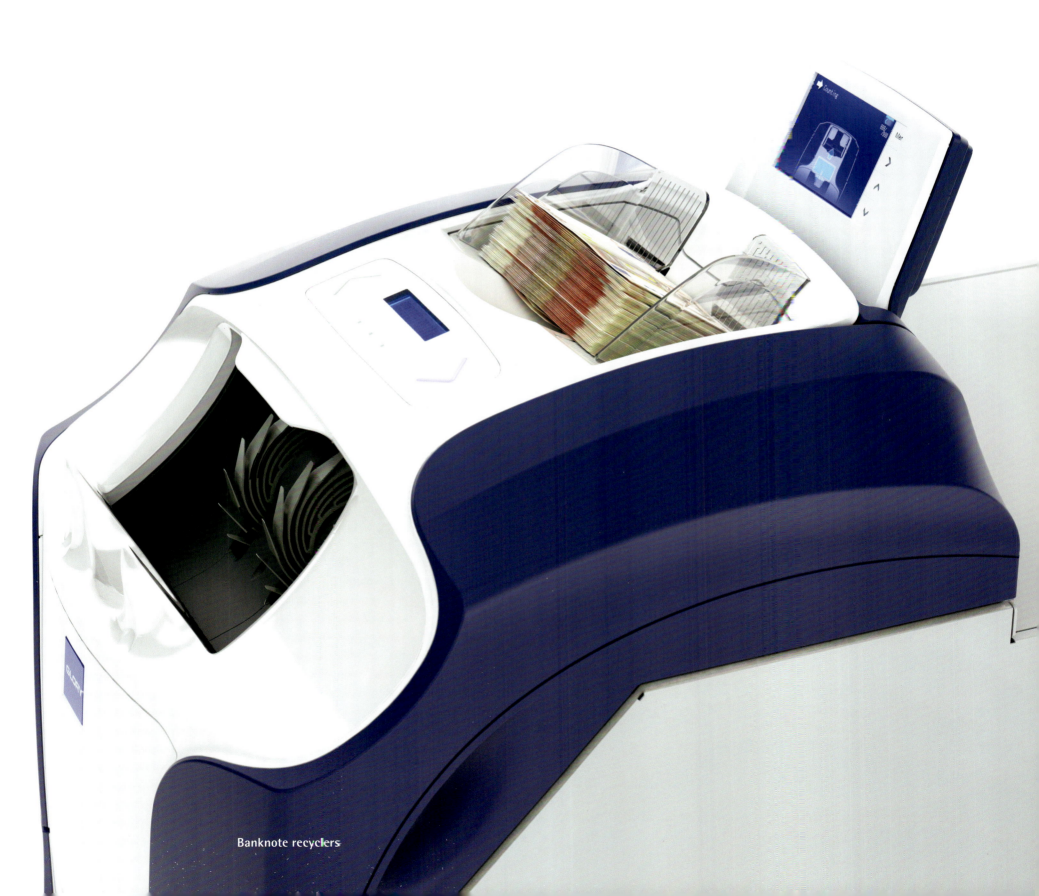

Banknote recyclers

RBG-100

Intelligent Banknote Recycler
Intelligenter Banknoten-Recycler

Manufacturer
GLORY LTD.,
Himeji, Japan
In-house design
Hirofumi Tougo,
Takayuki Kuroda
Web
www.glory-global.com

Despite compact measurements that make it small enough to be installed under the high counters of financial institutions, the RGB-100 banknote recycler can store no less than 17,000 banknotes while assuring a very high counting speed. It includes a function that records the banknotes' serial numbers, making exact cash management possible. Equipped with various functions for accommodating large volume deposits and disbursements, the device works both efficiently and extremely accurately. In addition, the RGB-100 features a built-in collection box, making cash collection by cash-in-transit companies easier. Due to the visual presentations on the monitor, the interface is easy to understand.

Trotz seiner kompakten Abmessungen, die einen Einbau unter den hohen Kassenschaltern von Geldinstituten ermöglicht, kann der Banknoten-Recycler RGB-100 bis zu 17.000 Banknoten aufnehmen und zeigt dabei eine sehr hohe Zählgeschwindigkeit. Er besitzt eine Funktion zur Aufzeichnung der Seriennummern der Banknoten und ermöglicht somit ein genaues Bargeldmanagement. Verschiedene Funktionen erlauben die Durchführung von Ein- und Auszahlung großer Mengen, dabei arbeitet das Gerät sowohl effizient als auch extrem genau. Außerdem verfügt der RGB-100 über eine eingebaute Sammelkassette, die die Bargeldentnahme durch Werttransportunternehmen erleichtert. Das Interface ist durch die grafische Darstellung auf dem Monitor verständlich gestaltet.

Banknote recyclers

Signalling Light
Signalleuchte

Manufacturer
Yellowtec GmbH,
Monheim/Rhein, Germany
In-house design
Hanno Mahr
Web
www.yellowtec.com

The litt signalling light features homogeneous and dense light output from the programmable lighting segments. It thus offers a pleasant sense of light combined with the maximum signalling effect. The fact that the unit is manufactured from solid aluminium makes a crucial contribution to its value and sustainability. Thanks to the use of high-end LED technology, there is virtually no wear and tear and the light is maintenance-free.

Die Signalleuchte litt weist eine homogene, dichte Lichtabgabe der programmierbaren Leuchtsegmente auf. Dadurch wird das Licht als angenehm wahrgenommen, bei maximaler Signalwirkung. Die Fertigung aus massivem Aluminium trägt entscheidend zur Wertigkeit und Nachhaltigkeit der Leuchte bei. Durch den Einsatz von High-End-LED-Technik ist sie praktisch verschleißfrei und bedarf keiner Wartung.

Statement by the jury
The aluminium material gives this signalling light a premium look. The regular distance between the three lighting segments fosters a particularly harmonious impression.

Begründung der Jury
Das Aluminium verleiht der Signalleuchte ein hochwertiges Aussehen. Die gleichmäßigen Abstände zwischen den drei Leuchtsegmenten muten sehr harmonisch an.

MAGIC Motion Detector
MAGIC Bewegungsmelder

Manufacturer
Siemens Building Technologies,
Zug, Switzerland
Design
Teams Design
(Klaus Baumgartner),
Esslingen, Germany
Web
www.siemens.com
www.teamsdesign.com

The Magic motion detector features a patented detection mirror, which homogeneously images the room to be monitored within the inside of the detector housing, achieving improved intruder detection. In addition, the mirror technology allows for a uniquely flat and thin motion detector design. The cubic form blends perfectly into modern architecture, such as offices, museums and banks.

Statement by the jury
The Magic motion detector combines a discreet and minimalist appearance with advanced detector technology.

Der Magic-Bewegungsmelder ist mit einem patentierten Detektionsspiegel ausgestattet, der den zu überwachenden Raum homogen in das Gehäuseinnere abbildet, dadurch kann z. B. ein Einbruchsversuch besser erkannt werden. Durch die Spiegeltechnologie ist der Melder außerdem einzigartig flach und dünn. Die kubische Form fügt sich hervorragend in die moderne Architektur von Büros, Museen oder Banken ein.

Begründung der Jury
Der Magic-Bewegungsmelder verbindet eine zurückhaltende und minimalistische Erscheinung mit einer fortschrittlichen Detektionstechnologie.

deTec4 Core
Light Beam Safety Device
Sicherheitslichtvorhang

Manufacturer
SICK AG,
Waldkirch, Germany
Design
2ND WEST
(Michael Thurnherr, Manuel Gamper),
Rapperswil, Switzerland
Web
www.sick.com
www.2ndwest.ch

The deTec4 Core light beam safety device safeguards people working around machines and equipment by implementing a high safety standard. In modular steps of 15 cm, a protective field is covered at heights ranging from 30 cm to 2.10 metres, with complete protection provided across the entire profile length. The shape of the profile allows for very quick and intuitive installation. The universal brackets can be installed in various places and rotated by 15 degrees.

Statement by the jury
The housing of the deTec4 Core is very robust and impact-resistant. The semi-circular contour with flexible brackets enables easy installation.

Der Sicherheitslichtvorhang deTec4 Core sichert auf hohem Niveau Personen an Maschinen und Anlagen ab. In modularen Schritten von 15 cm wird eine Schutzfeldhöhe von 30 cm bis zu 2,10 Metern abgedeckt. Zudem bietet der Sicherheitslichtvorhang blindzonenfreien Schutz über die gesamte Profillänge. Die Form des Profils ermöglicht eine sehr schnelle und intuitive Installation. Die universellen Halter lassen sich an beliebiger Stellen platzieren und sind um 15 Grad drehbar.

Begründung der Jury
Das Gehäuse des deTec4 Core ist sehr robust und schlagfest. Die halbrunde Kontur mit den flexiblen Halterungen erleichtert die schnelle Montage.

Life science and medicine
Life Science und Medizin

Medical equipment and devices, laboratory technology and furniture, medical furniture and sanitary equipment, furnishings for rehabilitation centres and hospitals, rehabilitation, mobility, care and communication aids, orthopaedic aids
Medizinische Geräte und Ausrüstung, Labortechnik und -mobiliar, Medizinische Möbel und Sanitärausstattung, Praxis- und Krankenhausausstattung, Rehabilitation, Mobilitäts-, Pflege- und Kommunikationshilfen, Orthopädische Hilfsmittel

3

Dräger Perseus A500
Anaesthesia Workstation
Anästhesiearbeitsplatz

Manufacturer
Dräger Medical GmbH,
Lübeck, Germany

Design
Corpus-C Design Agentur GmbH
(Alexander Müller,
Sebastian Maier),
Fürth, Germany

Web
www.draeger.com
www.corpus-c.de

reddot design award
best of the best 2013

Optimal overview

Working in anaesthetics is multifaceted and constantly changing. Besides the daily work routines in the operating room, it includes a number of diagnostic and therapeutic procedures, and anaesthetists also need to be able to reliably respond to emergencies. Embodying this complexity, the Dräger Perseus A500 anaesthesia workstation was designed with close focus on the needs of its target group. With newly adjusted, improved ergonomics, it provides an even safer and more convenient work environment. And thanks to its sophisticated and self-explanatory functionality, the workstation highly efficiently adapts to work routines in anaesthetics. The new design concept focuses on individual configurations, allowing versatile set-ups and adjustments. A large worktable, an additional stow-away writing table and generous storage shelves provide a variety of set-up options. In addition, the working and documentation surfaces are equipped with illumination that can be dimmed, while well thought-out cable management solutions allow for quick conversions of the workspace even in the periphery. With more than 120 individual workplace set-ups, the Dräger Perseus A500 is a highly flexible system that adjusts to any kind of work situation – it optimises and supports work routines in anaesthetics.

Alles im Blick

Die Arbeit in der Anästhesie ist sehr vielfältig und in stetigem Wandel begriffen. Sie beinhaltet neben der täglichen Arbeit im OP zahlreiche Diagnose- und Therapieverfahren ebenso wie Notfälle, die zuverlässig abgedeckt werden müssen. Diese Komplexität verinnerlichend, wurde der Anästhesiearbeitsplatz Dräger Perseus A500 sehr nahe an der Zielgruppe gestaltet. Seine Ergonomie wurde neu abgestimmt, um ein noch sichereres und angenehmeres Arbeiten zu ermöglichen. Auch durch seine gut durchdachte und selbsterklärende Funktionalität kann sich dieser Arbeitsplatz den Abläufen in der Anästhesie sehr gut anpassen. Im Mittelpunkt steht dabei ein neues Konzept der individuellen Konfiguration, welches vielfältige Anpassungen und Variationen ermöglicht. So bieten ein großer Arbeitstisch, eine zusätzlich ausziehbare Schreibplatte und großzügige Ablageflächen unzählige Variationsmöglichkeiten. Die Arbeits- und Dokumentationsflächen sind zusätzlich mit einer dimmbaren Beleuchtung ausgestattet, und schlüssige Kabelmanagementlösungen ermöglichen auch in der Peripherie ein schnelles Umbauen des Arbeitsplatzes. Der Dräger Perseus A500 stellt mit mehr als 120 Konfigurationsmöglichkeiten ein sehr variables System für eine Anpassung an die jeweilige Arbeitssituation dar – die Abläufe in der Anästhesie werden einfacher und besser unterstützt.

Statement by the jury

The Dräger Perseus A500 anaesthesia workstation provides physicians and nursing staff with versatile possibilities for configuring their workspaces according to individual needs. The ergonomics and functionality of this product promote teamwork and create a comfortable working environment. The Dräger Perseus A500 showcases a distinctive and consistent design in every respect.

Begründung der Jury

Den Ärzten und Pflegekräften bietet der Anästhesiearbeitsplatz Dräger Perseus A500 vielfältige Möglichkeiten, sich den Arbeitsplatz ihren Bedürfnissen entsprechend zu konfigurieren. Die Ergonomie und die Funktionalität dieses Produktes begünstigen das Teamwork und schaffen ein angenehmes Arbeitsumfeld. Der Dräger Perseus A500 ist in jeder Hinsicht konsequent gestaltet und prägnant in seiner Linienführung.

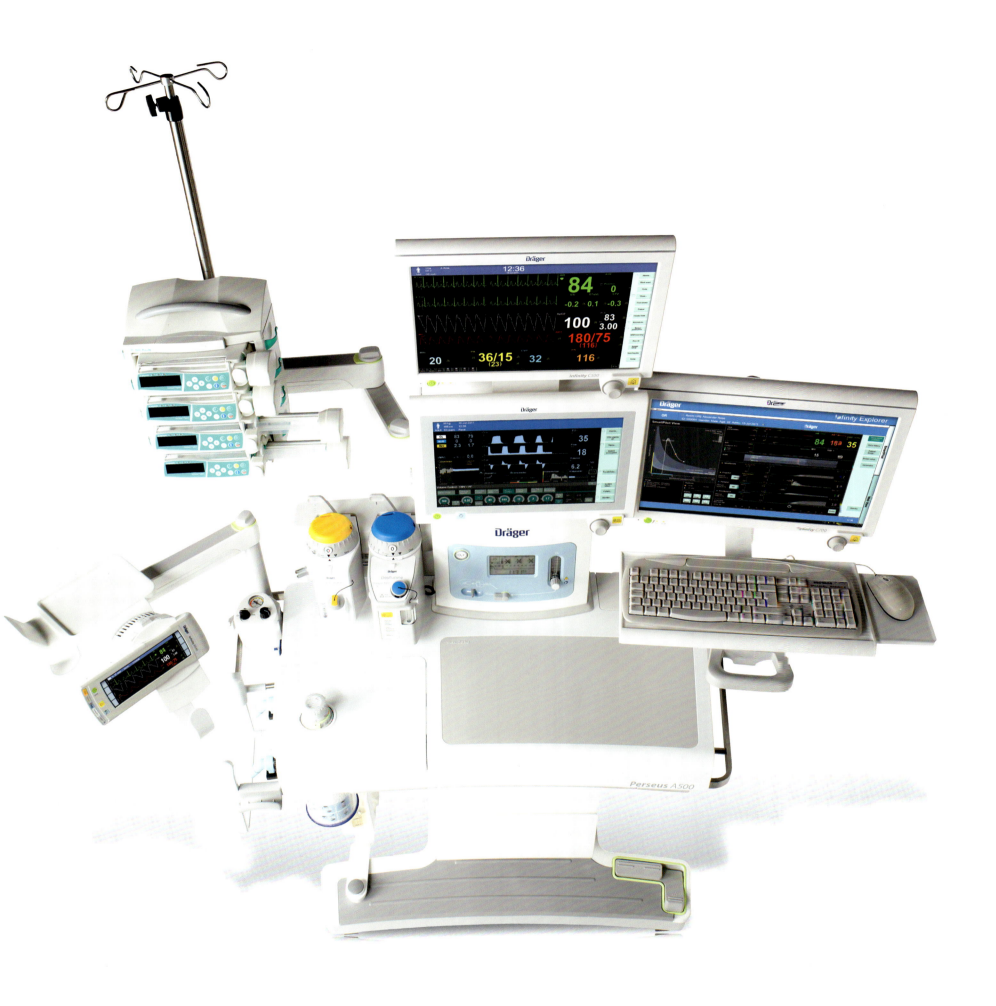

EVA
System for Vitreoretinal and Cataract Surgery
OP-System für die Vitreoretinal- und Kataraktchirurgie

Manufacturer
D.O.R.C. International B.V.,
Zuidland, Netherlands
Design
D'Andrea & Evers Design,
Enter, Netherlands
Web
www.dorc.nl
www.de-design.nl

EVA is a combined system for vitreoretinal and cataract surgery. At the heart of EVA is a revolutionary fluidcontrol system, the so-called VacuFlow VTi with timing intelligence valve control. This gives the surgeon control over flow and vacuum during eye surgery. The use of high-quality materials reflects the instruments high precision. Thanks to its modular structure, the system is easy to maintain and a variety of customer specific configurations are possible. The intuitive user interface makes all relevant information readily available for doctors and nurses.

Bei EVA handelt es sich um ein Operationssystem für die Vitreoretinal- und Kataraktchirurgie. Das Herzstück bildet ein Fluidsteuerungssystem, das VacuFlow VTi mit intelligenter Ventilsteuerung. Dadurch behält der Augenchirurg während des gesamten Eingriffs die Kontrolle über das Absaugen und die Vakuumbildung. Der Einsatz von hochwertigen Materialien spiegelt die hohe Präzision der Instrumente wider. Durch den modularen Aufbau des Systems ist die Wartung einfach, und es können verschiedene kundenspezifische Konfigurationen erstellt werden. Die intuitive Benutzeroberfläche sorgt für sofortigen Zugriff auf alle relevanten Informationen für Ärzte und Krankenschwestern.

Statement by the jury
Due to the slightly conical form and immaculately smooth front, EVA exhibits a particularly modern and high-quality appearance.

Begründung der Jury
Durch die leicht konische Form und die makellos glatte Front besitzt EVA ein besonders modernes und hochwertiges Erscheinungsbild.

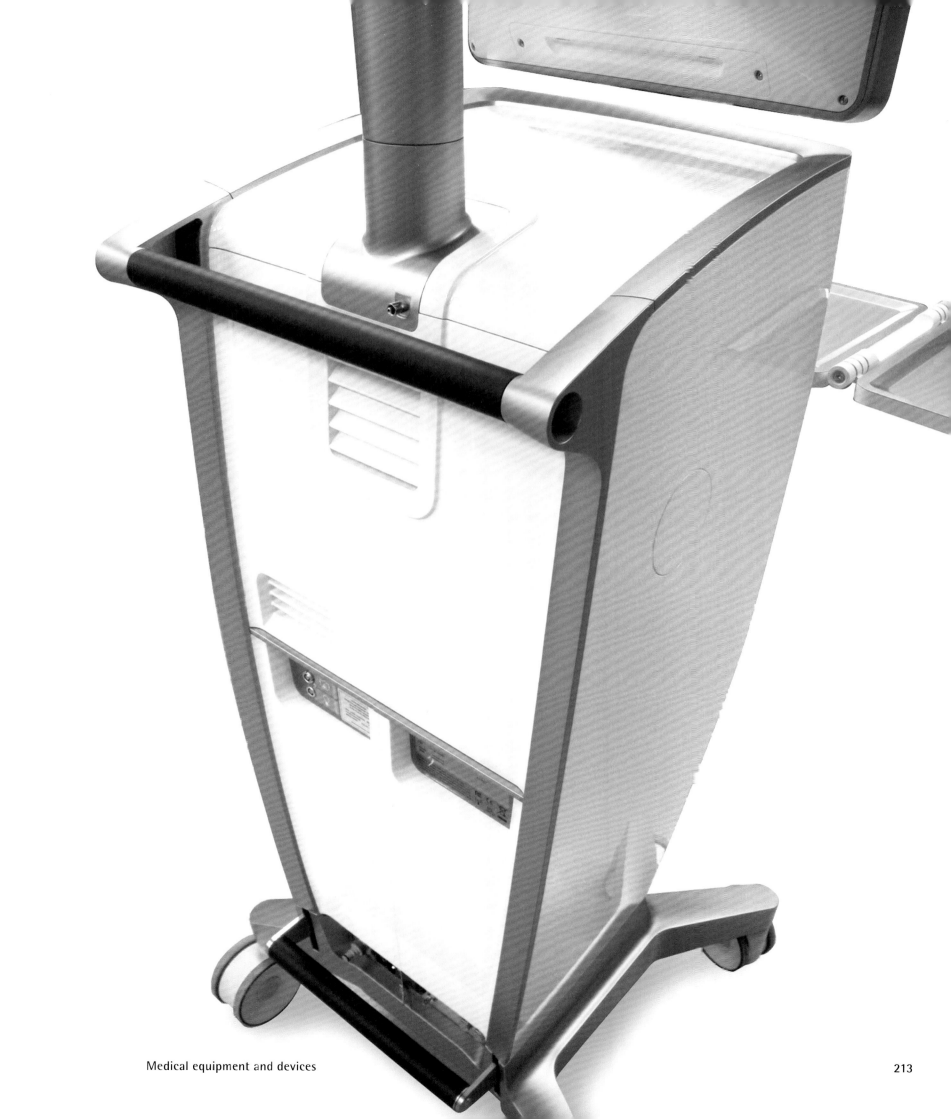

Ambient Experience Electro Physiology Lab

Electrophysiology Catheterisation Suite

Elektrophysiologisches Katheterlabor

Manufacturer
Royal Philips Electronics,
Eindhoven, Netherlands
In-house design
Philips Design Healthcare Team
Web
www.philips.com

This electrophysiology catheterisation suite is one of the first of its kind with Ambient Experience, a multimedia solution that can reduce patients' physical and mental stress during operations. Dynamic lighting, video and audio recordings, which the patients can choose themselves, foster a comforting atmosphere. All of the suite's treatment systems are integrated in a single user interface, which is displayed on a 55" LCD colour monitor.

Dieses elektrophysiologische Katheterlabor ist eines der ersten seiner Art mit Ambient Experience, einer Multimedialösung, die die physische und mentale Belastung von Patienten während der Operation reduzieren kann. Dynamische Licht-, Video- und Audioeinspielungen, die der Patient selbst auswählen kann, sorgen dabei für eine beruhigende Atmosphäre. Alle Systeme des Labors sind in einer einzigen Benutzeroberfläche integriert und werden auf einem 55"-LCD-Farbdisplay dargestellt.

Statement by the jury
The Ambient Experience System merges with the technical equipment in this electrophysiology catheterisation suite to form a seamless unit.

Begründung der Jury
In dem elektrophysiologischen Katheterlabor verschmilzt das Ambient-Experience-System mit dem technischen Equipment zu einer Einheit.

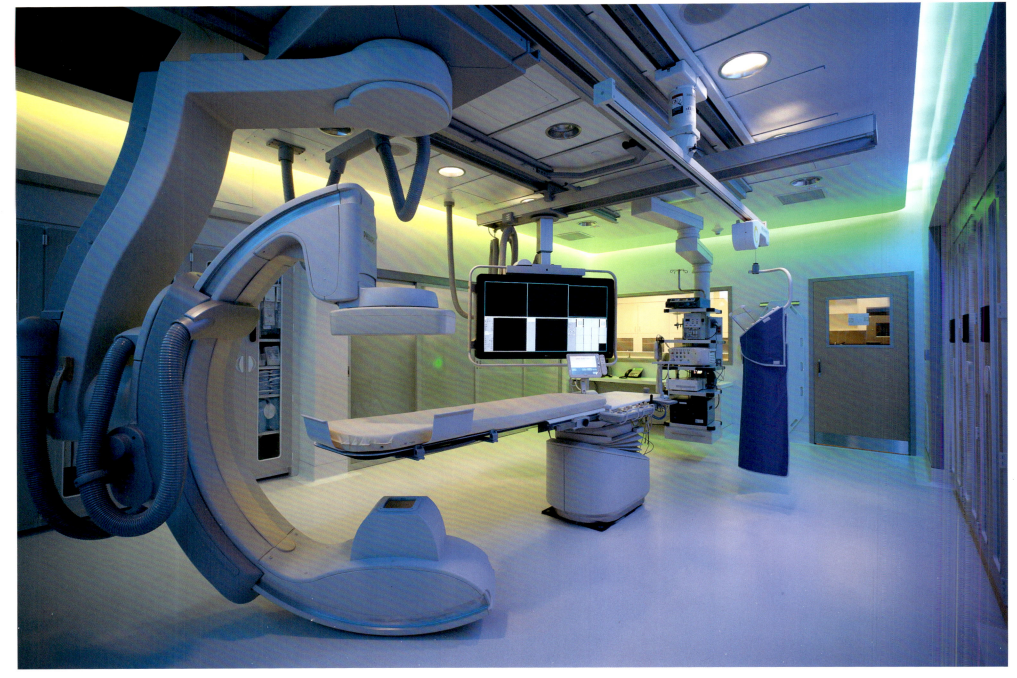

Ambient Experience
Hybrid Operating Room
Surgical Suite
Hybrid-OP

Manufacturer
Royal Philips Electronics,
Eindhoven, Netherlands
In-house design
Philips Design Healthcare Team
Web
www.philips.com

This surgical suite allows the switch from a minimally invasive procedure to a surgical intervention in critical situations. The Ambient Experience solution contributes to calming the patient by integrating interactive media, along with dynamic lighting and audio settings. An angiographic imaging device is attached to the ceiling and is not immediately visible, thus giving the environment a calm, tidy and well-organised impression.

In diesem Hybrid-OP kann in kritischen Situationen von einem minimal-invasiven zu einem chirurgischen Eingriff gewechselt werden. Das Ambient-Experience-System trägt mit interaktiven Medien und dynamischer Licht- und Tontechnik zur Beruhigung des Patienten bei. Eine bildgebende Angiographie-Einheit ist an der Zimmerdecke befestigt und fällt nicht direkt ins Auge, sodass das Umfeld ruhig, ordentlich und gut organisiert wirkt.

Statement by the jury
The surgical suite's futuristic appearance with minimalist, clear lines and different multimedia options highlights its ultra-modern character.

Begründung der Jury
Durch seine futuristische Erscheinung, die puristisch-klaren Linien und verschiedenen Multimediaangebote erhält der Hybrid-OP einen ultramodernen Charakter.

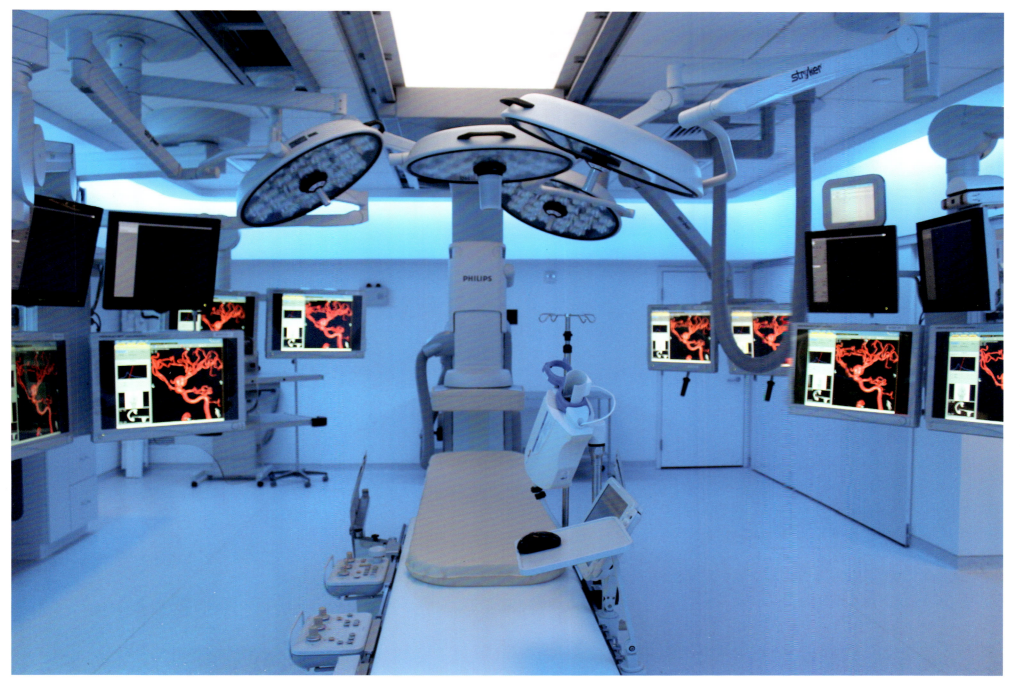

EndoVR (Endoscopic) and LapVR (Laparoscopic)
Surgical Simulators
Chirurgische Simulatoren

Manufacturer
CAE Healthcare,
Saint-Laurent, Canada
Design
Alto Design Inc
(Patrick Mainville, Benoit Orban),
Montreal, Canada
Web
www.cae.com
www.alto-design.com

The cohesive design of the surgical simulators EndoVR and LapVR enables medical students to quickly understand and operate the different models. The operating unit is height-adjustable. Both of the 21" LCD screens can be adjusted according to individual requirements thanks to the aluminium joints which are attached to a column. Furthermore, the devices are lightweight and compact, which makes them easy to move and store away.

Der einheitliche Aufbau der chirurgischen Simulatoren EndoVR und LapVR ermöglicht es Medizinstudenten, die verschiedenen Modelle schnell zu verstehen und zu bedienen. Die Steuerungskonsole ist höhenverstellbar. Die beiden 21"-LCD-Bildschirme können dank der sich auf einem Mast befindlichen Gelenke aus Aluminium wunschgemäß ausgerichtet werden. Die Apparate sind zudem leicht und kompakt, damit sie einfach zu bewegen und zu verstauen sind.

Statement by the jury
The design of these surgical simulators has impressively succeeded in expressing robustness, simplicity and durability.

Begründung der Jury
Bei der Gestaltung dieser chirurgischen Simulatoren ist es eindrucksvoll gelungen, Robustheit, Einfachheit und Langlebigkeit zum Ausdruck zu bringen.

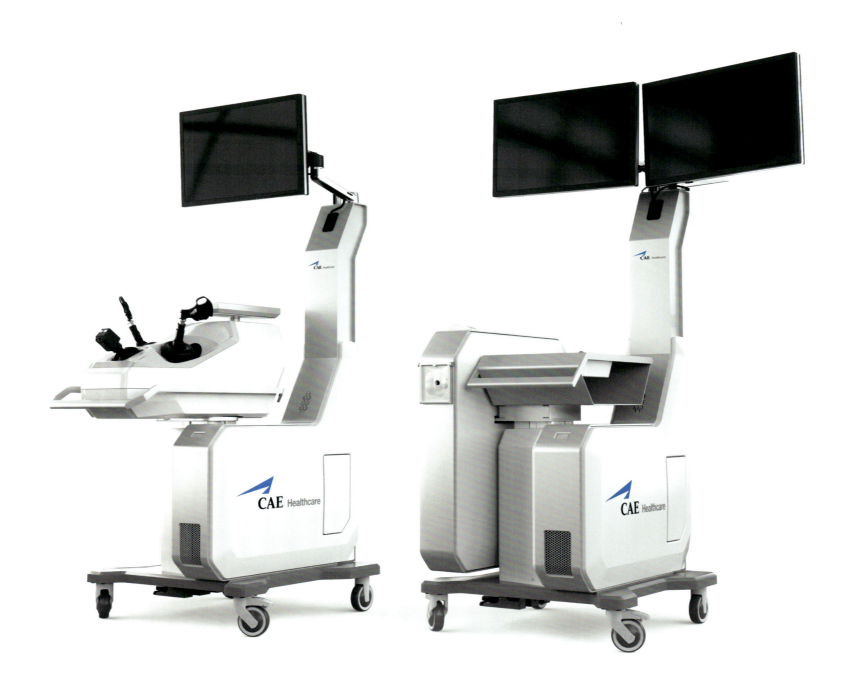

Mobilett Mira
Mobile X-ray System
Mobiles Röntgensystem

Manufacturer
Siemens AG,
Munich, Germany
Design
designaffairs GmbH,
Erlangen, Germany
Web
www.siemens.com
www.designaffairs.com

Mobilett Mira is a digital mobile X-ray system with high-resolution imaging quality comparable to that of stationary X-ray imaging systems. Depending on customer requirements and budget, Mobilett Mira is available with either a wireless or wired detector. The 180-degree rotating swivel arm provides flexible access to the patient. Thanks to an integrated motor, the unit is particularly easy to manoeuvre.

Bei Mobilett Mira handelt es sich um ein digitales mobiles Röntgensystem, dessen hochauflösende Bildqualität mit der von stationären Röntgensystemen vergleichbar ist. Je nach Kundenanforderungen und Budget ist Mobilett Mira entweder mit einem kabellosen oder einem kabelgebundenen Detektor erhältlich. Der um 180 Grad schwenkbare Dreharm erlaubt einen flexiblen Zugang zum Patienten. Dank des integrierten Motors lässt sich das System zudem besonders leicht manövrieren.

Statement by the jury
Since the electric cables are concealed within the swivel arm, the device provides excellent manoeuvrability and flexibility.

Begründung der Jury
Da die elektrischen Kabel innerhalb des Schwenkarms verlaufen, beeindruckt das Gerät durch seine Wendigkeit und Flexibilität.

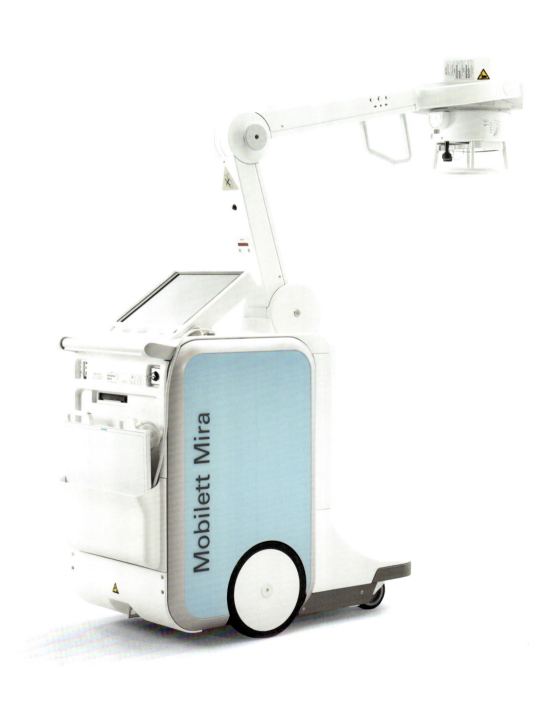

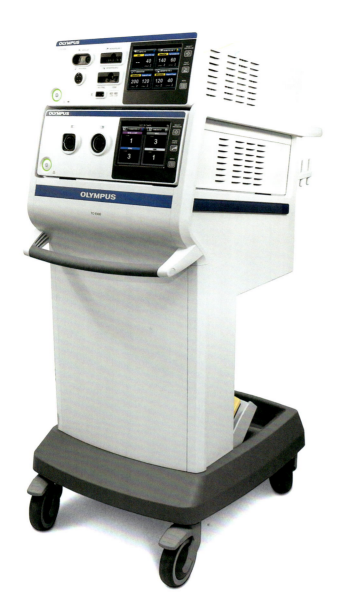

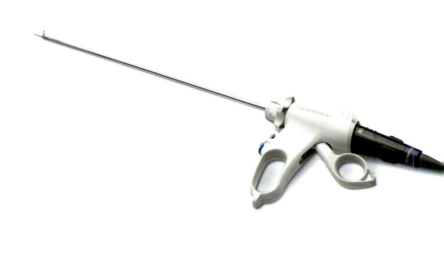

Thunderbeat
Tissue Management System
Gewebemanagementsystem

Manufacturer
Olympus Corporation,
Tokyo, Japan
In-house design
Hirotoshi Amano,
Kenji Tajima
Web
www.olympus.com

The Thunderbeat tissue management system is one of the first devices in the world to combine ultrasonic and bipolar technologies for the fast cutting and reliable sealing of vessels in a single instrument. All components are compact and incorporated into a rollable unit. The fine design of the instrument tip ensures precise dissection and reduced mist generation. With two grip types – pistol and inline handle – and four working lengths, the system provides high operating comfort.

Das Gewebemanagementsystem Thunderbeat ist eines der weltweit ersten Instrumente, das Ultraschalltechnologie und Bipolartechnologie für schnelles Schneiden und zuverlässige Gefäßversiegelung in einem einzigen Instrument vereint. Alle Komponenten sind kompakt in einem rollbaren Element zusammengefasst. Die Feinheit der Instrumentenspitze sorgt für eine präzise Dissektion und eine geringe Dampfentwicklung. Mit zwei Handgrifftypen, dem Pistolen- und dem Inline-Griff, sowie vier verschiedenen Arbeitslängen bietet das System hohen Anwenderkomfort.

Statement by the jury
Thunderbeat's combination of bipolar and ultrasonic energy is excellently communicated through its coherent design language.

Begründung der Jury
Die Zusammenführung von bipolarer und Ultraschall-Energie im Thunderbeat wird durch die in sich geschlossene Formsprache ausgezeichnet herausgearbeitet.

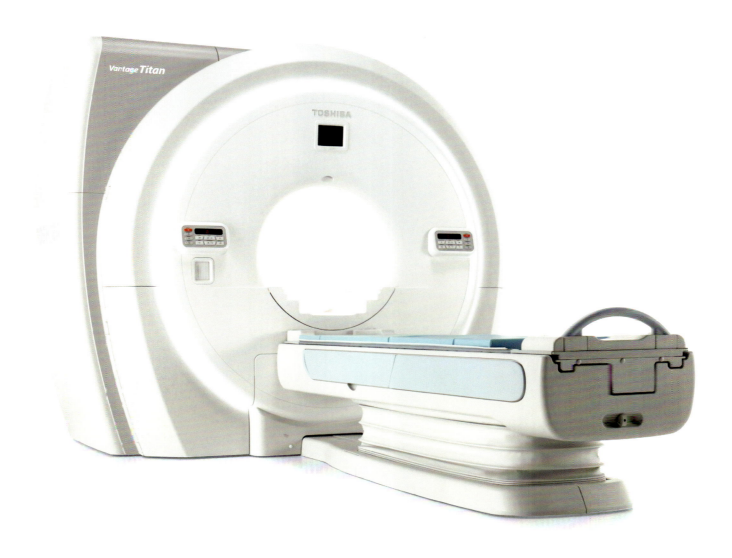

Toshiba MRI
Vantage Titan MRI-2004
Magnetic Resonance
Imaging System
Magnetresonanztomograph

Manufacturer
Toshiba Medical Systems Corporation,
Otawara City, Tochigi Prefecture, Japan
In-house design
Toshiba Corporation Design Center
(Atsunobu Banryu, Yasuhiro Shiino)
Web
www.toshiba-medical.co.jp
www.toshiba.co.jp

With its frontal, ring-shaped LED lighting, the MRI Vantage Titan magnetic resonance imaging system fosters a relaxed atmosphere, which reduces anxiety in the patient. This lighting also lends the large device a more compact appearance, which likewise serves to comfort the patient. Other stress-reducing elements include the 71-cm-wide examination bed, a noise-reduction mechanism and an illuminated tunnel.

Der MRI Vantage Titan kreiert durch seine ringförmige LED-Beleuchtung an der Vorderseite eine entspannte Atmosphäre, die Angstgefühle beim Patienten reduziert. Durch die Beleuchtung wirkt das große Gehäuse zudem kompakter, was den Patienten ebenfalls beruhigt. Weitere stressreduzierende Elemente sind das 71 cm große Untersuchungsbett, der geräuscharme Mechanismus und die Beleuchtung im Tunnel.

Statement by the jury
The concept behind this MRI scanner makes an innovative contribution to providing patients with a more stress-free examination situation.

Begründung der Jury
Der MRT-Scanner trägt durch seine Konzeption dazu bei, dass Patienten die Untersuchung stressfreier erleben können.

XDR3
Digital X-Ray System
Digitales Röntgengerät

Manufacturer
Veterinärmedizinisches
Dienstleistungszentrum GmbH (VetZ),
Isernhagen, Germany
Design
RpunktDESIGN Werbeagentur GmbH
(Martin Rinderknecht, Florian Langer),
Hannover, Germany
Web
www.vetz.de
www.rpunktdesign.de

XDR is a mobile, digital X-ray system for veterinarians. Constructed from carbon and aluminium, it is a lightweight device at just 14 kg. Its technical heart is a wireless flat-panel detector, which delivers digital X-ray images in seconds. The housing is extremely robust and conceived for hassle-free system maintenance. The broad, centrally mounted handle ensures optimal portability.

XDR3 ist ein mobiles digitales Röntgengerät für Tierärzte. Es besteht aus Carbon und Aluminium und ist mit nur 14 kg ein Leichtgewicht. Das technische Herzstück, ein Wireless-Flachdetektor, erzeugt sekundenschnell digitale Röntgenaufnahmen. Das Gehäuse ist außerordentlich stabil und so konzipiert, dass das System problemlos zu warten ist. Der breite Griff in der Mitte bietet optimalen Tragekomfort.

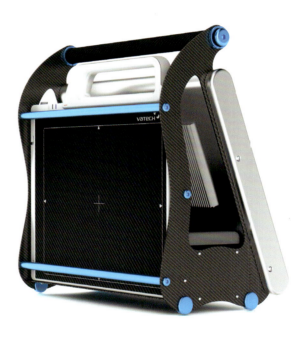

X-Mind unity
Intraoral X-Ray Unit
Intraorales Röntgengerät

Manufacturer
Acteon,
Merignac, France
In-house design
Claudio Giani
Design
frog
(Mariano Cucchi, Andrea Besana),
Milan, Italy
Web
www.acteongroup.com
www.frogdesign.com

The intraoral X-Ray unit X-Mind unity is equipped with a digital sensor based on so-called ACE technology. This causes the automatic closure of the sensor as soon as optimal image quality has been achieved. Radiation exposure is thus reduced by 52 per cent. The ultra light arm can be smoothly navigated and positioned with one hand thanks to its ergonomic design and intuitive handle. The digital sensor's connecting wires are concealed within the generator arm.

Statement by the jury
The telescopic arm of the X-Mind unity offers outstanding flexibility and precision. Its geometric form is a true highlight in terms of design.

Das intraorale Röntgengerät X-Mind unity verfügt über einen Digitalsensor mit sogenannter ACE-Technologie. Dadurch wird die automatische Schließung des Sensors verursacht, sobald eine optimale Bildqualität erreicht ist. Die Strahlenbelastung kann so um bis zu 52 Prozent reduziert werden. Der ultraleichte Arm lässt sich durch sein ergonomisches Design und die intuitive Griffgestaltung mühelos mit einer Hand bewegen und positionieren. Die Anschlusskabel des Digitalsensors sind im Generatorarm verborgen.

Begründung der Jury
Der Teleskoparm des X-Mind unity bietet außergewöhnliche Flexibilität und Präzision. Seine Geometrie ist auch architektonisch ein Highlight.

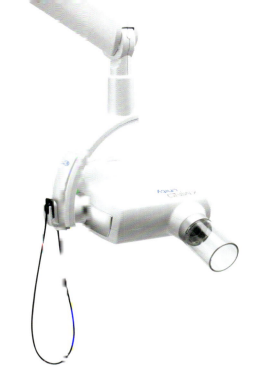

MyRay Hyperion X9
Dental Imaging Platform
Dentale Bildgebungsplattform

Manufacturer
Cefla s.c. Dental Group,
Imola, Italy
Design
Mechanema
(Giulio Mattiuzzo),
Rimini, Italy
Web
www.cefla.com
www.mechanema.it

Hyperion X9 is an upgradeable multi-level platform that provides a variety of imaging solutions, ranging from 2D panoramic imaging to cone beam 3D tomography, as well as true full arch volumetric scan capability. The device can be fitted either with a large or a small image detector and is compatible with software by other manufacturers.

Statement by the jury
Hyperion X9 impresses with its friendly and fresh design, which is characterised by a rounded, curvaceous form and colourful accents.

Hyperion X9 ist eine nachrüstbare Multi-Level-Plattform, die verschiedene Möglichkeiten der Bildgebung zur Verfügung stellt. Dazu zählen die 2D-Panorama-Bildgebung, die 3D-Tomographie mit konischer Strahlung sowie die Ausführung von volumetrischen Scans der gesamten Zahnbögen. Das Gerät kann mit einem großen oder kleinen Bilddetektor ausgerüstet werden und ist mit der Software anderer Hersteller kompatibel.

Begründung der Jury
Hyperion X9 überzeugt durch seinen freundlichen und frischen Charakter, der durch die abgerundete kurvige Form und die farbigen Akzente geprägt ist.

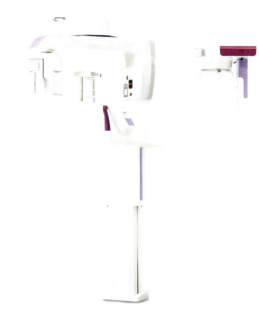

VistaRay 7
X-ray System
Röntgensystem

Manufacturer
Dürr Dental AG,
Bietigheim-Bissingen, Germany
In-house design
Andreas Kühfuß
Web
www.duerrdental.de

The VistaRay 7 X-ray system provides images with a high level of detail at low radiation. It is used in the control of implant fixation and also in endodontics. Due to its depiction of the finest grey scales, anomalies like D1 caries lesions can be reliably detected. Thanks to the latest sensor technology, the images are quickly available and can be transmitted directly to a screen via a USB port.

Statement by the jury
The VistaRay 7 houses state-of-the-art technology in a particularly discreet and slim body, which decreases feelings of anxiety in the patient.

Das Röntgensystem VistaRay 7 liefert sehr detaillierte Bildaufnahmen bei geringer Strahlendosis. Es findet Anwendung sowohl bei der Kontrolle von Implantatbefestigungen als auch in der Endodontie. Aufgrund der Darstellung von feinsten Graustufen können beispielsweise Karies-D1-Läsionen sicher erkannt werden. Dank modernster Sensortechnologie sind die Bilder schnell verfügbar und lassen sich über einen USB-Anschluss direkt auf einen Bildschirm übertragen.

Begründung der Jury
Modernste Technik wurde beim VistaRay 7 in einen besonders dezenten und schlanken Korpus überführt, der dem Patienten die Angst nimmt.

DC-N3
Color Doppler Diagnostic
Ultrasound Imaging System
Farbdoppler-Ultraschallsystem

Manufacturer
Shenzhen Mindray
Bio-Medical Electronics Co., Ltd.,
Shenzhen, China
In-house design
Qi Zhang, Jun Luo,
Xiang Zhou
Web
www.mindray.com

The DC-N3 colour Doppler diagnostic ultrasound imaging system is designed to meet the daily requirements of an interdisciplinary hospital environment. An impressive number of ten ultrasonic probes makes it easy to adjust to varying examination requirements. The height-adjustable, swivelling operating panel is complemented by a 180-degree-rotatable LCD monitor which, in combination with the two-level foot rest, makes the system comfortable to use in any situation.

Das Farbdoppler-Ultraschallsystem DC-N3 ist auf die täglichen Anforderungen im interdisziplinären Krankenhausumfeld abgestimmt. Gleich zehn Schallköpfe erlauben es problemlos, sich auf variierende Untersuchungsbedingungen einzustellen. Das höhenverstellbare und schwenkbare Bedienfeld wird durch einen um 180 Grad drehbaren LCD-Monitor ergänzt und ermöglicht in Verbindung mit der 2-Stufen-Fußstütze unter allen Umständen ein komfortables Arbeiten.

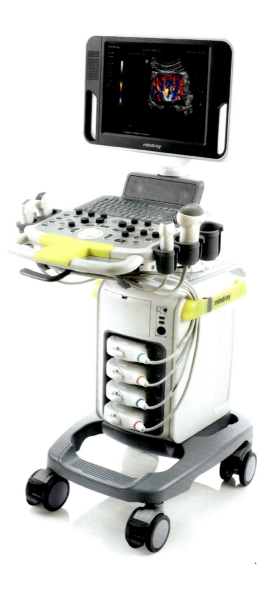

S-CAPE
Medical Glass Console
Medizinische Glaskonsole

Manufacturer
S-Cape GmbH,
Reichenbach-Heinsdorfergrund, Germany,
In-house design
Uwe Seidel, Matthias Lubkowitz
Web
www.s-cape.com

The S-Cape medical glass console is designed for use in operating theatres, their anterooms and other medically sensitive areas, such as intensive care or anaesthesiology. In designing this device, particular attention was paid to hygiene in the operating theatre. The front features a large safety glass surface, and the station does without any cables or elements that might prove detrimental to a hygienic environment.

Statement by the jury
The seamless design of this medical glass console conveys self-contained aesthetics. The elegant, high-gloss frame makes contaminations immediately visible.

Die medizinische Glaskonsole S-Cape ist für den Einsatz in Operationssälen und deren Vorräumen sowie in anderen medizinisch sensiblen Bereichen wie Intensivstation oder Anästhesie konzipiert. Bei der Gestaltung wurde besonderer Wert darauf gelegt, hygienische Belastungen für den OP-Saal zu vermeiden. Die Front besteht aus einer durchgängigen Sicherheitsglasscheibe. Nach außen führende Kabel oder andere hygienisch kritische Elemente sind nicht vorhanden.

Begründung der Jury
Die fugenlose Verarbeitung der Multikonsole ergibt eine in sich geschlossene Ästhetik. Der elegante hochglänzende Rahmen macht Verunreinigungen sofort sichtbar.

PadScan HD 5
Bladder Scanner
Blasenscanner

Manufacturer
Caresono Technology Co., Ltd.,
Shenzhen, China
In-house design
Anderson Bo Sun,
Yuanping Shu
Web
www.caresono.com

The PadScan HD 5 bladder scanner is a mobile ultrasound system for the non-invasive measurement of bladder volume. The device, which features an integrated thermal printer, is extremely lightweight and compact. It is operated via an 8" touchscreen, and there are no keys apart from the power button. The scanner charges automatically on the docking station, and the battery lasts for hours of scanning.

Statement by the jury
Beyond its puristic and functional design, this bladder scanner conveys freshness and friendliness with its colour accents and rounded contours.

Der Blasenscanner PadScan HD 5 ist ein mobiles Ultraschallsystem zur nichtinvasiven Messung des Blasenvolumens. Das Gerät mit integriertem Thermodrucker ist extrem leicht und kompakt konzipiert. Die Bedienung läuft über einen 8"-Touchscreen, außer dem An-/Aus-Knopf sind keine weiteren Tasten vorhanden. Der Scanner lädt sich automatisch über die Dockingstation auf, und der Akku hält mehrere Stunden lang.

Begründung der Jury
Über seine puristische Sachlichkeit hinaus wirkt der Blasenscanner durch seine Farbakzente und kurvigen Formen frisch und freundlich.

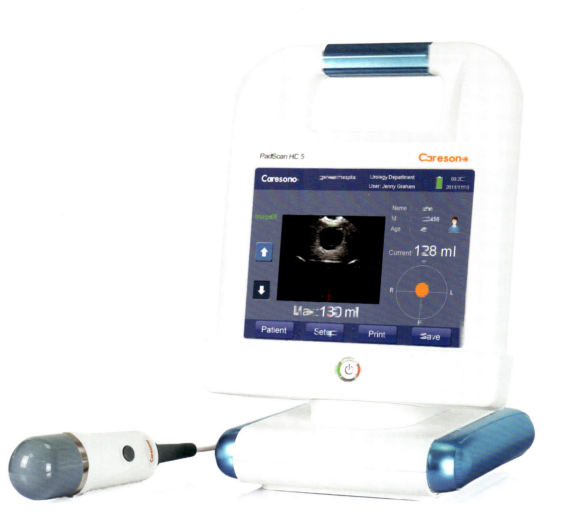

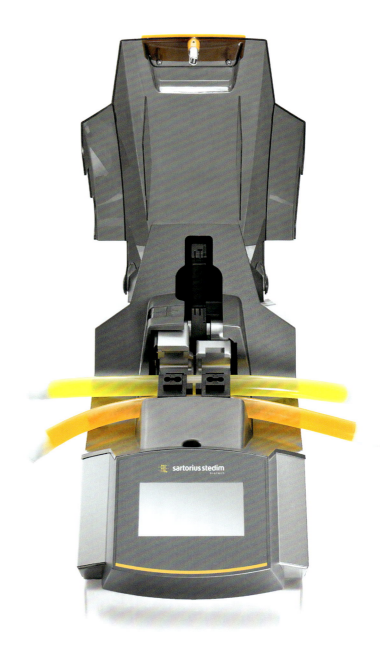

BioWelder® TC
Fully Automated Tube Welding Device
Vollautomatisiertes Schlauchschweißgerät

Manufacturer
Sartorius Stedim Switzerland AG,
Tagelswangen, Switzerland
In-house design
Sartorius Corporate Administration GmbH
Design
Corpus-C Design Agentur GmbH
(Alexander Müller, Sebastian Maier),
Fürth, Germany
formfabrik AG
(Christoph Jaun, Dominic Spiess),
Zwillikon, Switzerland
Web
www.sartorius-stedim.com
www.corpus-c.de

The BioWelder TC (Total Containment) is a fully automated device for the sterile welding of thermoplastic tubing. The system also permits the sterile connection of liquid-filled tubing. Its interchangeable tube holder allows for flexible adaptability to process requirements. The device is operated via a touchscreen, which ensures ease of use and promotes safety at all times.

Der BioWelder TC (Total Containment) ist ein vollautomatisiertes Gerät zum sterilen Verschweißen thermoplastischer Schläuche sowie zur sterilen Verbindung von flüssigkeitsgefüllten Schläuchen. Die auswechselbaren Schlauchhalter erlauben eine flexible Anpassung an die Prozessanforderungen. Das Gerät wird über einen Touchscreen bedient. Dies macht die Handhabung einfach und sicher.

Statement by the jury
The functionality of the BioWelder TC is reflected in its clear aesthetics, which convey the impression of strong reliability.

Begründung der Jury
Die Funktionalität des BioWelder TC spiegelt sich in einer klaren Ästhetik wider, die den Eindruck hoher Zuverlässigkeit vermittelt.

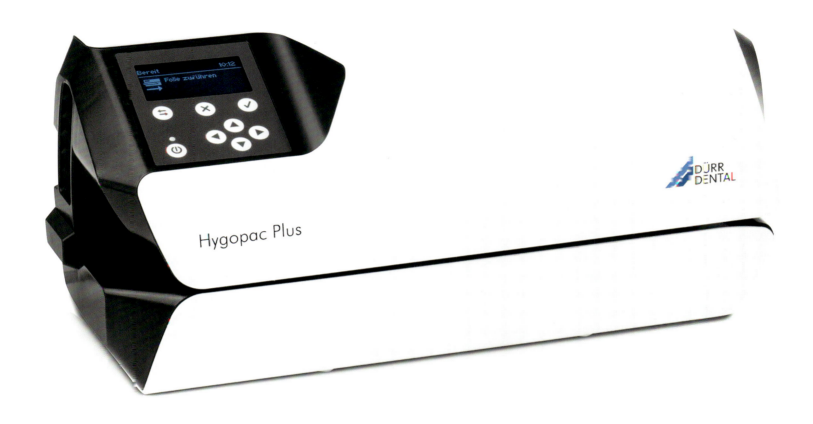

Hygopac Plus
Rotary Sealing Device
Durchlaufsiegelgerät

Manufacturer
Dürr Dental AG,
Bietigheim-Bissingen, Germany
Design
formstudio merkle park
(Ulrich Merkle, Min-Jung Park),
Stuttgart, Germany
Web
www.duerrdental.de
www.formstudio.com

This rotary sealing device for the sterile packaging of dental instruments supports quick and safe handling. The unit's display guides the operator through the validation process step by step. The sealing temperature, contact pressure and speed can be continually monitored and also saved onto an SD card, for example.

Das Durchlaufsiegelgerät zur sterilen Verpackung von dentalen Instrumenten unterstützt ein sicheres und schnelles Arbeiten. Der Anwender wird über das Display auf dem Gerät Schritt für Schritt durch den Validierungsprozess geführt. Siegeltemperatur, Anpressdruck und Siegelgeschwindigkeit lassen sich nachvollziehen und werden ständig erfasst, überprüft und gespeichert, zum Beispiel auf einer SD-Karte.

Statement by the jury
The Hygopac Plus device impresses with its compact and simple appearance, which displays a distinctive look through the use of bright metal parts and dark plastic housing components.

Begründung der Jury
Hygopac Plus besticht durch sein kompaktes, schlichtes Aussehen. Durch die hellen Metall- und dunkler Kunststoffgehäuseteile erhält es ein eigenständiges Profil.

Hoffmann 3
External Fixator
Externer Fixateur

Manufacturer
Stryker Osteosynthesis,
Selzach, Switzerland
In-house design
Beat Mürner
Web
www.osteosynthesis.stryker.com

This external fixator for the primary care of bone fractures combines three different connector diameters in one coupling. Quick fixation of the components using the disposable thumbwheel supports fracture reduction and indicates that the final tightening of the coupling is yet to be carried out. The convex connecting elements converge to a triangular shape, and the green colour scheme has a calming effect.

Dieser externe Fixateur für die Erstversorgung von Knochenfrakturen vereint drei unterschiedliche Konnektorendurchmesser in einer Kupplung. Die schnelle Fixation der Komponenten mithilfe des wegwerfbaren Drehknopfs erleichtert die Reduktion der Fraktur und signalisiert, ob das finale Festziehen der Spannschraube noch aussteht. Die Verbindungselemente laufen konvex in einer Dreiecksform zusammen. Die grüne Farbgebung wirkt beruhigend.

Pirol / Skua
Spatz / Specht
Medical Suction Regulator
Medizinische Absaugeinheiten

Manufacturer
Greggersen Gasetechnik GmbH,
Hamburg, Germany
Design
Bazarganzi | Design+Innovation
(Parviz Bazargani),
Hamburg, Germany
Web
www.greggersen.com
www.bazargani.de

Pirol and Skua have been designed for the safely controlled suction of body fluids in medical environments. The Pirol device is vacuum-powered, while the Skua operates with compressed air. Both units integrate high-quality aluminium and plastic components, which give the product line a cohesive appearance. In addition, they are easy to clean due to their precise, flush transitions and can also be recycled.

Statement by the jury
The medical suction regulators impress with their comfortable surface feel and coloured details, which cleverly direct attention to essential components like the indicator needle.

Spatz und Specht sind für das regulierbare Absaugen von Körperflüssigkeiten im medizinischen Umfeld konzipiert. Spatz ist ein Vakuumregler, Specht funktioniert mit Druckluft. Beide Geräte bestehen aus hochwertigen Aluminium- und Kunststoffteilen, so besitzt die gesamte Produktlinie ein einheitliches Erscheinungsbild. Durch die präzisen Übergänge sind sie leicht zu reinigen, und sie sind gut recycelbar.

Begründung der Jury
Die Absaugeinheiten begeistern durch eine angenehme Haptik sowie bunte Details, die geschickt auf entscheidende Stellen wie die Anzeigennadel hinweisen.

bee-system[3ad]
Surgical Power System
Chirurgisches Motorensystem

Manufacturer
medical bees GmbH,
Emmingen-Liptingen, Germany
In-house design
medical bees Design Team
Web
www.medical-bees.de
Honourable Mention

A particular innovation of the bee-system surgical power system is its battery-powered drive unit that recognises and adapts automatically to the five different attachments. The charging status of the machine is indicated on an illuminated display. At the same time, the surgical field is lit up by four LEDs. The drive unit and the handle are made of a lightweight aluminium alloy and are equipped with a high-quality, non-reflective coating. The handles are ergonomically well-balanced, and all edges of the machine are gently rounded.

Statement by the jury
The bee-system is defined as a true precision tool with its clear delineation of handle and drive unit, featuring distinctive colouring and material highlights.

Besonders innovativ beim chirurgischen Motorensystem bee-system ist die akkubetriebene Steuereinheit, welche die fünf unterschiedlichen Handstücke erkennt. Der Ladezustand ist über eine Leuchtanzeige einsehbar. Gleichzeitig wird durch vier Beleuchtungs-LEDs die Ausleuchtung des OP-Feldes unterstützt. Steuereinheit und Handstück bestehen aus einer leichten Aluminiumlegierung und sind mit einer nicht reflektierenden Beschichtung versehen. Der Schwerpunkt der Handstücke ist zentral austariert. Alle Kanten der Maschinen sind abgerundet.

Begründung der Jury
Die klare Unterteilung in Handstück und Aufsätze setzt markante Farb- und Materialakzente, die das bee-system als echtes Präzisionswerkzeug ausweisen.

Stryker System 7
Surgical Power Tools
Elektrowerkzeuge
für die Chirurgie

Manufacturer
Stryker Instruments,
Portage, USA
In house-design
Jose Calderon, Matt Jousma,
Brey Hansford
Stryker R & D Team
Design
Tekna
(Mike Rozewicz, Dave Veldkamp),
Kalamazoo, USA
Web
www.stryker.com
www.teknalink.com

This set of cordless power tools has been designed specifically for orthopaedic surgery, especially for use in preparing bones for joint implants. The handle of each tool is made of anodised aluminium and rests comfortably in the hand. Moreover, the control elements are ergonomically arranged, the user has an unobstructed view of the surgical site, and the weight distribution is optimised for one-handed operation.

Statement by the jury
The design of the Stryker System 7 appears strong and robust, which excellently reflects the tools' capacity for high performance.

Die kabellosen Elektrowerkzeuge wurden speziell für den Einsatz in der orthopädischen Chirurgie entwickelt. Mit ihnen werden Knochen für den Einsatz von Gelenk-Implantaten präpariert. Der Griff der Geräte besteht jeweils aus anodisiertem Aluminium und passt sich der Hand an, die Bedienelemente sind ergonomisch angeordnet, die Sicht auf das Operationsfeld ist frei, und die Gewichtsverteilung wurde für die Arbeit mit einer Hand optimiert.

Begründung der Jury
Die Gestaltung der Werkzeuge des Stryker System 7 wirkt kraftvoll und stabil, was ihre Leistungsfähigkeit hervorragend widerspiegelt.

SpeediCath Compact Set
Compact Catheter Set
Katheterset

Manufacturer
Coloplast A/S,
Humlebæk, Denmark
In-house design
Marlene Corydon,
Benny Matthiassen,
Bent Hagel
Design
Native Design,
London, GB
Web
www.coloplast.com
www.native.com

SpeediCath is a ready-to-use compact catheter set that supports an active lifestyle. Thanks to its discreet design language, this all-in-one solution becomes a non-stigmatising accessory that fits into virtually any bag due to its compact size. The catheter is intuitive, easy to use and does not attract any unnecessary attention. The technical functions have all been cleverly incorporated into its ergonomic design. The user is therefore assured of a smooth catheterisation process.

SpeediCath ist ein gebrauchsfertiges Set von Katheter und Beutel, das einen aktiven Lebensstil unterstützt. Aufgrund der diskreten Formensprache wird diese Komplettlösung zu einem Accessoire ohne Stigma, das durch seine Kompaktheit in praktisch jede Tasche passt. Dadurch zieht das Katheterset keine unnötige Aufmerksamkeit auf sich. Zudem ist es einfach in der Handhabung, die technischen Funktionen sind allesamt clever in die ergonomische Gestaltung integriert. Der Anwender muss daher keine Bedenken haben, bei der Katheterisierung etwas falsch zu machen.

PureWhite
Endoscopy Instruments
Endoskopie-Instrumente

Manufacturer
medwork medical products and services GmbH,
Höchstadt/Aisch, Germany
Design
at-design GbR
(Tobias Kreitschmann, Christoph Tomczak),
Fürth, Germany
Web
www.medwork.com
www.atdesign.de

The basis for the characteristic design language of the PureWhite endoscopy instruments is an emphasis on functional surfaces and operating elements as expressed through form. The enlarged handles and the rotatable thumb ring enhance ergonomics and functionality without needing to integrate extra material. Safety is increased by optimised tactile feedback. White high-gloss surfaces and clear lines convey cleanliness and sterility.

Die gestalterische Betonung der Funktions-flächen und Bedienelemente bildet die Basis für die charakteristische Formen-sprache der Endoskopie-Instrumente PureWhite. Die vergrößerten Griffbereiche und der drehbare Daumenring erhöhen die Ergonomie und Funktionalität, ohne dass mehr Material verwendet wurde. Die Sicherheit wird durch das optimierte taktile Feedback gesteigert. Weiße Hochglanz-oberflächen und klare Linien assoziieren Sauberkeit und Sterilität.

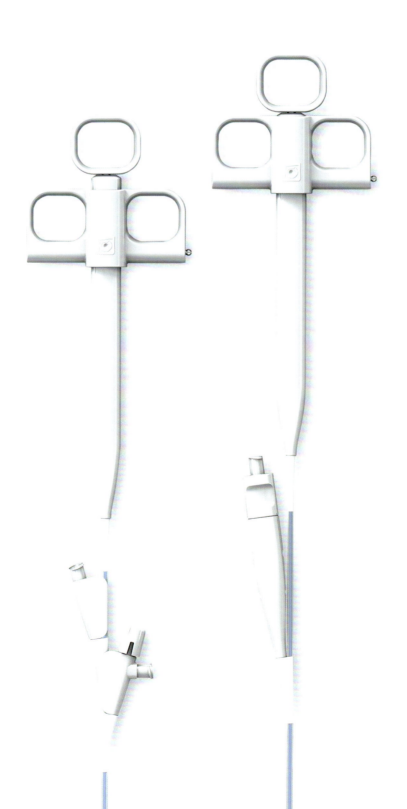

Multipette M4
Hand Dispenser
Handdispenser

Manufacturer
Eppendorf AG,
Hamburg, Germany
Design
Koop Industrial Design
(Norbert Koop),
Hamburg, Germany
Web
www.eppendorf.com
www.koopdesign.com

The Multipette M4 hand dispenser can repetitively dispense the aspirated liquid up to 100 times without refilling. With the integrated step counter, the dispensing procedure may be continued error-free even after an interruption. The emptied Combitip can be comfortably ejected by single hand control using the operating lever. An integrated sleep function turns the device off when not in use, thus reducing energy consumption and saving batteries.

Statement by the jury
This hand dispenser convinces with its constructive and functional clarity, complemented by outstanding ergonomics.

Der Handdispenser Multipette M4 kann die aufgenommene Flüssigkeit wiederholt bis zu 100 Mal abgeben, ohne neu befüllt werden zu müssen. Dank des integrierten Schrittzählers kann der Dispensiervorgang auch nach Unterbrechungen fehlerfrei fortgesetzt werden. Der entleerte Combitip wird in Einhandbedienung bequem über den Bedienhebel abgeworfen. Die Sleep-Funktion schaltet das Gerät bei Nichtbenutzung ab, was den Energieverbrauch reduziert und die Batterien schont.

Begründung der Jury
Der Handdispenser überzeugt durch seine konstruktive und funktionale Klarheit, die durch eine ausgezeichnete Ergonomie ergänzt wird.

Eppendorf Reference 2
Manual Pipette
Pipette

Manufacturer
Eppendorf AG,
Hamburg, Germany
Design
Koop Industrial Design
(Norbert Koop),
Hamburg, Germany
Web
www.eppendorf.com
www.koopdesign.com

The use of stainless-steel edging gives the Eppendorf Reference 2 its outstanding robustness. The round upper section, which does without a finger hook, makes it possible to work in any position. The pipette has a one-piece design, so that it is particularly easy to clean. The four-digit display is clearly visible from every angle. Colour coding and labelling visible on the volume area of the pipette enable quick and safe identification of volume and tip size.

Statement by the jury
The pipette's ergonomic design allows users to freely decide how they wish to hold the device. The striking colour coding ensures particularly intuitive handling.

Der Einsatz von Edelstahl an potenziellen Stoßkanten verleiht der Eppendorf Reference 2 ihre hohe Robustheit. Das runde Oberteil ohne Fingerhaken ermöglicht das Arbeiten in jeder Position. Es ist aus einem Stück gefertigt und so besonders einfach zu reinigen. Das vierstellige Display ist aus jedem Winkel gut sichtbar. Durch die Farbcodierung und die Aufschrift des Volumenbereichs auf der Pipette lässt sich die Spitzen- bzw. Volumengröße schnell und sicher identifizieren.

Begründung der Jury
Die ergonomische Gestaltung der Pipette stellt es dem Benutzer frei, wie er sie halten möchte. Die auffällige Farbcodierung macht die Bedienung besonders intuitiv.

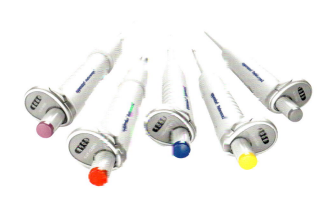

Eppendorf Tube 5.0 mL and Tube Rack 5.0 mL
Eppendorf Gefäß 5,0 mL
und Ständerplattform 5,0 mL

Manufacturer
Eppendorf AG,
Hamburg, Germany
In-house design
Detlef Schwarzwald,
Kathlen Gruner
Web
www.eppendorf.com

The Eppendorf Tube 5.0 mL has been designed for the simple and safe processing of medium-sized sample volumes. Up to now, when working with sample sizes from 0.5 to 2.0 ml, users had to handle large conical screw-cap tubes that were both impractical and prone to contamination. The newly developed tube version includes a well-conceived assortment of matching system accessories with which all lab procedures can be executed safely and ergonomically.

Statement by the jury
The Eppendorf Tube 5.0 mL fills a significant gap in the processing of samples. Its large lid moreover provides plenty of marking space.

Das Eppendorf Gefäß 5,0 mL ist für die einfache und sichere Bearbeitung von mittleren Probenvolumina konzipiert. Bisher mussten Anwender bei Proben im Bereich 0,5 bis 2,0 ml mit großen konischen Schraubdeckel-Gefäßen arbeiten, die sowohl unpraktisch als auch kontaminationsanfällig waren. Zu der neu entwickelten Gefäßvariante gehört ein durchdachtes Sortiment an passendem Systemzubehör, mit dem alle Laborabläufe sicher und ergonomisch durchgeführt werden können.

Begründung der Jury
Das Eppendorf Gefäß 5,0 mL schließt eine wichtige Lücke in der Probenverarbeitung. Der große Deckel bietet eine sehr gute Beschriftungsfläche.

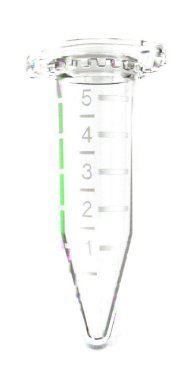

3M™ ESPE™
Dental Retraction Capsule
Retraktionskapsel
für die Zahnmedizin

Manufacturer
3M Deutschland GmbH,
3M ESPE Division,
Seefeld, Germany
In-house design
Andreas Boehm
Web
www.3mespe.com

This retraction capsule was developed for the artificial retraction of gingiva from the tooth neck. By using contrasts of matte and glossy surfaces, but also relief-like height differences, it renders the impression of a high-quality product at first glance. The cylindrical shape with rounded edges provides pleasant haptics and a well-controlled grip. The asymmetric fin underlines the friendly character of the capsule. Upon retraction, the tapered and elongated capsule tip can be easily inserted into the sulcus without injury. The conical geometry allows for easy mechanical displacement of the tissue. In addition, the attached orientation ring simplifies precise intra-oral handling.

Diese Retraktionskapsel wurde für die künstliche Entfernung des Zahnfleisches vom Zahnhals entwickelt. Sie vermittelt durch den Kontrast von matten und glänzenden Oberflächen sowie durch reliefartige Höhenunterschiede schon auf den ersten Blick den Eindruck eines hochwertigen Produktes. Die zylindrische Form mit abgerundeten Kanten sorgt für eine angenehme Haptik und einen kontrollierten Griff. Die asymmetrische Finne unterstreicht den freundlichen Charakter der Kapsel. Bei der Retraktion wird die taillierte und lang auslaufende Kapselspitze verletzungsfrei in die Zahnrinne eingeführt. Durch die konische Geometrie kann das Gewebe einfach mechanisch verdrängt werden. Der Orientierungsring vereinfacht die präzise intraorale Handhabung zusätzlich.

Statement by the jury
In designing this retraction capsule, the smallest details of use were considered for every single future working step.

Begründung der Jury
Bei der Gestaltung dieser Retraktionskapsel wurde jeder Arbeitsschritt in der späteren Anwendung bis ins kleinste Detail durchdacht und berücksichtigt.

Easypet® 3
Electronic Pipette Controller
Elektronische Pipettierhilfe

Manufacturer
Eppendorf AG,
Hamburg, Germany
Design
Koop Industrial Design
(Norbert Koop),
Hamburg, Germany
Web
www.eppendorf.com
www.koopdesign.com

The Easypet 3 electronic pipette controller offers great precision in the aspiration and dispensing of liquids. It is controlled by exerting different amounts of pressure on the ergonomically shaped buttons. The device sits comfortably in the hand with a form that supports fatigue-free pipetting. A powerful motor delivers efficient performance and speeds up the pipetting process. LEDs indicate the battery status.

Die elektronische Pipettierhilfe Easypet 3 bietet sehr hohe Präzision bei der Flüssigkeitsaufnahme und -abgabe. Die Geschwindigkeit wird durch unterschiedlich starken Druck auf die ergonomisch geformten Tasten reguliert. Das Gerät liegt bequem in der Hand und unterstützt durch seine Form ermüdungsfreies Arbeiten. Ein leistungsstarker Motor sorgt für eine effiziente Leistung. LEDs zeigen den Ladezustand der Batterie an.

Statement by the jury
Due to its sophisticated ergonomic form, the Easypet 3 is particularly user-friendly and also very comfortable to operate for extended periods of time.

Begründung der Jury
Easypet 3 ist aufgrund der durchdachten Ergonomie besonders anwenderfreundlich und auch bei längerem Gebrauch äußerst komfortabel zu benutzen.

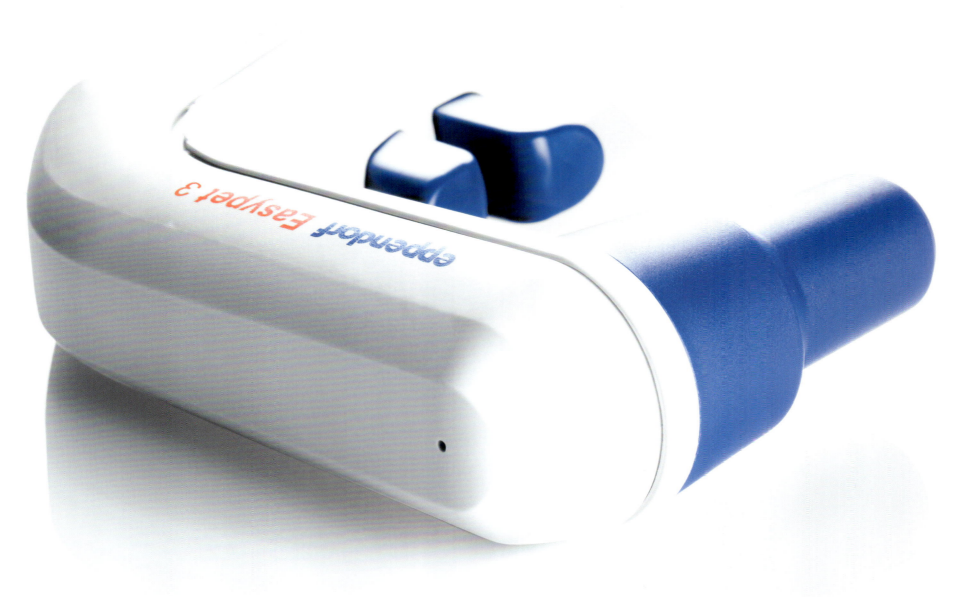

Broen Hand-held Eye Shower
Handaugendusche

Manufacturer
BROEN A/S, BROEN Lab Group,
Assens, Denmark
Design
Bjarne Frost Design Office
(Bjarne Frost),
Aarhus, Denmark
Web
www.broen.com
www.bjarnefrost.com

The Broen hand-held eye shower has been developed especially for emergency situations in modern laboratories and industrial environments. Grasping the eye shower firmly activates soft jets of water, which provide optimum flushing of the eyes. The consistent use of plastics has produced a hand-held shower that is both light and well-balanced.

Statement by the jury
This hand-held eye shower impresses with it ergonomic and functional form, which is complemented by a distinctive but calm colour scheme.

Die Broen Handaugendusche wurde speziell für Notfallsituationen in modernen Laboratorien und Industrieumgebungen entwickelt. Durch das Packen der Handaugendusche mit festem Griff wird der sanfte Strahl aktiviert, der die Augen optimal auswäscht. Die durchgehende Verwendung von Kunststoffmaterialien ist so konzipiert, dass das Gerät leicht und gut ausbalanciert in der Hand liegt.

Begründung der Jury
Diese Handaugendusche überzeugt durch ihre ergonomische und funktionelle Form, die durch eine markante, aber ruhige Farbgestaltung abgerundet wird.

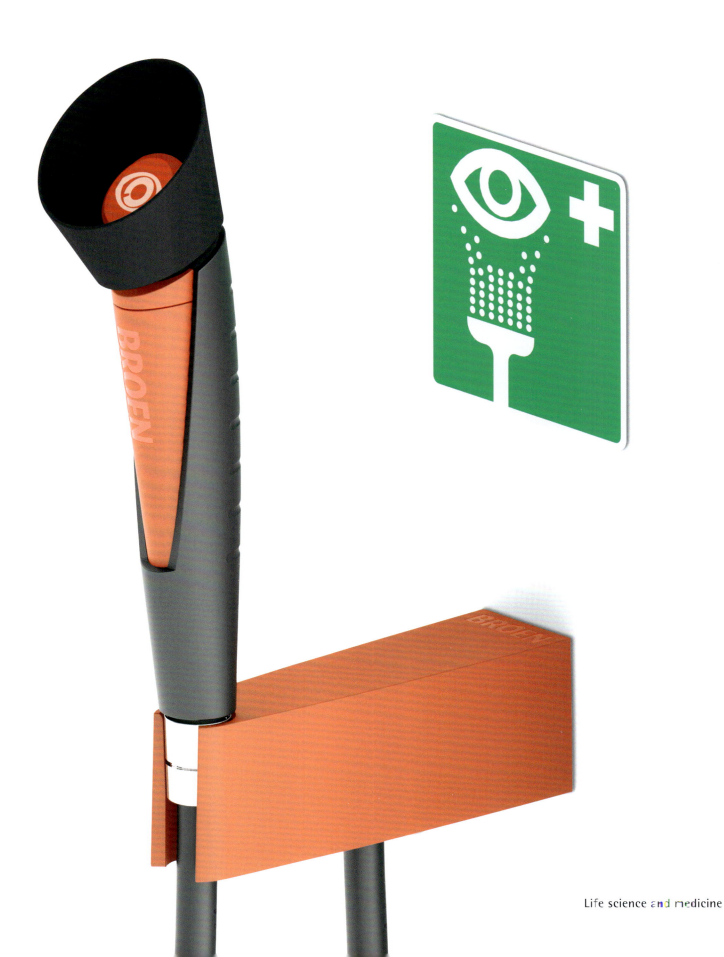

Tornado
Dental Compressor
with Silencer Hood
Dentalkompressor
mit Schallschutzhaube

Manufacturer
Dürr Dental AG,
Bietighe m-Bissingen, Germany
Design
Scala Design und technische
Produktentwicklung (Peter Theiss),
Böblingen, Germany
Web
www.duerrdental.de
www.scala-design.de

Tornado is a dental compressor with
a silencer hood that strongly reduces
noise emissions. This makes the device
very user-friendly when operated in the
vicinity of treatment rooms. Tank and
compressor are visually indicated by the
two-part design. With its asymmetrically
curved form, the housing symbolises
the route of the airflow to the outlet.

Statement by the jury
The low noise level of this dental compressor is stylishly communicated by its
smooth and dynamic design.

Tornado ist ein Dentalkompressor mit
Schallschutzhaube, welche die Geräusch-
emission stark reduziert. Dadurch ist das
Gerät bei der Verwendung in der Nähe
des Behandlungsraums sehr anwenderfreundlich. Tank- und Kompressionsbereich
werden durch den zweiteiligen Aufbau
optisch veranschaulicht. Mit seiner einsei-
tig geschwungenen Form visualisiert das
Gehäuse den Verlauf des Luftstroms bis
hin zur Auslassöffnung.

Begründung der Jury
Das geringe Arbeitsgeräusch des Dental-
kompressors wird durch die glatte und
dynamische Gestaltung besonders stilvoll
kommuniziert.

lay:art
System of Mixing Trays and Brushes
Anmischplatten- und Pinselsystem

Manufacturer
Renfert GmbH,
Hilzingen, Germany
Design
einmaleins Büro für Gestaltung
(Judith Tenzer),
Burgrieden, Germany
Web
www.renfert.com
www.einmaleins.net

lay:art consists of high-quality natural hair brushes and mixing trays for dentistry. Both the extended wrist rest and the completely detachable lid allow easy access to the tray. The brushes made from Kolinsky hair are characterised by their long life and high elasticity. The very lightweight and ergonomically shaped brush handle has a soft-touch surface, while the integrated protection against accidental roll-off provides additional customisation options with its different colours.

lay:art besteht aus optisch und qualitativ hochwertigen Naturhaarpinseln und Anmischplatten für Zahntechniker. Die verlängerte Handballenauflage ermöglicht ebenso wie der komplett abnehmbare Deckel einen optimalen Zugang zur Platte. Die Pinselspitze aus Kolinski-Haaren zeichnet sich durch ihre lange Lebensdauer und hohe Spannkraft aus. Der sehr leichte und ergonomisch geschnittene Pinselgriff besitzt eine Softtouch-Oberfläche, der integrierte Rollschutz lässt sich durch verschiedene Farben individualisieren.

Statement by the jury
With its ergonomic details lay:art provides strong support while working. The high-quality craftsmanship and materials ensure a long service life.

Begründung der Jury
lay:art sorgt durch ergonomische Details für eine sehr gute Unterstützung beim Arbeiten. Die hohe Qualität versorieht eine lange Haltbarkeit.

CEREC Omnicam
Intraoral Camera
Intraoralkamera

Manufacturer
Sirona Dental Systems,
Bensheim, Germany
Design
Puls Produktdesign
(Hendrik Breitbach, Andreas Ries),
Darmstadt, Germany
Web
www.sirona.com
www.puls-design.de

Cerec Omnicam is a CAD/CAM system
for personalised dental restorations. It
creates 3D images of teeth in natural
colours. Thanks to its ergonomic shape
and small camera head, the lateral tooth
area can also be easily scanned. The
camera head is moved over the teeth in
a flowing movement to capture images
while the high-precision 3D model is
gradually created.

Cerec Omnicam ist ein CAD/CAM-System
für individuelle Zahnrestaurationen. Es
erzeugt 3D-Aufnahmen von Zahnpartien
in natürlichen Farben. Dank der ergonomisch optimierten Form und des kleinen
Kamerakopfes kann auch der hintere
Seitenzahnbereich problemlos eingescannt
werden. Der Kamerakopf wird dabei in
einer flüssigen Bewegung über die Zähne
geführt, während sich das mit hoher Präzision vermessene 3D-Modell sukzessive
aufbaut.

Statement by the jury
The three harmoniously designed
material and colour transitions of
the Cerec Omnicam give the device
a premium, functional look.

Begründung der Jury
Die drei harmonisch verlaufenden
Material- und Farbübergänge der Cerec
Omnicam verleihen dem Gerät eine
hochwertige, sachliche Anmutung.

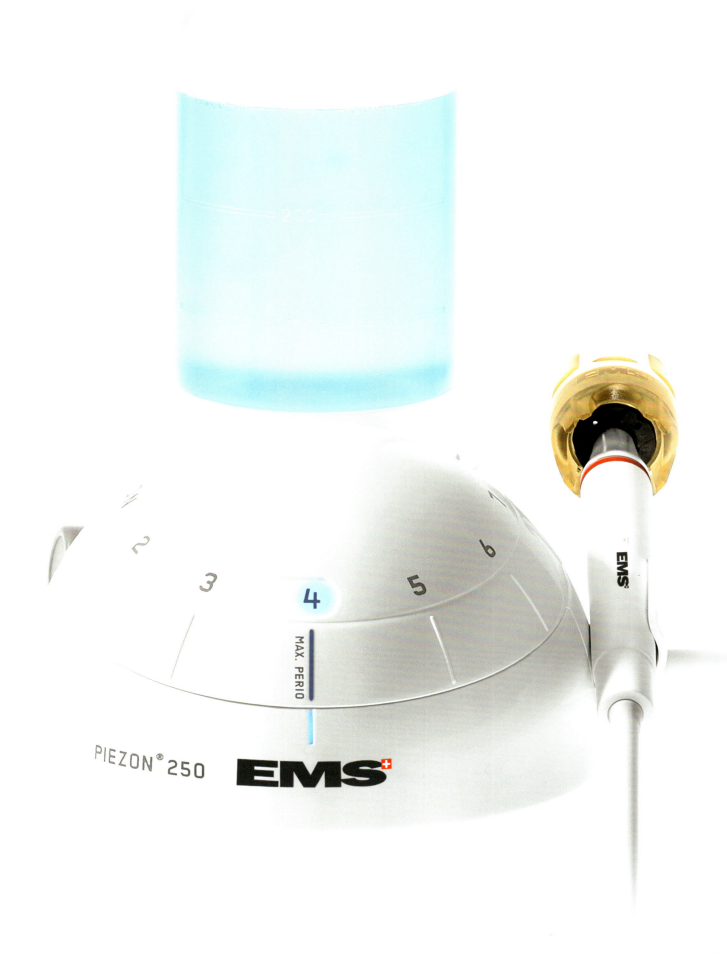

Piezon 250
Dental Scaler
Dentalscaler

Manufacturer
EMS Electro Medical Systems SA
Design
N+P Industrial Design GmbH
(Christiane Bausback),
Munich, Germany
Web
www.ems-company.com
www.neumeister-partner.com

The Piezon 250 is a new dental unit for the professional removal of tartar. Its compact form takes up less space than an A5-sized sheet of paper. The Piezon handpiece is thin, lightweight and robust. The most evident design element – the control sphere for regulating power – offers little resistance when touched, yet its rotation is precisely attuned to the pressure applied by the intuitive one-finger control element. When the foot pedal is released, an LED timer starts. At the same time, six LED lights located around the tip of the handpiece illuminate the mouth cavity for 20 seconds, thus making plaque clearly visible.

Der Piezon 250 ist eine neue Dentaleinheit zur professionellen Zahnsteinentfernung. Die platzsparende Form nimmt weniger als die Fläche eines DIN-A5-Blattes ein. Das Piezon-Handstück ist schlank, leicht und robust. Das augenscheinlichste Design-element, die Bedienkugel zur Leistungs-regulierung, leistet bei Berührung kaum Widerstand und rotiert dabei dennoch nur so stark wie Druck über das intuitive Ein-Finger-Bedienelement ausgeübt wird. Beim Loslassen des Fußpedals startet ein LED-Timer. Gleichzeitig beleuchten sechs LED-Lichter am Ende des Handstücks den Mundraum für 20 Sekunden, sodass der Zahnbelag gut sichtbar wird.

Statement by the jury
The spherical shape of the control unit excellently communicates the devices' timer function while, together with the handpiece, simultaneously conveying a compact and functional impression.

Begründung der Jury
Die runde Form des Leistungsreglers kommuniziert hervorragend seine Timerfunktion. Zusammen mit dem Handstück wirkt er kompakt und funktional.

3M™ ESPE™
Intra-Oral Syringe
Intraorale Einwegspritze

Manufacturer
3M Deutschland GmbH,
3M ESPE Division,
Seefeld, Germany
In-house design
Andreas Boehm
Web
www.3mespe.com

This single-use, intra-oral syringe is available in two variants: green for the impression material Express 2 and purple for the impression material Impregum. It can be used to obtain precise and high-quality prosthetic impressions in dentistry. The lean and functional design of the syringe body transitions into the transparent XXS mixing tip, ensuring optimum mixing quality with minimal waste. The pleasant tactile feel supports the precise guidance of the syringe. Two ergonomically arranged finger rests and the accompanying support rings allow for controlled extrusion independent of the syringe's fill level.

Diese intraorale Einwegspritze ist in zwei Varianten verfügbar: Grün für das Abformmaterial Express 2 und Lila für das Abformmaterial Impregum. Damit lassen sich prothetische Präzisionsabformungen in der Zahnheilkunde auf einem hohen Qualitätsniveau durchführen. Der schlanke und funktionale Spritzenkörper mündet in der glasklaren XXS-Mischkanüle, die eine optimale Mischqualität bei minimalem Materialverwurf gewährleistet. Die angenehme Haptik unterstützt die präzise Führung der Spritze. Zwei ergonomisch angeordnete Fingerplatten und davor platzierte Auflageringe ermöglichen ein kontrolliertes Ausbringen des Materials, unabhängig vom Füllstand der Spritze.

Statement by the jury
The special tactile construction of the 3M Espe intra-oral syringe fosters a sound visual overview during the filling and applying process.

Begründung der Jury
Die besonders taktile Bauweise der 3M Espe Intra-Oral Syringe macht das Befüllen und Applizieren visuell hervorragend erfassbar.

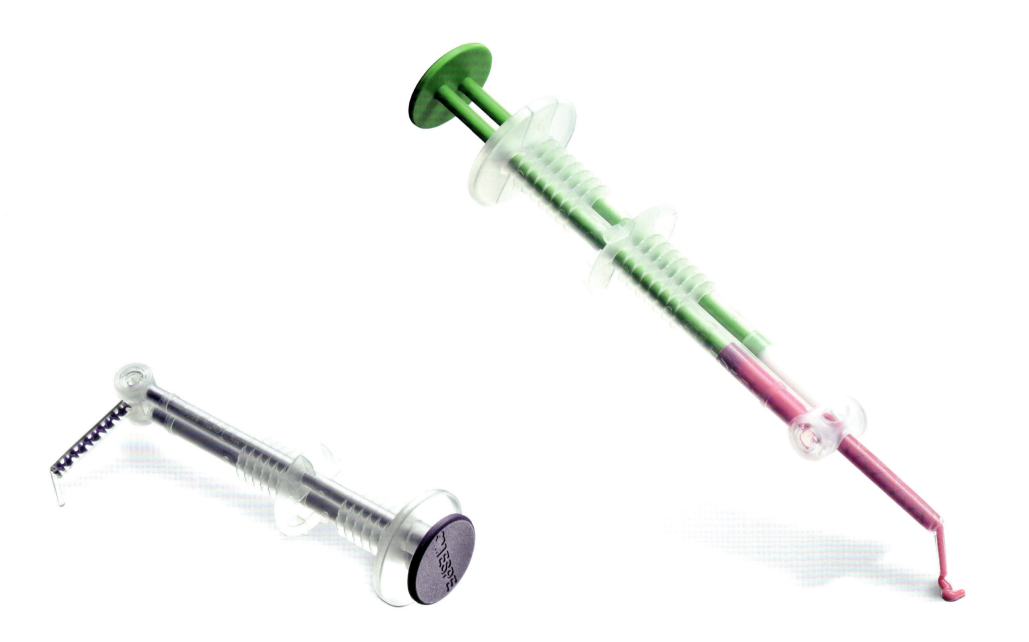

CS 8100
Dental X-Ray System
Dental-Röntgensystem

Manufacturer
Carestream Dental,
Marne-la-Vallée, France
In-house design
Yann Lecuyer,
Olivier Nesme
Web
www3.carestreamdental.com

The CS 8100 dental X-ray system combines advanced technologies in an elegant unit and enables simplified panoramic imaging. Providing digital images of the highest quality, it is optimally suited for standard panoramic images of the teeth and jaw. The system offers all of the technical functions required to improve workflow while providing the best possible patient care.

Statement by the jury
The open and particularly airy design of the CS 8100 provides the patient with maximum freedom of movement and high comfort during the examination.

Das Dental-Röntgensystem CS 8100 vereint fortschrittliche Technologien in einer eleganten Geräteform und ermöglicht eine vereinfachte Panorama-Bildgebung. Es eignet sich optimal für standardmäßige Übersichtsaufnahmen der Zähne und des Kiefers und liefert digitale Bildergebnisse in höchster Qualität. Es bietet sämtliche technische Möglichkeiten, um Arbeitsabläufe zu verbessern, und sorgt dabei für eine bestmögliche Patientenversorgung.

Begründung der Jury
Die offene, besonders luftige Gestaltung des CS 8100 bietet dem Patienten während der Untersuchung maximale Bewegungsfreiheit und hohen Komfort.

LiteTouch™
Dental Laser
Dentallaser

Manufacturer
Light Instruments Ltd.
(Syneron Dental Lasers),
Yokneam, Israel
In-house design
Yossi Weinberger
Web
www.synerondental.com

LiteTouch is one of the most versatile dental lasers for both hard and soft tissue treatments. The laser technology has been completely incorporated into the handpiece. This innovative solution mimics the handling of the turbine drill, allowing dentists to maintain complete freedom of movement and control. The LiteTouch offers all of the benefits of a laser in that it facilitates quicker healing after micro-surgical and minimally invasive treatments, as well as higher acceptance by the patients.

Statement by the jury
The dental laser impresses with its contrasting material and colour accents, while the technology is completely hidden inside the slim handpiece.

LiteTouch ist einer der vielfältigsten Dentallaser für Behandlungen im Hart- und Weichgewebe. Die Lasertechnologie wurde komplett in das Handstück integriert. Das System ahmt die Handhabung eines Turbinenbohrers nach, sodass dem Zahnarzt die uneingeschränkte Bewegungsfreiheit und Kontrolle erhalten bleibt. Gleichzeitig bietet LiteTouch alle Vorteile eines Lasers, indem es bei mikrochirurgischen und minimal-invasen Eingriffen eine schnellere Wundheilung und höherer Akzeptanz beim Patienten ermöglicht.

Begründung der Jury
Der Dentallaser besticht durch spannungsvolle Material- und Farbakzente, wobei die Technik absolut unsichtbar in das schlanke Handstück eingebaut ist.

Soprocare
Intraoral Camera
Intraoralkamera

Manufacturer
Sopro, Acteon Group,
La Ciotat, France
In-house design
Laurent Zenou
Design
EDDS (Eric Denis),
Lyon, France
Web
www.soprocare.com
www.edds-design.com

This intraoral camera offers three operating modes for checking teeth and gums: Day light mode, Cario mode and Perio mode. The latter uses an innovative fluorescence technology to highlight gum inflammation and is able to differentiate new from older plaque. The ergonomically shaped handle and the rounded camera head improve access to areas that are difficult to reach. The sensitive touch button that allows the practitioner to freeze the images is situated directly on the handpiece.

Statement by the jury
The components of the intraoral camera are precisely synchronised, lending the device a form that conveys professional flexibility and trust.

Diese Intraoralkamera bietet drei Betriebsarten für die Kontrolle von Zähnen und Zahnfleisch: den Tageslicht-Modus, den Karies-Modus und den Perio-Modus, der mithilfe einer innovativen Fluoreszenz-Technologie Zahnfleischentzündungen farblich markiert und neue von älteren Zahnbelägen unterscheiden kann. Durch den ergonomisch geformten Griff und den abgerundeten Kamerakopf sind auch schwer erreichbare Stellen gut zugänglich. Die Sensortaste, mit der sich die Bilder einfrieren lassen, befindet sich direkt am Handstück.

Begründung der Jury
Die Bauteile der Intraoralkamera sind präzise aufeinander abgestimmt. Die Form vermittelt professionelle Flexibilität und wirkt Vertrauen erweckend.

Lucid
Looking Model
Schaumodelle
für die Dentalprothetik

Manufacturer
DESIGN by TOM,
Münster, Germany
In-house design
Tom Kath
Web
www.db-tom.de

Lucid provides professional three-dimensional models for prosthetic dentistry, enabling the dentist to explain and clearly illustrate the different treatment options. High-quality materials in combination with Plexiglas foster visual transparency and enhance the patient's understanding of the different options available.

Lucid bietet professionelle Schaumodelle für die Dentalprothetik, mit denen der Zahnarzt seinen Patienten die Möglichkeiten der unterschiedlichen Versorgungsformen klar und verständlich vor Augen führen kann. Hochwertige Materialien in Kombination mit Plexiglas sorgen sowohl in der Optik als auch im Verständnis der verschiedenen Behandlungsformen für Transparenz.

Statement by the jury
The arrangement of these high-quality models is reminiscent of a slide tray, lending it a particularly structured and distinctive look.

Begründung der Jury
Die Anordnung dieser hochwertig anmutenden Schaumocelle erinnern an ein Diamagazin, so erscheinen sie ausgesprochen prägnant und aufgeräumt.

Micro Sprint
Orthodontic Brackets
Kieferorthopädische Brackets

Manufacturer
FORESTADENT Bernhard Förster GmbH,
Pforzheim, Germany
In-house design
Rolf Förster, Michael Wessinger
Web
www.forestadent.com

Micro Sprints are among the smallest orthodontic brackets in the world. They are manufactured as a single piece using a sophisticated metal injection moulding process. All bracket edges have been rounded to optimise wearing comfort. The base is anatomically curved, providing the brackets with a clean fit to the tooth surface. Tiny opposing hooks in the base ensure a secure attachment to the tooth. Laterally positioned slants allow for the safe placement of special pliers to remove the brackets at a later stage.

Statement by the jury
Micro Sprint's complex geometry is surprisingly robust. It lends the brackets a high-quality and unobtrusive appearance.

Micro Sprint gehört zu den kleinsten kieferorthopädischen Brackets weltweit. Die Brackets werden im aufwendigen MIM-Spritzgussverfahren in einem Stück gefertigt. Alle Kanten sind für den Tragekomfort optimal abgerundet. Die Basis ist anatomisch gewölbt, sodass die Brackets sauber auf der Zahnoberfläche sitzen. Winzige gegenläufige Häkchen in der Basis sorgen für sicheren Halt auf dem Zahn. Seitlich angebrachte Schrägen ermöglichen später das sichere Ansetzen der Entfernungszange.

Begründung der Jury
Die komplexe Geometrie von Micro Sprint ist überraschend stabil. Sie verleiht den Brackets ein hochwertiges und unauffälliges Aussehen.

AMS-003
Portable Steriliser
Tragbarer Sterilisator

Manufacturer
AcoMoTech,
Hsinchu, Taiwan
Design
Nova Design Co., Ltd. (Roger Lin),
New Taipei, Taiwan
Web
www.acomotech.com
www.e-novadesign.com

The portable steriliser AMS-003 incorporates the medical technology of UVC sterilisation and provides great versatility in positioning and placing dentures. The semi-transparent material filters UVC rays and allows the process of sterilisation to be observed at the same time. A soft light indicates that the device is ready for operation. Incorporating a cold cathode fluorescent lamp makes the steriliser more energy-saving and reduces its weight. The holder is made of environmentally friendly soft plastic so the lid may be opened comfortably and the dentures positioned easily.

Statement by the jury
The smoothly curved shape, paired with subtle colouring and soft material, lends this portable sterilizer an aesthetic appeal.

Der tragbare Sterilisator AMS-003 bedient sich des medizinischen Verfahrens der UVC-Entkeimung und erlaubt dem Benutzer vielfältige Positionierungsmöglichkeiten für den Zahnersatz. Das halbtransparente Material filtert die UVC-Strahlen und gestattet es dem Benutzer, den Entkeimungsprozess zu beobachten. Eine sanfte Beleuchtung zeigt an, dass der Sterilisator in Betrieb ist. Die Lichtquelle besteht aus einer kalten Kathoden-Leuchtstofflampe, die Energie spart und das Gewicht des Sterilisators reduziert. Der Behälter besteht aus einem umweltfreundlichen weichen Kunststoff, sodass der Deckel leicht zu öffnen und der Zahnersatz einfach hineinzulegen ist.

Begründung der Jury
Die sanft geschwungene Form, gepaart mit der dezenten Farbgebung und dem weichen Material, verleiht dem tragbaren Sterilisator eine ansprechende Ästhetik.

Etherena
IUD Insertion Device
Einführungsinstrument
für Intrauterinpessar

Manufacturer
Pregna International Ltd,
Mumbai, India
Design
Ticket Design Pvt Ltd,
Pune, India
Web
www.pregna.com
www.ticketdesign.com

Etherena is a loader and insertion device for the CopperT 380A contraceptive intrauterine device (IUD). The loading process is entirely aseptic, as the physician's hand touches only the grip area. The system also assists in accurate fundal placement of the IUD, as it features an integrated uterine depth scale. The form makes it easy to insert since the unit follows the natural curvature of the female body. The device is also safer for the client since the chances of perforation of uterus are lesser.

Etherena ist ein Lade- und Einführungsinstrument für das Intrauterinpessar CopperT 380A. Der Ladevorgang ist vollständig steril, da die Hand mit keinem anderen Bereich des Geräts außer dem Griff in Berührung kommt. Das System hilft auch bei der exakten Platzierung der Spirale, da es über eine integrierte Messskala für die Bestimmung der Tiefe des Uterus verfügt. Die Form erleichtert die Einführung, da sie der natürlichen Krümmung des Körpers angepasst ist. Das System ist zudem sicherer für die Patientin, da das Perforationsrisiko im Uterus geringer ist.

ORCHID SPEC
Speculum
Spekulum

Manufacturer
Bridea Medical bv,
Amsterdam, Netherlands
In-house design
Bob Roeloffs
Web
www.brideamedical.com

This disposable speculum for gynaeco-logical examinations is optimised for patient comfort and for usability by the doctor. The beak of the Orchid Spec fea-tures inwardly folded edges, allowing for a larger radius on its outer edges. This reduces friction against vaginal tissue and supports the cervix, thus improving functionality. The uniquely placed part-ing lines on the beak are seated deep inside to avoid any sharp outer edges. The locking function allows for single-handed operation. The front handle is angled away from the patient, giving the physician better control. The high-gloss white inner surface of the beak reflects the examination light, to provide optimal viewing conditions. In addition, patients regard the white plastic as more professional and calming.

Dieses Einwegspekulum für gynäkolo-gische Untersuchungen bietet optimierter Patientenkomfort und Anwenderfreund-lichkeit. Das Schnabelteil des Orchid Spec besitzt nach innen gefaltete Ränder, die einen vergrößerten Außenradius ermög-lichen. Das führt zu weniger Reibung gegen das Vaginalgewebe und den Gebär-mutterhals, was auch die Funktionalität verbessert. Die Trennlinie des Schnabels sitzt besonders tief, um jegliche scharfen Kanten zu vermeiden. Durch den Verrie-gelungsmechanismus kann das Spekulum mit nur einer Hand benutzt werden. Der vordere Griff ist nach hinten abgewinkelt, also von der Patientin weg. Dadurch ergibt sich eine bessere Kontrolle für den Arzt. Die weiße Hochglanzinnenfläche des Schnabels reflektiert das Untersuchungs-licht, um optimale Sichtverhältnisse zu schaffen. Aufgrund des Materials und seiner Farbe wird das Spekulum von den Patientinnen als professionell und beruhi-gend empfunden.

Freestyle
Soother
Beruhigungssauger

Manufacturer
Mapa GmbH,
Zeven, Germany
In-house design
Bodo Warden
Web
www.mapa.de

The balance between ergonomic and functional aspects was key to the development of the Freestyle soother. Its mouth shield is characterised by large ventilation openings, and a minimised contact surface ensures that sufficient air reaches the baby's face when sucking on the soother. The flattened shape of the baglet provides more space for the tongue, and the soother only remains in the mouth when the baby is actively sucking. This assures that muscles are exercised and deformities avoided.

Statement by the jury
With its open design, Freestyle is visually appealing while offering tangible freedom for the baby. Moreover, the large selection of different colours is inspiring.

Die Balance zwischen ergonomischer und funktionalen Aspekten war entscheidend bei der Entwicklung des Beruhigungssaugers Freestyle. Seine Mundplatte zeichnet sich durch ihre großen Belüftungsöffnungen aus. Die geringe Auflagefläche sorgt dafür, dass auch während des Schnullerns genug Luft an die Gesichtshaut kommt. Das Lutschteil bietet viel Platz für die Zunge und verbleibt durch die flache Form nur im Mund, wenn das Baby aktiv saugt. So wird die Muskulatur trainiert und Fehlstellungen werden vermieden.

Begründung der Jury
Durch seine offene und ansprechende Gestaltung bietet Freestyle dem Baby spürbare Freiheit und begeistert durch seine große Farbauswahl.

Optiflow Junior
Nasal Cannula
Nasenkanüle

Manufacturer
Fisher & Paykel Healthcare,
Auckland, New Zealand
In-house design
Optiflow Product Design Team
Web
www.fphcare.com

The Optiflow Junior nasal cannula is a ventilation cannula for infant and child patients, which helps to provide them with humidified oxygen in situations of respiratory distress. The kink-proof, breathable tubing system reduces condensation by 87 per cent as compared to conventional tube systems. Easy to apply and remove, the cannula is likewise distinguished by the way it gently fits to the patient's face. Colour coding enables quick identification of the anatomically correct size.

Statement by the jury
With its gentle wing shape and soft padding, Optiflow Junior provides exceptional support and wearing comfort for the smallest of patients.

Die Nasenkanüle Optiflow Junior ist eine Beatmungskanüle, um Patienten im Säuglings- und Kindesalter, die unter Atemschwierigkeiten leiden, befeuchteten Sauerstoff zuzuführen. Das knickfreie und atmungsaktive Schlauch-System reduziert die Atemkondensation um 87 Prozent gegenüber konventionellen Schlauch-Systemen. Es lässt sich einfach anbringen und entfernen und zeichnet sich darüber hinaus durch eine sanfte Anpassung an das Gesicht aus. Durch eine Farbcodierung wird die anatomisch passende Größe schnell gefunden.

Begründung der Jury
Optiflow Junior bietet durch die sanfte Flügelform und weiche Polsterung besonders guten Halt und Tragekomfort für die kleinsten Patienten.

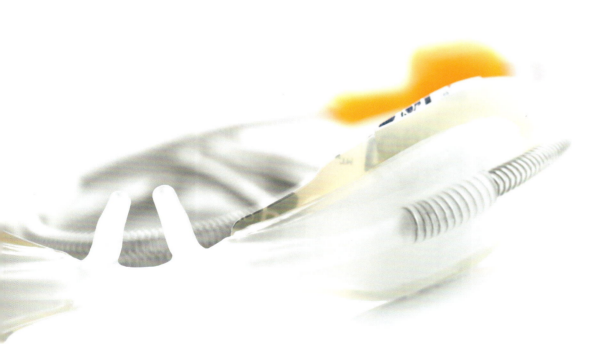

NUK First Choice Clinic
Baby Hospital
Bottle Range
Babyflaschenserie
für die Klinik

Manufacturer
Mapa GmbH,
Zeven, Germany
In-house design
Bodo Warden
Web
www.mapa.de

Nuk First Choice Clinic aims at sustainable production. The gasket geometry of the TPE teats and the smoothly structured screw ring are manufactured in an energy-saving way from one single piece of material. Moreover, the design allows for easy opening of the bottles. A soft part on the upper side of the teat takes the papilla incisiva, a sensitive nerve centre on roof of babies' mouths, into consideration. The small container for colostrum features gradation down to ⁻ ml in the spherical base, significant for mothers in a psychological sense, and can be emptied residue-free by syringe.

This baby bottle range for clinical facilities excels in particular due to its uncomplicated handling and positive environmental properties.

Nuk First Choice Clinic ist auf eine nachhaltige Produktion ausgerichtet. Der TPE-Trinksauger und der sanft strukturierte Schraubring sind energiesparend einteilig produziert und ermöglichen durch ihre Gestaltung ein griffiges, leichtes Öffnen der Flaschen. Eine Weichstelle auf der Oberseite des Saugers berücksichtigt die Papilla incisiva, ein bei Babys empfindliches Nervenzentrum im Gaumen. Der kleine Behälter für die Erstmilch weist eine für Mütter psychologisch wichtige Skala bis 1 ml im halbkugelförmigen Boden auf und kann mittels Spritze rückstandsfrei gelehrt werden.

Diese Babyflaschen für die Klinik stechen vor allem durch ihre unkomplizierte Handhabung und ihre positiven Umwelteigenschaften hervor.

Philips AVENT SCF330/20
Manual Breast Pump
Handmilchpumpe

Manufacturer
Royal Philips Electronics,
Eindhoven, Netherlands
In-house design
Philips Design Consumer Lifestyle Team
Web
www.philips.com

The Philips Avent SCF330/20 manual breast pump makes expressing milk particularly convenient. The blossom-shaped cushion can be softly positioned, where it gently massages the breast. The soft and warm texture helps support milk flow. The special shape of the pump allows for upright posture, and the ergonomically-shaped handle makes it comfortable to hold while expressing milk.

The concept of this manual breast pump is based on the natural nursing process. The pump is compact, light and easy to operate, making it perfect for use outside of the home.

Mit der Handmilchpumpe Philips Avent SCF330/20 ist das Abpumpen besonders praktisch. Das blütenförmige Kissen kann leicht an die Brust angelegt werden und massiert sie sanft, das warme und weiche Gewebe regt den Milchfluss an. Durch die besondere Form der Pumpe ist es möglich, bequem aufrecht sitzend die Milch abzupumpen, statt sich unbequem nach vorne lehnen zu müsse. Der ergonomisch geformte Griff bietet angenehmen und guten Halt.

Bei der Konzeption der Handmilchpumpe war das natürliche Stillen Vorbild. Zudem ist die Pumpe kompakt, handlich und leicht und dadurch ideal für unterwegs.

Philips AVENT SCF690/693
Baby Bottle
Babyflasche

Manufacturer
Royal Philips Electronics,
Eindhoven, Netherlands
In-house design
Philips Design Consumer Lifestyle Team
Web
www.philips.com

The Philips Avent SCF690/693 natural bottle range is ideal for mothers who combine breast and bottle feeding. The naturally shaped teat with petal design is particularly soft and flexible at the same time, so that the teat does not collapse. The innovative twin-valve system prevents colic. The bottles have a slightly curved design, making them more intuitive to hold for both mother and baby, while their wide necks make them easier to fill and clean.

With a comfort cushion in the teat, the baby bottle range makes the transition from breastfeeding to bottle feeding a particularly natural experience.

Die Naturnah-Flaschen Philips Avent SCF690/693 sind ideal für Mütter, die Stillen und Füttern mit der Flasche kombinieren. Der natürlich geformte Sauger in Blütenform ist besonders weich und gleichzeitig flexibel, ohne sich beim Saugen zusammenzuziehen. Das innovative Doppelventilsystem beugt Koliken vor. Die Flaschen sind leicht gebogen, damit Mutter und Kind sie besser halten können, der breite Hals erleichtert das Füllen und Reinigen.

Durch die weichen Komfortkissen im Sauger ist mit den Babyflaschen der Übergang vom Stillen zur Ernährung mit der Flasche besonders natürlich.

HeartStart FR3
Defibrillator

Manufacturer
Royal Philips Electronics,
Eindhoven, Netherlands
In-house design
Philips Design Healthcare Team
Web
www.philips.com

The HeartStart FR3 defibrillator automatically turns on when the case is opened, which leaves the user free to fully focus on attaching the pads. Pre-connected, immediately attachable electrodes that are not foil-wrapped reduce the preparation time. With its weight of only 1.6 kg, the device is particularly lightweight. In addition to audio instructions, the high-resolution LCD display also provides image- and text-based directions.

Der Defibrillator HeartStart FR3 schaltet sich beim Öffnen des Koffers selbst ein. So kann sich der Anwender auf das Anbringen der Pads konzentrieren. Vorab angeschlossene, sofort abzieh- und aufklebbare Elektroden ohne Folienverpackung verkürzen die Vorbereitungszeit. Das Gerät ist mit einem Gewicht von nur 1,6 kg besonders leicht. Das hochauflösende LCD-Display gibt Sprach-, Bild- und Textanweisungen.

Statement by the jury
The design of the HeartStart FR3 with its signal red case successfully places the fast and intuitive use of the defibrillator at centre stage.

Begründung der Jury
Die Gestaltung des HeartStart FR3 im signalroten Koffer rückt den schnellen und intuitiven Gebrauch des Defibrillators in den Mittelpunkt.

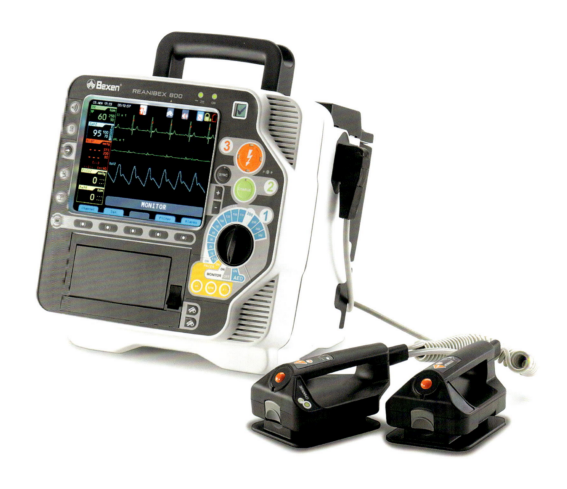

Reanibex 800
Multi-parameter Defibrillator

Manufacturer
Osatu S.Coop,
Ermua, Spain
Design
Grandesign
(Jorge Inacio, Vanda Amor),
Marinha Grande, Portugal
Web
www.bexencardio.com
www.grandesign.pt

The Reanibex 800 is a multi-parameter defibrillator designed to integrate a modular multi-purpose system and thus increase its versatility. The modularity of the device helps to retain a compact shape and volume, while simultaneously making it easier to use and transport. All interfaces are designed to optimise readability, with each key associated to its function through colour and a specific symbol, thus enhancing user-friendliness.

Reanibex 800 ist ein Defibrillator mit Multiparameteranzeige. Durch sein modulares Mehrzwecksystem wurde die Funktionsvielfalt erhöht. Die Modularität hilft dabei, die Form und das Volumen des Geräts kompakt zu halten und gleichzeitig die Benutzung und den Transport zu erleichtern. Alle Schnittstellen sind auf optimierte Lesbarkeit ausgerichtet, jeder Taste ist durch Farbe und Symbole eindeutig eine Funktion zugeordnet, was die Benutzerfreundlichkeit verbessert.

Statement by the jury
The clear, calm design of the Reanibex 800 conveys the impression of competence and safety. It directs the eye to the keypad in an optimal way.

Begründung der Jury
Die aufgeräumte, ruhige Gestaltung des Reanibex 800 vermittelt den Eindruck von Kompetenz und Sicherheit. Sie lenkt den Blick optimal auf die Funktionstasten.

5aver
Emergency Flashlight and Mask
Notfall-Taschenlampe und Maske

Manufacturer
GemVax & KAEL,
South Korea

In-house design
Eung Seok Kim,
Sang Hoon Lee

Web
www.5aver.com

reddot design award
best of the best 2013

First rescue

When a fire breaks out, every minute counts. Within a very short time, the fire produces toxic smoke and gases, which harm the lungs and can be extremely dangerous for humans. This is why a common rule says that persons in danger have five minutes to evacuate the danger area. The concept of the 5aver integrates a flash light and a respirator in one functional unit and thus is perfectly adjusted for a sensible and quick response in the case of fire. The design of this rescue system is highly self-explanatory, it is compact and allows for an intuitive use even in extreme emergency situations. In only five seconds, users can activate the respirator, which filters 92 per cent of the toxic COs out of the air. The way the 5aver integrates all the components in a compact and aesthetic appearance is innovative. This system has a fresh appeal and also immediately illustrates its purpose through colouring; its clear design vocabulary enriches almost any interior. To comply with a wide variety of locations and scenarios, such as in public buildings, the design of the 5aver follows a modular concept that allows for different configurations. This rescue system presents a highly successful reinterpretation, which both saves lives and enriches its environment through an aesthetic appearance.

Erste Rettung

Wenn es anfängt zu brennen, geht es um Minuten. Innerhalb kürzester Zeit entwickeln sich im Feuer gefährlicher Rauch und auch Gase, die die Lungen so angreifen, dass es für den Menschen extrem gefährlich wird. Die allgemeingültige Regel besagt deshalb, dass gefährdete Personen fünf Minuten Zeit haben, um außerhalb des Gefahrenbereichs zu gelangen. Das Konzept von 5aver integriert eine Taschenlampe sowie eine Atemschutzmaske in einer funktionalen Einheit und ist damit exakt auf ein möglichst sinnvolles Handeln im Brandfall abgestimmt. Dieses Rettungssystem ist gut selbsterklärend gestaltet, es ist handlich und lässt sich auch im extremen Notfall intuitiv bedienen. Innerhalb von fünf Sekunden kann der Nutzer die Atemschutzmaske aktivieren, die für fünf Minuten 92 Prozent des giftigen COs aus der Atemluft herausfiltert. Innovativ ist die Art und Weise, wie sich alle Komponenten des 5aver in eine kompakte und ästhetische Gestaltung einfügen. Dieses System mutet frisch an und visualisiert auch durch seine Farbgebung auf direkte Weise seine Bestimmung; seine klare Formensprache bereichert das jeweilige Interieur. Um den unterschiedlichen Einsatzbereichen und Szenarien, etwa in öffentlichen Gebäuden, zu entsprechen, folgt die Gestaltung von 5aver zudem einem modularen Konzept, welches unterschiedliche Konfigurationen erlaubt. Durch eine sehr gelungene Neuinterpretation entstand hier ein Schutzsystem, dass durch seine Gestaltung Leben retten kann und zugleich sein Umfeld ästhetisiert.

Emergency devices

CyFlow Cube 6
Flow Cytometry System
Durchflusszytometrie-System

Manufacturer
PARTEC GmbH,
Görlitz, Germany

Design
formfreun.de
Gestaltungsgesellschaft
(Jens Kaschlik),
Berlin, Germany

Web
www.partec.com
www.formfreun.de

A distinctive cell analysis station

Flow cytometry is the complex principle of automated analysing and counting up to 100,000 cells per second, e.g. used as standard method in medical diagnostics. The key technology is based on the analysis of fluoresent and scatter signals emitted by specially labeled cells when passing one or more laser beams at high speed. In the late 1960s, the German developer and manufacturer Partec was a pioneer in this field with the ICP 11, the first commercially available device. An impressive new concept for this technology is presented with the design of the CyFlow Cube 6. The cubic design vocabulary of this device immediately fascinates the beholder. It is defined by flowing lines and has a distinctive appearance. The CyFlow Cube 6 is small and compact, and thus easily fits into a laboratory. Thanks to its well-balanced construction and high stability, it can be used also for mobile laboratories, without having to compromise on quality. It allows high precision measurements for a wide variety of applications, such as analysing blood samples for immune status measurements, e.g. for leukaemia, lymphoma, and HIV/AIDS. Since it features a highly organised, easy-to-read interface, it can also be intuitively used by semi-qualified users. The design of the CyFlow Cube 6 lends a measuring system in the field of flow cytometry a new aesthetics – while at the same time enhancing its range of applications.

Markante Zellanalysestation

Die Durchflusszytometrie ist ein komplexes Prinzip der automatischen Zellanalyse und -zählung, das u.a. als Standardmethode in der medizinischen Diagnostik genutzt wird. Die Schlüsseltechnologie beruht auf einer Analyse von Fluoreszenz- und Streulichtsignalen, die speziell markierte Zellen aussenden, wenn sie in hohem Tempo einen oder mehrere Laserstrahlen passieren. Ein Pionier auf dem Gebiet der Durchflusszytometrie war der deutsche Entwickler und Herstellers Partec Ende der 1960er Jahre mit dem ersten kommerziell einsetzbaren Gerät ICP 11. Ein beeindruckend neues Konzept für diese Technologie bietet die Gestaltung des CyFlow Cube 6. Verblüffend ist auf den ersten Blick die kubisch anmutende Formensprache dieses Gerätes. Es ist mit fließenden Linien gestaltet und wirkt sehr markant. Das CyFlow Cube 6 ist klein und kompakt, weshalb es in jedes Labor passt. Da es zudem gut austariert ist und über eine gute Stabilität verfügt, kann es ohne qualitative Einbußen auch in mobilen Laborkonzepten betrieben werden. Es ermöglicht dabei hochpräzise Messungen und kann flexibel eingesetzt werden, um z. B. Blutproben für die Diagnostik im Bereich Leukämie, Lymphoma sowie HIV/AIDS zu analysieren. Da es über ein sehr übersichtlich gestaltetes Interface verfügt, kann es auch von wenig trainierten Nutzern intuitiv bedient werden. Die Gestaltung des CyFlow Cube 6 verleiht einem Messsystem aus dem Bereich der Durchflusszytometrie eine neue Ästhetik – sie erweitert zugleich seine Einsatzmöglichkeiten.

Statement by the jury

The design of the CyFlow Cube 6 combines highly precise advanced analysis with a new user concept of a mobile application in an impressively new way. Its minimalist and cubic design vocabulary is both distinctive and elegant, inviting users wanting to touch it. A visionary design sets new standards for the future of laboratory medicine.

Begründung der Jury

Die Gestaltung des CyFlow Cube 6 verbindet auf beeindruckend neue Weise hochpräzise Hightech-Messung mit dem neuen Nutzerkonzept eines mobilen Einsatzes. Seine minimalistisch-kubische Formensprache ist ebenso markant wie elegant, man möchte dieses Gerät spontan berühren. Ein visionäres Design zeigt hier neue Wege für die Zukunft der Labormedizin auf.

CyFox
Gel Electrophoresis Device
Gel-Elektrophorese-Gerät

Manufacturer
Partec GmbH,
Görlitz, Germany
Design
formfreun.de Gestaltungsgesellschaft
(Jens Kaschlik),
Berlin, Germany
Web
www.partec.com
www.formfreun.de

The CyFox gel electrophoresis device is a fully integrated, all-in-one solution for DNA analysis. The migration speed of the molecules located in the gel is illustrated and measured in real time during the complete measuring process. With only five buttons, the device can even be operated by a semi-skilled worker. Thanks to an integrated micro computer, automatic data analysis is possible both on the device itself and on an external computer.

Statement by the jury
CyFox impresses with is geometric, cubic shape, which conveys both functionality and minimalism while also saving considerable space.

Mit dem Gel-Elektrophorese-Gerät CyFox steht eine der ersten voll integrierten Komplettlösungen für die DNA-Analyse zur Verfügung. Die Wanderungsgeschwindigkeit der im Gel befindlichen Moleküle wird während des gesamten Messvorgangs in Echtzeit dargestellt und gemessen. Mit nur fünf Tasten ist sogar ein angelernter Laie imstande, das System zu steuern. Die Daten können dank eines integrierten Mini-Computers direkt ausgewertet, aber auch auf einen externen Computer übertragen werden.

Begründung der Jury
CyFox beeindruckt durch seine geometrische Kubusform, die sowohl Funktionalität als auch Minimalismus ausstrahlt und zudem Platz spart.

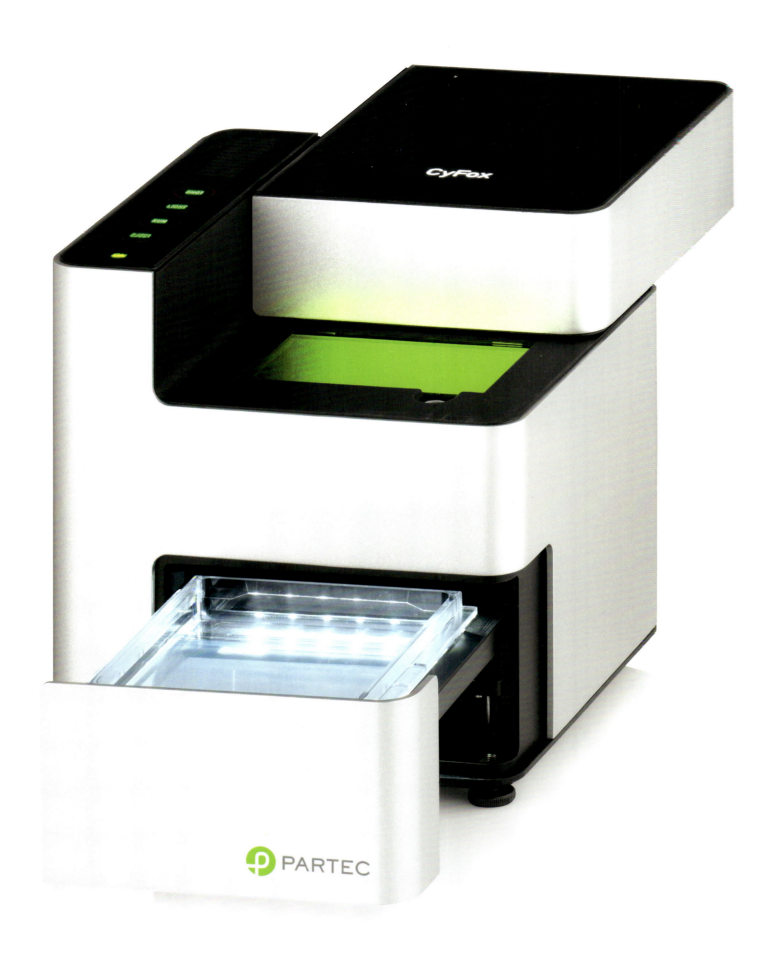

rotarus®
Peristaltic Pump
Schlauchpumpe

Manufacturer
Hirschmann Laborgeräte GmbH & Co. KG,
Eberstadt, Germany
In-house design
Hans-Jürgen Bigus
Design
Phoenix Design GmbH + Co. KG,
Stuttgart, Germany
Web
www.hirschmannlab.com
www.phoenixdesign.com

rotarus is a product family of continuous dispensing pumps. Different motors, varying housing safety classes and intelligent control of delivery volumes cover a broad spectrum of application areas in the laboratory and industry. The pumps can also accurately dispense high-viscosity media. Radio-frequency identification (RFID) technology is used to detect the pump head and the tubes.

Bei rotarus handelt es sich um eine Serie von kontinuierlich fördernden Schlauchpumpen. Verschiedene Motoren, Gehäuse unterschiedlicher Schutzklassen und intelligente Steuerungstechnik der Fördermengen decken ein breites Spektrum an Anwendungsbereichen im Labor und in der Industrie ab. Mit den Pumpen können auch Medien hoher Viskosität exakt dosiert werden. Zur Erkennung des Pumpenkopfes und der Schläuche kommt die RFID-Technologie zum Einsatz.

Statement by the jury
The rotarus peristaltic pumps are convincing with a compact design and functions that are reduced to the essentials, thus enabling intuitive use.

Begründung der Jury
Die Schlauchpumpen überzeugen durch ihr kompaktes Auftreten. Ihre Funktionen sind auf das Wesentliche reduziert und erlauben dadurch ein intuitives Handling.

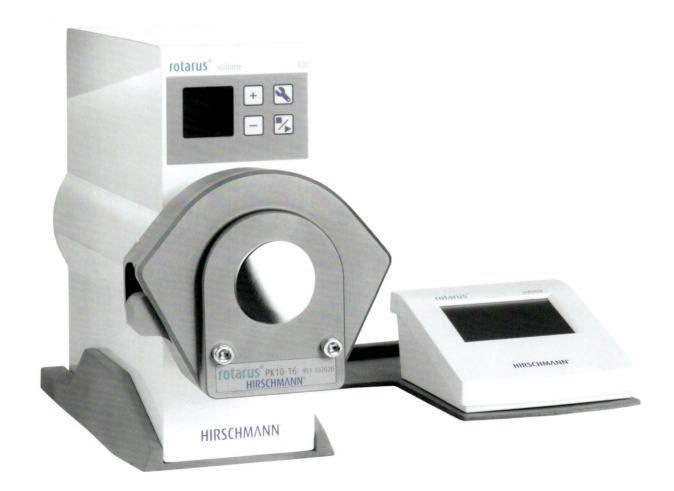

BIOSTAT® B

Bioreactor
Bioreaktor

Manufacturer
Sartorius Stedim Systems GmbH,
Guxhagen, Germany
In-house design
Sartorius Corporate Administration GmbH
Design
Corpus-C Design Agentur GmbH
(Alexander Müller, Sebastian Maier),
Fürth, Germany
Web
www.sartorius-stedim.com
www.corpus-c.de

The Biostat B bioreactor has been specifically designed to meet the requirements of biotechnological and biopharmaceutical research and development. There are different configurations available for animal, plant and insect cell cultures, as well as microbial fermentation. The control unit can be combined with various glass containers and disposable bioreactors made of polycarbonate.

Der Bioreaktor Biostat B wurde speziell für die Anforderungen in der biotechnologischen und biopharmazeutischen Forschung und Entwicklung konstruiert. Es sind verschiedene Konfigurationen für tierische, pflanzliche und Insektenzellkulturen sowie für die mikrobielle Fermentation verfügbar. Die Steuerungseinheit kann mit autoklavierbaren Glasgefäßen und Einwegkulturgefäßen aus Polykarbonat kombiniert werden.

Statement by the jury
Biostat B is convincing with its clear, geometrical design concept which harmoniously integrates into the work environment.

Begründung der Jury
Biostat B überzeugt durch sein aufgeräumtes geometrisches Gestaltungskonzept, das sich harmonisch in das Arbeitsumfeld einfügt.

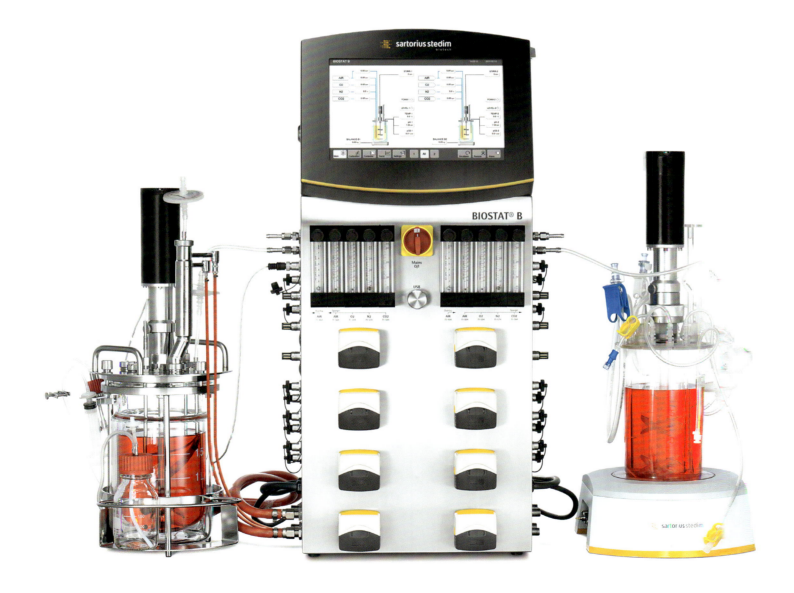

Thrombodynamics Analyser T2
Analyser for Coagulation Diagnostics
Analysegerät für die Gerinnungsdiagnostik

Manufacturer
HemaCore LLC,
Moscow, Russia
Design
WILDDESIGN GmbH & Co. KG
(Roland Wulftange),
Gelsenkirchen, Germany
velixX GmbH (Manfred Augstein),
Mannheim, Germany
Web
www.hemacore.com
www.wilddesign.de
www.velixx.com

In the medical laboratory, the Thrombodynamics Analyser T2 can be used to optimally simulate the blood-clotting process and to show anti-coagulant changes in the haemostatic balance. This enables the risk of blood clots, for instance, to be diagnosed. The design focuses on the optical measurement path in the form of the cylindrical measuring chamber. A readily visible ring of light displays the current process status.

Im medizintechnischen Labor kann man mit dem Thrombodynamics Analyser T2 den Blutgerinnungsprozess optimal simulieren und gerinnungshemmende Veränderungen im hämostatischen Gleichgewicht anzeigen. Dadurch lässt sich beispielsweise das Thromboserisiko bestimmen. Der optische Messweg wird in Form der zylindrischen Messkammer in den Mittelpunkt gerückt. Ein weithin sichtbarer Leuchtring zeigt den aktuellen Prozessstatus an.

Statement by the jury
The reduced form of the Thrombodynamics Analyser T2, which places a clear focus on the light ring, signifies successful concentration on the process of analysis.

Begründung der Jury
Die reduzierte Form des Thrombodynamics Analyser T2 mit klarem Fokus auf den Leuchtring stellt eine gelungene Konzentration auf den Analyseprozess dar.

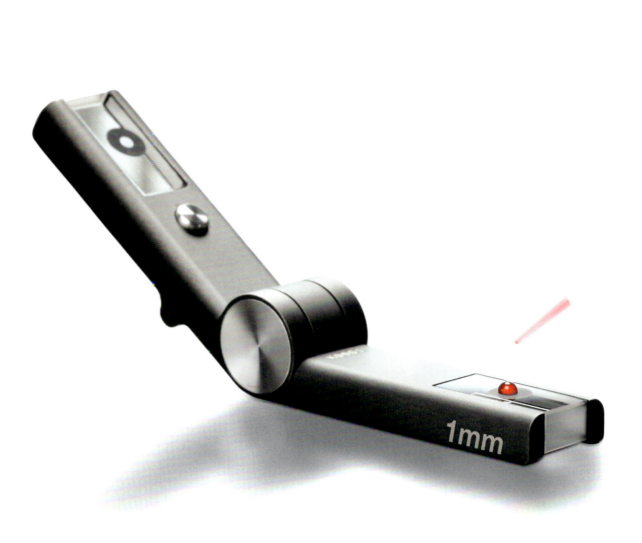

µCuvette G1.0
Cuvette
Küvette

Manufacturer
Eppendorf AG,
Hamburg, Germany
In-house design
Kathlen Gruner
Web
www.eppendorf.com

The µCuvette G1.0 is a microlitre measuring cell that is particularly well suited for the photometric measurement of high concentrations in very small volumes. It features an optical thickness of only 1 mm, which equates to a light path ten times shorter than that of standard cuvettes. This allows, for instance, nucleic acid concentrations to be measured with high reproducibility in a much higher concentration range.

Die µCuvette G1.0 ist eine Mikroliter-Mess-zelle, die sich besonders für die photome-trische Bestimmung hoher Konzentra-tionen in kleinsten Volumina eignet. Sie weist eine optische Schichtdicke von nur 1 mm auf, was einem zehnfach kürzeren Lichtweg im Vergleich zu Standardküvet-ten entspricht. Somit können z. B. Nukle-insäurekonzentrationen reproduzierbar in einem weit höheren Konzentrationsbereich gemessen werden.

Statement by the jury
The cuvette's housing made of coated aluminium, along with an innovative slide-bearing joint, conveys a high degree of robustness and dependability.

Begründung der Jury
Das Rahmengehäuse aus beschichtetem Aluminium und das innovative Gleitlager-gelenk der Küvette vermitteln ein hohes Maß an Robustheit und Sicherheit.

Berner Claire
Safety Cabinet
for Laboratories
Sicherheitswerkbank
für das Labor

Manufacturer
BERNER International GmbH,
Elmshorn, Germany
Design
Neomind
(Matthias Fischer, Mirko Kiesel),
Munich, Germany
Web
www.berner-international.de
www.neomind.eu

The Claire safety cabinet features a clear design language. Its most prominent feature is the arched front casing – the protection shield – featuring a harmoniously integrated touchscreen. There are recessed light-bands in both vertical side panels, which use colour coding to inform about the current state of operations and warn the user early of potential dangers. Moreover, the intelligent illumination of the front window's lower edge uses visual cues to immediately alert the user to any safety problems.

Statement by the jury
With its self-contained and minimalist design, in combination with innovative lighting technology, the Claire safety cabinet sets new standards.

Die Sicherheitswerkbank Claire weist eine klare Formensprache auf. Ihr markantes Merkmal ist die überwölbte Gehäuse-front – das Protection Shield – mit einem harmonisch eingepassten Touchscreen. In die beiden seitlichen Vertikalen ist ein Lichtband eingelassen, das durch seine Farbcodierung über den aktuellen Betriebszustand aufklärt und frühzeitig vor Gefahren warnt. Zusätzlich informiert eine intelligente Beleuchtung der Frontschei-ben-Unterkante den Benutzer unmittelbar in seinem Blickfeld über seine Sicherheit.

Begründung der Jury
Die Sicherheitswerkbank Claire setzt durch ihre in sich geschlossene reduzierte Bauweise in Kombination mit innovativer Lichttechnik neue Maßstäbe.

Secura®
Laboratory Balance
Laborwaage

Manufacturer
Sartorius Weighing Technology GmbH,
Göttingen, Germany
In-house design
Sartorius Corporate Administration GmbH
Design
Corpus-C Design Agentur GmbH
(Alexander Müller, Sebastian Maier),
Fürth, Germany
Web
www.sartorius.com
www.corpus-c.de

The Secura laboratory balance supports users in regulated areas of the pharmaceutical industry with functions that simplify compliance with documentation requirements and standardised workflows. The functional design and the materials used facilitate convenient operation. Moreover, the balance is very easy to clean, which is vital in pharmaceutical environments.

Die Laborwaage Secura unterstützt insbesondere Anwender in regulierten Bereichen der pharmazeutischen Industrie mit Funktionen, die die Erfüllung von Dokumentationspflichten und die Einhaltung standardisierter Arbeitsabläufe erleichtern. Die funktionale Gestaltung erlaubt eine einfache Handhabung. Darüber hinaus lässt sich die Waage sehr gut reinigen, was in einem pharmazeutischen Umfeld unabdingbar ist.

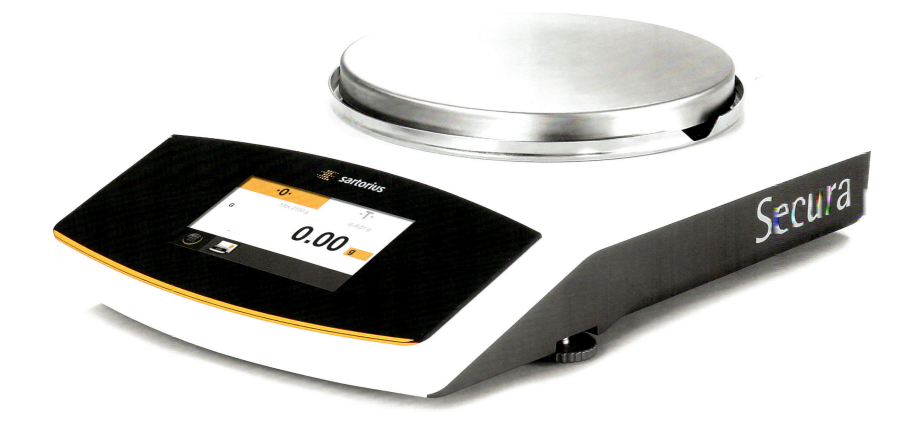

Axio Zoom.V16
Fluorescence Zoom Microscope
Fluoreszenz-Zoom-Mikroskop

Manufacturer
Carl Zeiss Microscopy GmbH,
Jena, Germany
Design
Henssler und Schultheiss
Fullservice Productdesign GmbH,
Schwäbisch Gmünd, Germany
Web
www.zeiss.com
www.henssler-schultheiss.de

The Axio Zoom.V16 fluorescence zoom microscope for large sample fields offers diverse expansion options thanks to various motor- or hand-operated stands. This high degree of modularity results from the systematic design of the individual components. At each stage of expansion the components form a self-contained device. This effect is further enhanced by the outer shape and the colour concept.

Das Fluoreszenz-Zoom-Mikroskop Axio Zoom.V16 für große Probenfelder besitzt vielfältige Ausbaumöglichkeiten durch unterschiedliche motorische oder handbetriebene Stative. Der hohe Modularitätsgrad wird durch eine systematische Gestaltung der Einzelkomponenten getragen. Bei jeder Ausbaustufe bilden die Komponenten ein in sich abgeschlossenes Gerät. Dieser Effekt wird durch die Außengeometrie und Farbgebung zusätzlich unterstützt.

eVO 500D
Slit Lamp
Spaltlampe

Manufacturer
Labo America,
Fremont, USA
In-house design
Neeraj Jain
Web
www.laboamerica.com

The inside of the eVO 500D slit lamp with touchscreen display consists of robust die-cast aluminium components, while its outside is completely covered in plastic. This prevents the housing from heating up and thus enables the ophthalmologist to safely touch the device in order to adjust the light angle. All controls are integrated in the base which enhances user convenience. The width of the light slit can be marked via a scale on the respective control knob. The image capture function is controlled via a joystick integrated in the base.

Die Spaltlampe eVO 500D mit Touchscreen besteht im Inneren aus robusten Aluminiumdruckgussteilen und ist außen vollständig mit Kunststoff verkleidet. Dadurch erhitzt das Gehäuse nicht, und der Augenarzt kann das Gerät zum Einstellen des Lichtwinkels problemlos anfassen. Alle Bedienelemente sind besonders benutzerfreundlich in den Basiskorpus integriert. Die Breite des Lichtschlitzes lässt sich über eine Skala am entsprechenden Einstellknopf markieren. Die Bildaufnahme wird über einen Joystick in der Plattform gesteuert.

LABterminal
Laboratory Furniture System
Laboreinrichtungssystem

Manufacturer
Wesemann GmbH,
Syke, Germany
Design
h&th design GmbH
(Günter Hartmann),
Hagen, Germany
Web
www.wesemann.com
www.designandconcept.de

The LABterminal modular and mobile laboratory furniture system is an ideal alternative or addition to the classic laboratory. It offers maximum planning security for process and equipment analysis or for serial experiments in instrumental analysis, because it can also be retrospectively adjusted with ease to changes in the operational environment. Electrical devices and computer systems can be reconfigured again and again according to project or task. Even during the day-to-day running of the laboratory, the system can be quite effortlessly configured and rearranged to suit the user's new requirements.

Das modulare und mobile Laboreinrichtungssystem LABterminal stellt eine ideale Alternative oder Ergänzung zur klassischen Laborzeile dar. Es bietet für die Prozess- und Geräteanalytik oder für Serienuntersuchungen in der instrumentellen Analyse maximale Planungssicherheit, weil es auch nachträglich ganz einfach an veränderte Einsatzumgebungen angepasst werden kann. Elektronische Geräte und Rechnersysteme können je nach Projekt oder Aufgabe immer wieder neu zusammengestellt werden. Auch im Laboralltag kann der Nutzer das System flexibel und ohne großen Aufwand nach seinen Wünschen konfigurieren und verschieben.

A-dec LED Dental Light
A-dec LED Dentalleuchte

Manufacturer
A-dec, Newberg, USA
In-house design
Jason Alvarez
Web
www.a-dec.com

The A-dec LED dental light offers the dentist a very efficient light with three brightness modes and a yellow cure-safe mode, which is used during the application of light-sensitive composite fillings. The current light status is indicated via a mode display, which is situated on a curved surface above the luminaire head. This gives users a clear view of the display even from an oblique angle.

Die A-dec LED Dentalleuchte bietet dem Zahnarzt ein sehr effizientes Licht mit drei Helligkeitsstufen und einem gelben „cure-safe"-Modus der während der Arbeit mit lichtempfindlichen Kompositfüllungen benutzt wird. Der aktuelle Lichtstatus wird über eine Modusanzeige dargestellt, die sich auf einer gekrümmten Fläche über dem Lichtkopf befindet. Dadurch haben die Benutzer auch aus einem schrägen Winkel immer freie Sicht auf die Anzeige.

XLED®
Surgical Lighting Systems
Operationslampe

Manufacturer
Steris Surgical Technologies,
Le Haillan, France
In-house design
Jean-Marie L'Hégarat,
Mathieu Besnard
Design
MBD Design
(Stéphane Pottier),
Paris, France
Web
www.steris-st.com
www.mbd-design.fr

The elegant contours of the XLED surgical light are not only visually appealing but also very practical in use. The LED lights are arranged so that all areas of the surgical operating table can be optimally illuminated without creating shadows of any kind. In addition, following a modular concept, the various possibilities in number and arrangement of spotlights helps achieve the ideal intensity and quality of light for each type of surgical application.

Statement by the jury
The unusual shape and dynamic appearance of the spotlights lends the XLED surgical light both a sculptural character and optimal functionality.

Die eleganten Konturen der Operationslampe XLED sind nicht nur optisch ansprechend, sondern verfolgen auch einen praktischen Nutzen. Die LED-Lichter sind dadurch so angeordnet, dass jeder Bereich des Arbeitsfeldes optimal ausgeleuchtet wird, ohne dass Schatten entstehen können. Darüber hinaus erlaubt es der modulare Aufbau des Systems, durch die Anzahl und Anordnung der Lichtbögen die Intensität und Beschaffenheit des Lichts individuell an die Operationsanforderungen anzupassen.

Begründung der Jury
Die außergewöhnliche Form der dynamisch anmutenden Lichtbögen verleiht der Operationslampe XLED einen skulpturalen Charakter und hohe Funktionalität.

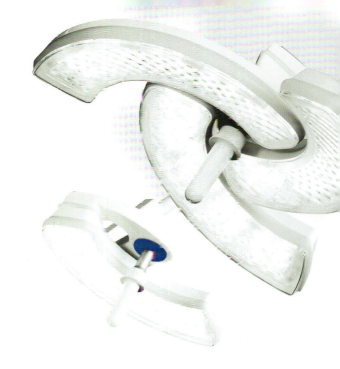

Triop Volista
Surgical Light System
OP-Leuchten-System

Manufacturer
Maquet SAS,
Orléans, France
Design
Designers Associés (James Cole),
Ivry-sur-Seine, France
Barré & Associés (Frédéric Alfonsi),
Lyon, France
Web
www.maquet.com
www.agence-da.com
www.barre-design.com

Triop Volista is a surgical light system with a modular design. The Triop suspension arm system facilitates a variety of different configurations for use in the operating theatre. The attached Volista surgical light enables surgeons to adapt lighting to their individual needs during various procedures. Thanks to a choice of three colour temperatures, they can select the lighting required for each specific operation.

Statement by the jury
The dynamic aesthetics of the Triop Volista elegantly incorporates the flexibility and movability of this surgical light system.

Triop Volista ist ein OP-Leuchten-System mit einem modularen Aufbau. Das Tragesystem Triop bietet zahlreiche Konfigurationsmöglichkeiten für die verschiedenen Einsätze in einem Operationssaal. Die daran befestigte OP-Leuchte Volista ermöglicht es den Chirurgen, die Beleuchtung den Anforderungen unterschiedlicher Eingriffe anzupassen. Dank der drei Stufen von Farbtemperaturen können sie das Licht nach den individuellen Erfordernissen einer Operation auswählen.

Begründung der Jury
Die dynamisch anmutende Ästhetik von Triop Volista greift die Flexibilität und Beweglichkeit des OP-Leuchten-Systems auf exzellente Weise auf.

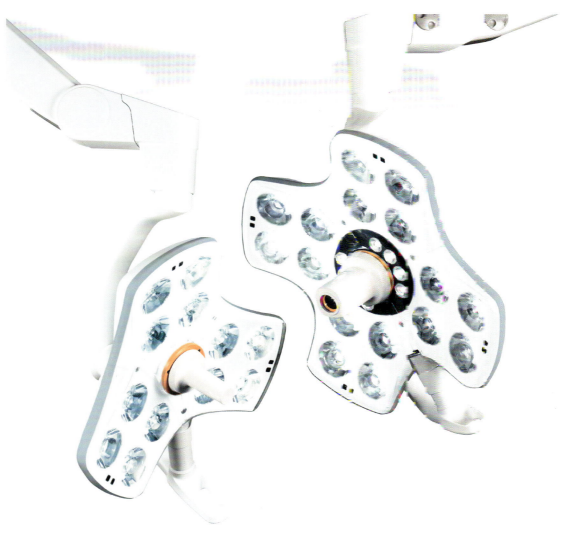

Solithera
Therapy Table
Therapieliege

Manufacturer
FREI AG,
Kirchzarten, Germany
Design
N+P Industrial Design
(Christiane Bausback),
Munich, Germany
Web
www.frei-ag.de
www.np-id.com

The Solithera therapy table, which appears to float in mid-air, is extremely robust while providing maximum legroom for the practitioner. Its electric height adjustment opens up a wide range of treatment options for all age groups. The fully integrated, invisible reel-lifting device in particular lends the therapy table its high mobility and flexibility.

Die Therapieliege Solithera scheint zu schweben und bietet dennoch eine enorme Belastbarkeit bei größtmöglicher Beinfreiheit für den Therapeuten. Sie ist elektrisch höhenverstellbar und eröffnet dadurch ein breites Spektrum an Behandlungsmöglichkeiten bei Patienten aller Altersklassen. Vor allem die unsichtbare, voll integrierte elektrische Rollenhebevorrichtung macht die Liege mobil und flexibel in der Anwendung.

Statement by the jury
The base and treatment surface of this therapy table form a harmonious whole. The central pillar gives it an extraordinarily light character.

Begründung der Jury
Sockel und Lagerungsfläche dieser Therapieliege bilden eine harmonische Einheit. Die Säule in der Mitte verleiht ihr eine außergewöhnliche Leichtigkeit.

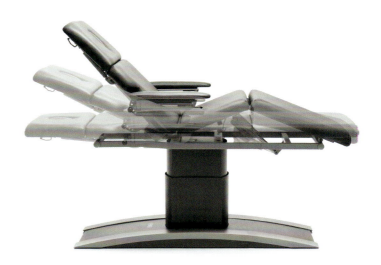

NTQ 7
Hydraulically
Emergency Stretcher
Hydraulische
Notfallkrankentrage

Manufacturer
Gökler A.Ş. / Nitrocare,
Sivas, Turkey
In-house design
Göksel Aras
Design
DESCENT, s.r.o. (Jiři Španihel),
Kopřivnice, Czech Republic
Web
www.nitrocare.com.tr
www.spanihel.cz

The NTQ 7 emergency stretcher meets the highest hygiene requirements. It is especially easy and fast to clean. For these purposes, a single-piece plastic body was designed to cover the backrest, the sitting surface and the leg rest. In addition, the body design hides the support legs that hold the compact surfaces up. By clearing the connection points, the stretcher prevents the build-up of microbes. The metal frame inside is also entirely covered so that fluids cannot enter the stretcher. In addition to this, the bodywork design has no sharp edges and does without connections to other apparatuses, a feature that makes both cleaning and handling it even easier and safer. The single-piece mattress features a special surface that is liquid-resistant and thus ensures longer durability. The hydraulic mechanism parts, which lift and lower the patient platform, are hidden in the telescopic aluminium columns, and the stretcher is equipped with sturdy bumpers.

Die Notfallkrankentrage NTQ 7 erfüllt höchste hygienische Ansprüche. Sie ist besonders einfach und schnell zu reinigen. Zu diesem Zweck wurde ein einteiliges Kunststoffgestell entworfen, das die Rückenlehne, Sitzfläche und Beinablage umschließt. Das Gestell verdeckt zudem die zusätzlichen Stützbeine, welche die kompakte Liegefläche tragen. Dadurch verringert sich die Gefahr, dass sich Mikroben an diesen Verbindungspunkten festsetzen können. Auch der Metallrahmen wird komplett von dem Gestell umschlossen, sodass keine Flüssigkeiten eindringen können. Zudem gibt es keine spitzen Ecken und Kanten oder Verbindungsstücke zu anderen Geräten, was die Reinigung zusätzlich erleichtert und die Benutzung für das Personal sicherer gestaltet. Die einteilige Matratze verfügt über eine spezielle Oberfläche, die Flüssigkeit abweist, was sie langlebiger macht. Die Hydraulik, mit der die Liegefläche angehoben und abgesenkt wird, ist in der teleskopischen Aluminiumsäule verborgen. Zudem besitzt die Trage robuste Stoßstangen.

Statement by the jury
The NTQ 7 emergency stretcher features appealing, contemporary colouring and a single-piece plastic body design that is extraordinarily robust and easy to clean.

Begründung der Jury
Die Notfalltrage NTQ 7 gefällt durch eine zeitgemäße Farbgestaltung und das durchgängige Kunststoffgestell, welches leicht zu reinigen und ausgesprochen robust ist.

PET/CT Uptake Room
PET/CT-Vorbereitungsraum

Manufacturer
Royal Philips Electronics,
Eindhoven, Netherlands
In-house design
Philips Design Healthcare Team
Web
www.philips.com

In this PET/CT uptake room, cancer
patients are injected with a radioactive
fluid which is required for the ensuing
imaging process. They wait for an hour
in the room until the examination starts.
The uptake room equipped with the
Ambient Experience system consists of a
stylishly designed ceiling that generates
specific mood settings. Moreover, there
is a practical procedure light to facilitate
an ideal working environment.

Statement by the jury
The softly illuminated wall and ceiling
panels give this uptake room a premium
lounge character and create a private
refuge.

In dem PET/CT-Vorbereitungsraum wird
Krebspatienten eine radioaktive Flüssigkeit
injiziert, die für die spätere Bildgebung
nötig ist. Hier halten sie sich dann bis zu
Beginn der Untersuchung eine Stunde
lang auf. Der Raum ist mit dem Ambient-
Experience-System ausgestattet, das aus
einer stilvollen Deckeneinheit mit Einstell-
möglichkeiten für spezielle Stimmungen
besteht. Außerdem steht praktisches
Arbeitslicht zur Verfügung, um für die
verschiedenen Aufgaben ein optimales
Umfeld zu schaffen.

Begründung der Jury
Die sanft hinterleuchtete Wand- und
Deckenpaneele verleiht diesem Vorberei-
tungsraum einen edlen Lounge-Charakter
und schafft einen privaten Rückzugsort.

Bedside Cabinet
Nachtschrank

Manufacturer
KBB-Tautmann,
Ankara, Turkey
Design
ipdd GmbH & Co. KG
(Özkan Isik),
Stuttgart, Germany
Web
www.tautmann.com
www.ipdd.com

This bedside cabinet looks like an elegant piece of furniture for the home but fulfils all the requirements of hospital furniture. It is made of easy-to-disinfect high pressure laminate (HPL), while the scratch-resistant profiles are crafted from anodised aluminium. Thanks to the double castor wheels, in addition to an overall lightweight construction, the cabinet is easy and safe to manoeuvre. It is available in different storage sizes and a range of colour and surface options.

Statement by the jury
In the design of this bedside cabinet, decorative versatility and durable materials are combined to create a convincing product in terms of both aesthetics and practicality.

Dieser Nachtschrank sieht aus wie ein elegantes Möbelstück für zu Hause, erfüllt aber alle Anforderungen an ein Krankenhausmöbel. Er besteht aus dem einfach zu desinfizierenden Werkstoff HPL, die kratzresistenten Profile sind aus anodisiertem Aluminium hergestellt. Durch die Leichtbaukonstruktion in Kombination mit den Doppelrollen lässt sich der Schrank leicht und sicher bewegen. Er ist mit variablem Stauraum und in vielen Farb- und Oberflächenvarianten erhältlich.

Begründung der Jury
Dekorative Wandelbarkeit und strapazierfähige Materialien machen diesen Nachtschrank zu einem überzeugenden Produkt sowohl in ästhetischer als auch in praktischer Hinsicht.

AQUABUDDY
Bed Shower
Bettdusche

Manufacturer
Pollution Group,
Budrio, Italy
Design
Mechanema
(Giulio Mattiuzzo),
Rimini, Italy
Web
www.pollution.it
www.mechanema.it

The circulation of water in the Aquabuddy bed shower is obtained without the use of in-line pumps. This helps to minimise the risk of bacterial infections for patients. The shower is suitable for use in any bed, both in hospitals and at home. The wheels and the horizontal handles make it easy to transport. A liquid crystal temperature indicator signals if temperatures rise too high, while built-in compartments offer storage space. The system does not require maintenance.

Statement by the jury
The purist and compact design of this bed shower skilfully underlines its straightforwardness in terms of assembly and use.

Die Wasserzirkulation der Bettdusche Aquabuddy funktioniert ganz ohne Inlinepumpen, dadurch wird das Risiko bakterieller Infektionen minimiert. Die Dusche ist sowohl für das Krankenhaus als auch für den Gebrauch zu Hause geeignet. Durch die Räder und die horizontalen Griffe ist sie einfach zu transportieren. Eine Temperaturanzeige aus Flüssigkristall warnt vor zu hohen Temperaturen. Eingebaute Fächer sorgen für Stauraum. Eine Wartung des Systems ist nicht nötig.

Begründung der Jury
Die puristische und kompakte Gestaltung der Bettdusche unterstreicht gekonnt ihre Unkompliziertheit im Zusammenbau und in der Anwendung.

Drawer Tray System
Schubladensystem

Manufacturer
Stryker Trauma GmbH,
Schönkirchen, Germany
In-house design
Andreas Heede
Design
Brennwald Design
(Jörg Brennwald, Chang-Chin Hwang),
Kiel, Germany
Web
www.osteosynthesis.stryker.com
www.brennwald-design.de

This drawer tray system is dedicated to the storage and sterilisation of medical implants and instruments in operating theatres. Its modular design facilitates use at variable heights and with different equipment. In all critical functional areas, the main body made of corrosion-resistant stainless steel is complemented by durable plastic parts, which have a standard size and can be used multiple times.

Dieses Schubladensystem dient zur Aufbewahrung und Sterilisation medizinischer Implantate und Instrumente im OP-Bereich. Der modulare Aufbau schafft die Voraussetzung für variable Bauhöhen und Ausstattungen. Der Hauptkorpus aus korrosionsfestem Edelstahl wird durch strapazierfähige Kunststoffteile ergänzt, die eine Standardgröße aufweisen und flexibel verwendet werden können.

Statement by the jury
The drawer tray system exhibits a striking, robust design of high-quality. The materials selected provide it with enormous durability.

Begründung der Jury
Dieses Schubladensystem überzeugt durch eine hochwertige, robuste Gestaltung. Die verwendeten Materialien verleihen ihm enorme Langlebigkeit.

Wireless Body Analysis Scale
Kabellose Körperanalysewaage

Manufacturer
iHealth Lab Inc.,
Mountain View, USA
In-house design
Cong Ming
Design
Beijing FromD Design Consultancy Ltd.
(Ma Ming),
Beijing, China
Web
www.ihealthlabs.com
www.fromd.net

The Wireless Body Analysis Scale communicates via Wi-Fi with the accompanying free iPhone and iPad app. A total of nine key measurements are taken, including weight, body fat, muscle mass and body mass index. Using the app, all measurement values and results can be administered, monitored and turned into easy-to-understand charts to facilitate smarter health planning. Moreover, the practical logging of daily calorie intake makes it easier to achieve weight loss goals.

Die Körperanalysewaage kommuniziert über Wi-Fi mit der dazugehörigen kostenlosen App für iPhone und iPad. Es werden insgesamt neun Messwerte wie z. B. Gewicht, Körperfettanteil, Muskelmasse und Body-Mass-Index erfasst. Sämtliche Werte und Ergebnisse können in der App verwaltet, überwacht, in anschaulichen Diagrammen dargestellt und für eine intelligente Gesundheitsplanung genutzt werden. Über die Verwaltung der täglichen Kalorienzufuhr fällt es leichter, gesetzte Ziele zu erreichen.

Statement by the jury
With its dynamic square shape and plainly designed weighing surface, this body analysis scale conveys distinctive and minimalist elegance.

Begründung der Jury
Durch ihre dynamische Quadratform und die schlichte Wiegefläche wirkt die Körperanalysewaage minimalistisch und auffallend elegant.

One Touch Select Simple
Blood Glucose Meter
Blutzuckermessgerät

Manufacturer
LifeScan Inc.,
Milpitas, USA
In-house design
Johnson & Johnson
Global Strategic Design Office
(Rochelle Kleinberg, Miya Osaki)
LifeScan Inc.
(Viru Parlikar, Neil Roberts,
Manoj Sharma, Grace Sheen, Ruth Yap)
Design
Shore Design Consultancy Ltd.,
Edinburgh, GB
Web
www.lifescan.com
www.shore-design.co.uk

One Touch Select Simple is a straight-forward blood glucose meter designed specifically for the needs of recently diagnosed diabetics without much experience. It does not require any buttons or menu navigation. The patient simply inserts the test strip, applies a drop of blood and receives the test result within a few seconds. The screen features large blood glucose numbers and an icon-driven interface. If a patient's blood sugar level deviates from the normal range, a flashing arrow appears, pointing towards the corresponding icons on the housing. Simultaneously, an audio alarm alerts the user to the blood sugar data results.

One Touch Select Simple ist ein einfaches Blutzuckermessgerät, das speziell auf die Bedürfnisse neu diagnostizierter Diabetiker ohne viel Erfahrung zugeschnitten ist. Es kommt ohne Knöpfe und Menüführung aus. Der Patient führt einfach den Test-streifen ein, gibt einen Tropfen Blut darauf und erhält das Ergebnis innerhalb weniger Sekunden. Die Messdaten erscheinen groß und deutlich auf dem Bildschirm, und bei einem zu hohen oder zu niedrigen Blutzuckerspiegel, wird ein blinkender Pfeil eingeblendet, der auf entsprechende Sym-bole auf dem Gehäuse deutet. Gleichzeitig weist ein Audioalarm auf die Bedeutung der Blutzuckerdaten hin.

Statement by the jury
The clear design of this blood glucose meter is exemplary. Its use can be grasped intuitively, and the measurement results are easy to understand.

Begründung der Jury
Die übersichtliche Gestaltung dieses Blutzuckermessgerätes ist vorbildlich. Die Anwendung ist intuitiv zu erfassen und Messergebnisse sind schnell zu verstehen.

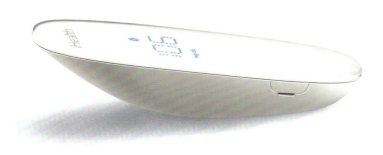

Wireless Smart Glucose Monitoring Device
Kabelloses intelligentes Blutzuckermessgerät

Manufacturer
iHealth Lab Inc.,
Mountain View, USA
In-house design
Li Zhenwei
Design
Beijing FromD Design Consultancy Ltd.
(Ma Ming, Li Duo),
Beijing, China
Web
www.ihealthlabs.com
www.fromd.net

This wireless smart glucose monitoring device was developed as an aid for integrating diabetes control into the everyday life of diabetics. It allows easy and convenient transmission of the blood glucose results to portable iOS devices. Users can monitor tendencies over longer periods, set reminder functions for medication intake times and automatically check on the expiry dates of test strips. In addition, the device contains a video guide function that can show how to correctly test blood glucose by it.

Das kabellose intelligente Blutzucker-messgerät wurde für Diabetiker entwickelt, die ihr Blutzuckermanagement in den Alltag integrieren möchten. Das Gerät ermöglicht es, den Blutzuckerspiegel bequem und einfach an mobile iOS-Geräte zu übermitteln. Dabei können Benutzer Entwicklungen über längere Zeiträume verfolgen, Erinnerungsfunktionen zur Medikamenteneinnahme einrichten und automatisiert die Ablaufdaten von Test-streifen überprüfen. Darüber hinaus verfügt das Gerät über eine Videofunktion, die zeigt wie der Blutzuckertest korrekt ausgeführt wird.

mylife™ Unio™
**Blood Glucose
Monitoring System**
Blutzuckermesssystem

Manufacturer
Ypsomed Distribution AG,
Burgdorf, Switzerland
Design
Process Design AG,
Lucerne, Switzerland
Web
www.mylife-diabetescare.com
www.process-group.com

mylife Unio is a practical, compact set for discreet and fast blood glucose monitoring at home and on the go. The main component is a blood glucose measuring device, which meets the highest standards. With its automatic load and release function, the precise mylife AutoLance lancing device provides constant pricking pressure and thus virtually painless blood sampling. Its flat storage container for test strips significantly reduces the bulk of the whole set.

mylife Unio ist ein praktisches Kompakt-Set zur diskreten und schnellen Blutzucker-messung zu Hause und unterwegs. Das Herzstück bildet ein Blutzuckermessgerät, das höchste Ansprüche erfüllt. Das präzise Lanzettengerät mylife AutoLance sorgt mit einer automatischen Lade- und Aus-lösefunktion für stets gleichbleibenden Einstechdruck und somit für eine schmerz-arme Blutentnahme. Durch eine flache Aufbewahrungsbox für die Teststreifen konnte die Dicke des gesamten Sets deut-lich reduziert werden.

Gmate™ Smart
**Blood Glucose
Monitoring System**
Blutzuckermesssystem

Manufacturer
philosys Co., Ltd.,
Gunsan City, Jeolabuk Province, South Korea
In-house design
Jeongin Bag, Soonwan Jeong
Web
www.philosys.com
Honourable Mention

This blood glucose monitoring system was designed especially for use in combination with the iPhone, iPad and iPod Touch. It is inserted into the earphone jack, and the blood glucose readings are transmitted directly to the terminal device where the data is presented in graphic form. It features a minimal configuration in order to be lightweight and easily portable. An integrated LED light makes it easier to find the test-strip port.

Dieses Blutzuckermesssystem wurde speziell für die Benutzung in Verbindung mit iPhone, iPad und iPod Touch entwickelt. Es wird in den Kopfhöreranschluss gesteckt, die Blutzuckermesswerte werden übertragen und in Form von Grafiken auf dem Endgerät dargestellt. Die Ausstattung beschränkt sich auf das Nötigste, damit es leicht und einfach zu transportieren ist. Ein LED-Licht weist auf den Port für den Teststreifen hin.

Statement by the jury
Gmate Smart combines advanced technologies with an ultra-small form, enabling the system to be easily taken anywhere.

Begründung der Jury
Gmate Smart vereint zukunftsweisende Technologien mit einer ultrakleinen Form, sodass es sich überallhin mitnehmen lässt.

Finger Pulse Oximeter
Fingerpulsoximeter

Manufacturer
iHealth Lab Inc.,
Mountain View, USA
In-house design
Han Qiubo
Design
Beijing FromD Design Consultancy Ltd.
(Ma Ming, Li Duo),
Beijing, China
Web
www.ihealthlabs.com
www.fromd.net

This finger pulse oximeter measures blood oxygen saturation and the pulse rate. The highly lightweight device carries out non-invasive measurements in just a few seconds when attached to the index finger. It is thus particularly suited for personal use and sports activities, but also for homecare and medical use. The pulse oximeter can communicate with the iPhone and iPad via a Bluetooth connection.

Der Fingerpulsoximeter zeigt die Sauerstoffsättigung im Blut und den Puls an. Das sehr leichte Gerät führt die nichtinvasive Messung in nur wenigen Sekunden aus, dazu muss es lediglich an den Zeigefinger gesteckt werden. Es ist daher insbesondere für den privaten Gebrauch, für sportliche Aktivitäten, aber auch für die häusliche Pflege und den medizinischen Bereich geeignet. Der Pulsoximeter kann über Bluetooth mit iPhone und iPad kommunizieren.

Statement by the jury
This elegant finger pulse oximeter conveys purist aesthetics that harmonise with the compatible Apple products.

Begründung der Jury
Der Fingerpulsoximeter präsentiert sich in einer puristischen Ästhetik und passt damit zu den Apple-Produkten, mit denen er kompatibel ist.

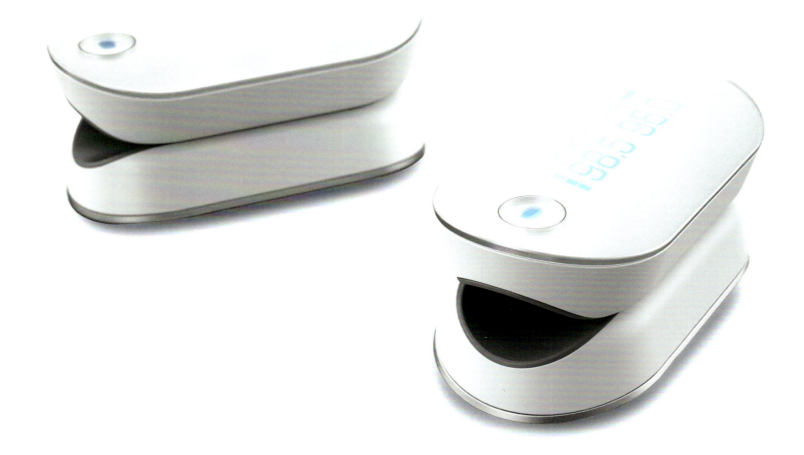

Microlife A6PC
Blood Pressure Monitor with Stroke Risk Detection
Blutdruckmessgerät mit Erkennung des Schlaganfallrisikos

Manufacturer
Microlife Corp,
Taipei, Taiwan
Design
Process Design AG (Peter Wirtz),
Lucerne, Switzerland
Web
www.microlife.com
www.process-group.com

Microlife A6PC is an innovative blood pressure monitoring device that can detect atrial fibrillation. The colour-coded cuff socket is one design feature that optimises usability. The wide range conical shape cuff is suitable for all common arm sizes from 22–42 cm and ensures a high level of comfort. Its flat form makes the device an ideal travel companion. The optional device stand provides a satisfactory viewing angle when using the product at home.

Statement by the jury
A simple and self-explanatory design turns the Microlife A6PC into an easy-to-understand solution that is also notably space-saving.

Microlife A6PC ist ein innovatives Blutdruckmessgerät, das Vorhofflimmern erkennt. Die farbig gekennzeichnete Buchse für den Anschluss der Manschette verbessert die Benutzerfreundlichkeit. Die anpassbare Manschette wurde für Oberarmumfänge von 22–42 cm entwickelt und sitzt sehr bequem. Durch seine flache Form ist das Gerät der ideale Reisebegleiter. Der optionale Ständer macht die Bedienung bei der Anwendung zu Hause besonders komfortabel.

Begründung der Jury
Durch die simple und selbsterklärende Gestaltung ist Microlife A6PC leicht anzuwenden und zudem sehr platzsparend.

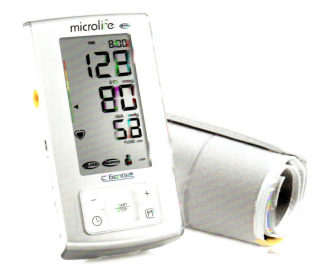

ME 80
Mobile ECG Device
Mobiles EKG-Gerät

Manufacturer
Beurer GmbH,
Ulm, Germany
Design
MD Biomedical, Inc. (Carl Yu),
Taipei, Taiwan
Web
www.beurer.com
www.mdbiomedical.com

With the touch of a button, this one-channel mobile ECG device offers straightforward self-monitoring by allowing users to reliably measure their heart rhythm. The data can be recorded spontaneously, for example when feeling unwell, and discussed during a consultation with a doctor at a later time. For precise evaluation and a detailed representation of the measurement data, the ME 80 may be simply connected to a PC using the integrated USB plug.

Statement by the jury
This elegant and very easy-to-use mobile ECG device is aligned to the smart aesthetics of modern MP3 players.

Mit nur einem Knopfdruck bietet das mobile 1-Kanal-EKG-Gerät eine unkomplizierte Selbstkontrolle, um den Herzrhythmus zuverlässig zu ermitteln. Die Werte können spontan, z. B. im Moment des Unwohlseins, aufgezeichnet und bei einem späteren Arztbesuch besprochen werden. Für die genaue Auswertung und detaillierte Darstellung der Messdaten wird das ME 80 einfach über den integrierten USB-Plug an den PC angeschlossen.

Begründung der Jury
Das elegante und besonders handliche EKG-Gerät passt sich auf smarte Weise der Ästhetik von modernen MP3-Playern an.

SBM 42
Blood Pressure Monitor
Blutdruckmessgerät

Manufacturer
Beurer GmbH,
Ulm, Germany
Design
uli schade industriedesign
(Matthias Kolb),
Neu-Ulm, Germany
Web
www.beurer.com
www.schadedesign.de

The prominent design elements of this fully automatic blood pressure monitor are its extra-large display and the operating knob, both of which are backlit in blue. The knob allows comfortable and easy use when taking the pulse and measuring blood pressure. In addition, the device features integrated arrhythmia detection. The coloured LED scale to the left of the display ensures quick classification of the measurements.

Statement by the jury
The soft lines of this blood pressure monitor and its clear focus on a large display foster a state-of-the-art character.

Hervorstechende Gestaltungselemente dieses vollautomatischen Blutdruckmessgeräts sind das extra große Display mit blauer Hintergrundbeleuchtung und der ebenfalls blau beleuchtete Bedienknopf. Dieser bietet eine komfortable und einfache Handhabung, um Puls und Blutdruck zuverlässig zu ermitteln. Das Gerät verfügt außerdem über eine integrierte Arrhythmie-Erkennung. Die farbige LED-Skala links neben dem Display sorgt für eine schnelle Einordnung der Werte.

Begründung der Jury
Die sanfte Linienführung und der klare Fokus auf das großzügige Display verleihen dem Blutdruckmesser einen modernen Charakter.

Mobile Health Mini-Lab
Mobiles Mini-Labor

Manufacturer
MIR – Medical International Research,
Rome, Italy
Design
Creanova srl (Matteo Moroni),
Como, Italy
Web
www.spirometry.com
www.creanova.it

Spirotel integrates four devices in one: a spirometer for measuring lung volume, a 3D oximeter, a triaxial accelerometer with motion analysis, which calculates physical activity and body posture, and an e-diary. The device is controlled using a touchscreen, and data may be transmitted via an embedded GSM module, USB or Bluetooth. The detachable turbine ensures safe use and can be easily removed for disinfection.

Spirotel vereint vier Geräte in einem: den Spirometer für die Messung des Lungenvolumens, ein 3D-Pulsoximeter, einen dreiachsigen Bewegungssensor, der die körperliche Aktivität und Körperhaltung berechnet, sowie ein E-Tagebuch. Das Gerät wird über einen Touchscreen bedient, Messdaten können über ein integriertes GSM-Modul, USB oder Bluetooth übertragen werden. Das Atemrohr mit der Turbine garantiert eine sichere Anwendung. Die Turbine lässt sich für die Desinfizierung leicht abnehmen.

Statement by the jury
Spirotel has a compact design that is very distinctive due to its contrasting orange elements.

Begründung der Jury
Spirotel gefällt durch sein kompaktes Äußeres, das durch die markant abgesetzten orangefarbenen Komponenten einen hohen Wiedererkennungswert hat.

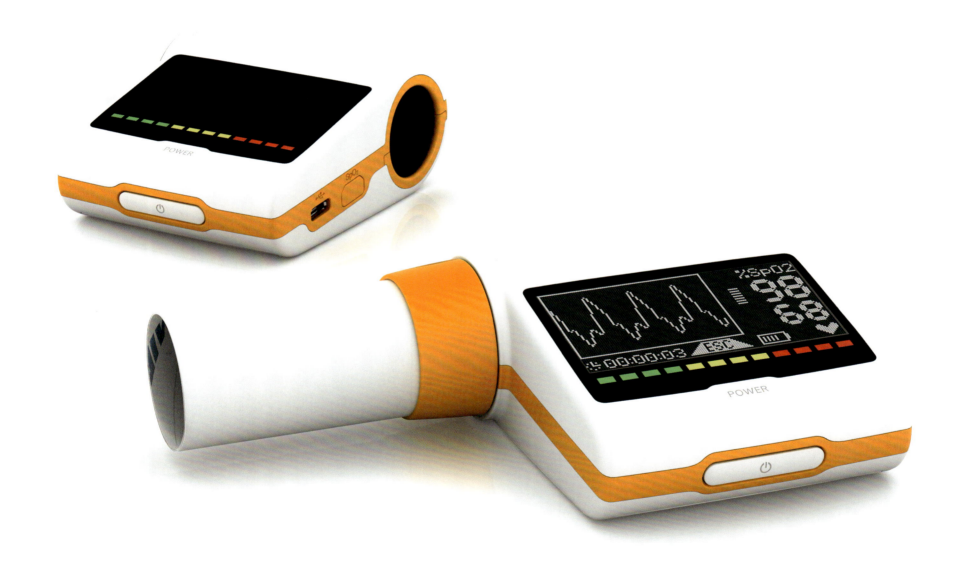

SurfLink Mobile
Assistive Listening Device
Multifunktionshörhilfe

Manufacturer
Starkey Hearing Technologies,
Eden Prairie, USA
In-house design
Jerry Ruzicka, Tim Trine
Web
www.starkeypro.com

SurfLink Mobile is a multifunctional assistive listening device, which in combination with a wireless hearing aid can be used as a mobile phone transmitter, media streamer and hearing aid remote control. The system enables hands-free use of a mobile phone. The call is transmitted directly to the wireless hearing aid. Audio signals from televisions or stereos can also be transmitted directly to the hearing device via Bluetooth.

Statement by the jury
The slanted surface at the front of the SurfLink Mobile device conveys an inviting impression, while its matte-black elegance is reminiscent of a modern smartphone.

SurfLink Mobile ist eine Multifunktionshörhilfe, die in Verbindung mit einem Drahtlos-Hörgerät als Mobiltelefon-Sender, Medien-Übertragungssystem und Hörgerät-Fernbedienung benutzt werden kann. Das System ermöglicht das Telefonieren mit einem Mobiltelefon über eine Freisprecheinrichtung. Der Anruf wird per Funk direkt an das Hörgerät gesendet. Audiosignale von Fernseher oder Musikanlage lassen sich per Bluetooth ebenfalls direkt auf das Hörsystem übertragen.

Begründung der Jury
Durch die Neigung der Grifffläche wirkt SurfLink Mobile einladend. In seiner mattschwarzen Eleganz erinnert es an ein modernes Smartphone.

MySign S
Patient Monitor
Patientenmonitor

Manufacturer
EnviteC-Wismar GmbH,
a Honeywell Company,
Wismar, Germany
In-house design
Michael Schiffner
Design
Koop Industrial Design
(Norbert Koop, Andreas Hildebrandt),
Hamburg, Germany
Web
www.envitec.com
www.koopdesign.com

The MySign S patient monitor is designed for monitoring the blood's functional oxygen saturation as well as for measuring the pulse rate. It can be used as both a portable handheld monitor and a stationary bedside device. Particularly practical features are the mounting concept for easy attachment to standard rail systems, beds and tables and the fold-out stand. The rubber-coated housing provides impact resistance and protects the device from streams of water and dust.

Statement by the jury
MySign S inspires with its clean and clear aesthetics. The digital colour display features large, easy-to-read numbers.

Der Patientenmonitor MySign S dient zur Überwachung und Kontrolle der funktionellen Sauerstoffsättigung im Blut sowie zur Messung der Pulsfrequenz. Es ist sowohl als mobiler Handmonitor als auch stationär am Patientenbett einsetzbar. Besonders praktisch sind das Halterungskonzept zur einfachen Befestigung an Standard-Schienensystemen, Betten oder Tischen sowie der ausklappbare Standfuß. Die Gummierung des Gehäuses macht es stoßsicher und schützt das Gerät vor Strahlwasser und Staub.

Begründung der Jury
MySign S begeistert durch seine saubere und klare Ästhetik. Das digitale Farbdisplay ist übersichtlich gestaltet, die großformatigen Zahlen sind sehr gut zu erkennen.

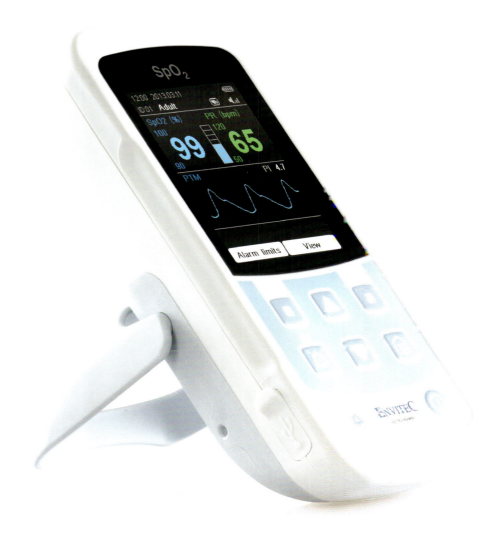

Manufacturer
LVI Low Vision International AB,
Växjö, Sweden
In-house design
Joachim Schill
Design
Myra Industriell Design AB
(Stina Juhlin),
Stockholm, Sweden
Web
www.lvi.se
www.myra.se

The MagniLink S video magnifier supports people with visual impairments. Images and text are transferred to a computer screen and enlarged via the camera lens for easier viewing. The system can be taken anywhere and is quickly connected to a laptop or desktop computer. Its foldable legs minimise its footprint, allowing the magnifier to be placed closely to a laptop. Orange components indicate how to set up and adjust the device. In addition, the system meets all the latest technical standards, including HDMI and USB 3.0. It does not require any batteries, since it is powered by the connected computer.

Die Videolupe MagniLink S unterstützt Menschen mit Sehschwäche, indem über das Kameraobjektiv Bilder oder Texte auf einen Computerbildschirm übertragen und dort vergrößert gelesen werden können. Das System kann überallhin mitgenommen werden und lässt sich einfach an einen Laptop oder Tischcomputer anschließen. Die Beine sind faltbar, sodass das Lesegerät dicht neben dem Laptop platziert werden kann. Die Komponenten in Orange zeigen an, wie das Gerät auf- und eingestellt wird. Das System verfügt über die neuesten technischen Standards, z. B. HDMI und USB 3.0, und benötigt keine Batterien, da es vom Computer aus mit Strom versorgt wird.

Charismo 2^C
Hearing Aid
Hörgerät

Manufacturer
Siemens Audiologische Technik GmbH,
Erlangen, Germany
Design
Brandis Industrial Design,
Nuremberg, Germany
Web
www.rexton.com
www.brandis-design.com

The Charismo 2^C hearing aid is among the world's smallest behind-the-ear hearing devices. Gently flowing contours foster an elegant impression and harmoniously integrate the haptically optimised controls. The form is finely adjusted to the ear's sensitive anatomy, thus providing high wearing comfort in combination with an excellent natural sound.

Statement by the jury
With its clear, unobtrusive appearance and monochrome colour scheme, this hearing device offers a strong degree of discretion for the wearer.

Das Hörgerät Charismo 2^C ist eines der kleinsten Hinterohr-Hörsysteme der Welt. Sanft verlaufende Kanten sorgen für eine elegante Erscheinung und binden die haptisch optimierten Bedienelemente harmonisch ein. Seine Form ist auf die anatomischen Gegebenheiten im Ohrbereich abgestimmt, dadurch ist das Hörgerät sehr komfortabel zu tragen. Zudem besitzt es eine exzellente, natürliche Klangqualität.

Begründung der Jury
Durch seine klare, zurückhaltende Gestaltung und monochrome Farbgebung bietet das Hörgerät ein hohes Maß an Diskretion für den Träger.

VSB QuickCheck
Tester for Hearing Implant Systems
Prüfgerät für Hörimplantatsysteme

Manufacturer
Vibrant Med-El,
Innsbruck, Austria
Design
Hyve AG
(Andreas Beer, Tobias Neutz),
Munich, Germany
Web
www.medel.com
www.hyve-design.de

The VSB QuickCheck tester has been designed to easily and quickly test the functioning of active hearing implants. This device meets a need formerly fulfilled by highly qualified technicians who carried out time-consuming tests during the surgical intervention. The highly sensitive measuring instrument provides all necessary information at the touch of a button and gives the ENT surgeon immediate certainty as to whether the implant is working.

Statement by the jury
The special feature of the VSB Quick-Check's design is its segmented silhouette, which distinctly highlights the front measuring sensor.

Mit dem Prüfgerät VSB QuickCheck kann die Funktionsfähigkeit von aktiven Hörimplantaten einfach und schnell getestet werden. So muss während der operativen Implantierung eines Hörgeräts nicht länger ein speziell ausgebildeter Techniker zugegen sein und aufwendige Tests durchführen. Das hochsensible Messinstrument liefert auf Knopfdruck alle notwendigen Informationen und gibt dem HNO-Chirurgen sofortige Gewissheit darüber, ob das Implantat funktioniert.

Begründung der Jury
Die Besonderheit der Gestaltung des VSB QuickCheck liegt in der segmentierten Silhouette, die den vorderen Messsensor markant in Szene setzt.

Aerosure™
Respiratory Trainer
Atemtrainer

Manufacturer
Actegy Health,
Ascot, GB
Design
Team Consulting Ltd
(Ben Turner, Paul Greenhalgh),
Cambridge, GB
Web
www.actegy.com
www.team-consulting.com

Aerosure helps alleviate the breathlessness associated with conditions such as asthma, cystic fibrosis and chronic obstructive pulmonary disease (COPD). It also assists with the clearing of mucus from the throat or lungs. The device displays a modular design, which makes it easy to clean and enables the quick replacement of components. Moreover, the respiratory trainer is rechargeable.

Statement by the jury
The colour accents, rounded edges and pleasant, easy-to-grip size of this respiratory trainer convey comfort and friendliness.

Aerosure hilft Atembeschwerden zu lindern, die bei Erkrankungen wie Asthma, zystischer Fibrose oder chronisch obstruktiver Lungenerkrankung (COPD) vorkommen. Ebenso wird die Schleimlösung in Rachen und Lunge unterstützt. Durch seinen modularen Aufbau ist das Gerät einfach zu reinigen, und einzelne Komponenten können schnell ausgetauscht werden. Über einen Adapter ist es wieder aufladbar.

Begründung der Jury
Mit seinen bunten Farbakzenten, abgerundeten Kanten und dem angenehmen Greifvolumen versprüht der Atemtrainer Sympathie und Freundlichkeit.

NeoPAP
Respiratory Treatment System for Newborns and Infants
Beatmungssystem für Neugeborene und Säuglinge

Manufacturer
Royal Philips Electronics,
Eindhoven, Netherlands
In-house design
Philips Design Healthcare Team
Web
www.philips.com

NeoPAP is a sophisticated continuous positive airway pressure (CPAP) delivery system for treating newborns and infants with respiratory distress syndrome. The low-profile interface and bonnet design reduce the need for adjustments and allow a looser fit, thus helping to minimise pressure on the baby's face. The leak compensation technology known as Baby-Trak allows the device to measure pressure at the infant's nose and regulate the target CPAP level.

Statement by the jury
The flush appearance of the monitoring display and its integration into a frame with rounded contours serves to underline the gentle character of NeoPap.

NeoPAP ist ein hoch entwickeltes Beatmungssystem zur Behandlung von Neugeborenen und Säuglingen mit Atemnot-Syndrom. Das Low-Profile-Interface und die Haubenausführung reduzieren den Anpassungsbedarf und ermöglichen eine lockere Passform. Dadurch wird der Druck auf das Gesicht des Babys gemindert. Mit der als Baby-Trak bezeichneten Leckagekompensation kann das Gerät den Druck an der Nase des Neugeborenen messen und regulieren.

Begründung der Jury
Die flächenbündige Gestaltung des Überwachungsmonitors in einer Ebene mit dem abgerundeten Rahmen unterstreicht den behutsamen Charakter von NeoPap.

T7™
Point-of-Care Technology Cart
Mobiler Technologie-Gerätewagen

Manufacturer
Humanscale,
New York, USA
In-house design
Robert Volek,
Mesve Vardar,
Jane Abernethy
Web
www.humanscale.com

The T7 point-of-care technology cart encourages medical personnel to work ergonomically while sitting or standing. The position of the cart can be adjusted via the touchscreen to accommodate the height of each individual. The Power Track steering with a fifth wheel simplifies the manoeuvring and positioning of the cart. The completely encased wire management system, which ranges throughout the cart starting at the monitor, prevents cable clutter and hygienic risks.

Statement by the jury
With flexible height adjustment capability and dynamic mobility, T7 enables users to quickly custom-tailor the workplace according to individual needs.

Der mobile Technologie-Gerätewagen T7 unterstützt das medizinische Personal dabei, ergonomisch zu arbeiten. Über den Touchscreen lässt sich die Einstellung des Wagens individuell anpassen, sie wird aufgrund der eingegebenen Körpergröße berechnet. Die Power-Track-Lenkung mit einem fünften Rad vereinfacht das Manövrieren und Positionieren des Wagens. Das komplett verborgene Kabelführungssystem vom Monitor durch den Wagen vermeidet Kabelgewirr und hygienische Risiken.

Begründung der Jury
T7 erlaubt es durch seine flexible Höhenverstellbarkeit und dynamische Beweglichkeit, den Arbeitsplatz den eigenen Bedürfnissen schnell anzupassen.

3M™ Kind Removal Silicone Tape
3M™ hautschonend ablösbares Silikon-Pflaster

Manufacturer
3M Health Care,
St. Paul, USA
In-house design
Fang Zhou
Web
www.3m.com

The 3M Kind Removal Silicone Tape integrates 3M's adhesive technology with a silicone base to provide an optimal balance of securement and gentleness. The new technology allows this repositionable tape to be removed easily and harmlessly from patients' skin. The edge protectors maintain neat appearance over time. Thanks to the gentle blue colour this tape can be easily identified among patients and clinicians.

Das 3M hautschonend ablösbare Silikon-Pflaster ist mit einem innovativen Silikon-klebstoff ausgestattet, der ein optimales Zusammenspiel aus sicherem Halt und Schonung der Haut ermöglicht. Er bietet eine gleichbleibende Klebekraft über die gesamte Anwendungsdauer und lässt sich rückstandsfrei wieder entfernen, ohne dabei die Haut zu verletzen oder unnötigen Schmerz zu verursachen. Die Seitenschoner schützen die Pflasterrolle vor Verschmutzungen. Durch die Blaufärbung hat das Pflaster sowohl beim Patienten als auch beim Personal einen hohen Wiedererkennungswert.

Statement by the jury
With a decorative feather pattern in combination with a muted blue colour, the side protectors visualise the tape's extraordinary gentleness to the skin.

Begründung der Jury
Durch ihr dekoratives Federmuster in Kombination mit dem gedämpften Blauton visualisieren die Seitenschoner die besondere Sanftheit des Pflasters zur Haut.

TL 90
Brightlight
Tageslichtlampe

Manufacturer
Beurer GmbH,
Ulm, Germany
Design
uli schade industriedesign
(Matthias Kolb),
Neu-Ulm, Germany
Web
www.beurer.com
www.schadedesign.de

The TL 90 daylight lamp, designed for in-home light therapy, simulates sunlight using two fluorescent tubes with 36 watts each, generating a light intensity of up to 10,000 lux. It features continuous adjustment inclination, one-button operation and an LED timer that displays the treatment time. The daylight is produced by an extra-large 51 x 34 cm illumination surface and is emitted very evenly.

Statement by the jury
This daylight lamp displays a puristic and simultaneously luxurious appearance thanks to its minimalist design and broad illumination surface.

Die Tageslichtlampe TL 90 für die Lichttherapie zu Hause simuliert Sonnenlicht durch zwei Leuchtstoffröhren mit jeweils 36 Watt, die eine Lichtstärke von bis zu 10.000 Lux erreichen. Der Neigungswinkel ist stufenlos verstellbar, und es gibt nur einen Bedienungsknopf. Ein LED-Timer zeigt die Behandlungszeit an. Das Tageslicht wird über eine extra große Beleuchtungsfläche von 51 x 34 cm abgegeben und besonders gleichmäßig ausgestrahlt.

Begründung der Jury
Die Tageslichtlampe wirkt durch ihre zurückgenommene Gestaltung puristisch und gleichzeitig luxuriös durch die überbreite Beleuchtungsfläche.

JeNu
Skincare System
Hautpflegesystem

Manufacturer
JeNu Biosciences,
Seattle, USA
Design
Design Partners
(Kim MacKenzie, Blaithin Brennan),
Bray, Ireland
Web
www.myjenu.com

JeNu is a clinically tested skincare system with anti-ageing technology. It uses ultrasound waves to increase the effective application of specially formulated ingredients to the skin. The rear of the case is made of anodised aluminium, while the front features a soft, comfortable rubber surface A polished metal applicator head communicates the correct use of the device, which has been realised with no visible screws or fixings.

Statement by the jury
The plain, carefully crafted design of this skincare system imparts professionalism and possesses a luxurious appearance.

JeNu ist ein klinisch getestetes Hautpflegesystem mit Anti-Aging-Effekt. Mithilfe von Ultraschallwellen werden die Inhaltsstoffe der Pflegeprodukte besonders wirkungsvoll in die Haut eingearbeitet. Die Rückseite des Gehäuses besteht aus eloxiertem Aluminium, die Vorderseite aus angenehm weichem Gummi. Der Applikatorkopf aus poliertem Metall vermittelt die richtige Benutzung des Geräts, das ansonsten ohne sichtbare Nähte oder Schrauben gefertigt ist.

Begründung der Jury
Die schlichte und sorgfältig verarbeitete Gestaltung des Hautpflegesystems unterstreicht seine Professionalität und wirkt gleichzeitig luxuriös.

Iriver on
Fitness-Tracking Earbuds
Ohrhörer zur Vitaldatenmessung

Manufacturer
Iriver,
Seoul, South Korea
In-house design
Beak Jin Seong,
Jeongbeom Han
Web
www.iriver.com

The Iriver on is a Bluetooth audio headset with earbuds that measure different vital signs, such as heart rate and maximum oxygen absorption, while simultaneously providing high sound quality. A new type of sensor in the earbuds measures the blood supply in the capillaries. The data can be accessed via an app on the connected device. The headset features an ergonomically shaped, padded neckband that allows the wearer to engage in extreme sports without any limitations. All music and monitoring functions can be controlled comfortably via the neckband.

Iriver on ist ein Bluetooth-Audio-Headset mit Ohrhörern, die verschiedenste Vital-daten wie die Herzfrequenz und maximale Sauerstoffaufnahme messen und dabei gleichzeitig eine ausgezeichnete Klangqua-lität bieten. Ein neuartiger Sensor im Ohrhörer ermittelt die kapillare Blutver-sorgung. Die Daten sind über eine App auf einem entsprechenden Gerät abrufbar. Die Ohrhörer besitzen ein ergonomisches, weich gepolstertes Nackenband, das es dem Träger auch erlaubt, Extremsport zu betreiben, ohne ihn zu beeinträchtigen. Über das Nackenband sind alle Funktionen des Headsets bequem zu steuern.

AL1 Solid
Linear Lift

Manufacturer
AMF-Bruns GmbH & Co. KG,
Apen, Germany
In-house design
Gerit Bruns
Web
www.amf-bruns.de

The fully automated AL1 Solid linear lift for rear or side doors owes its specific design to the patented lifting arms made from aluminium profiles. Due to a construction that is accurate to the millimetre, the arms guarantee lasting and precise functionality. Despite a load-bearing capacity of 400 kg, the system is 25 per cent lighter than average lifts of its kind, which increases the vehicle's payload. The lip plate at the transition between lift platform and vehicle interior and an automatic roll-off barrier offer wheelchair users absolute dependability. A transparent platform design provides excellent rear visibility while driving.

Der vollautomatische Linearlift AL1 Solid für Heck- oder Seitentüren verdankt seine besondere Bauform den patentierten Hubarmen aus Aluminiumprofilen. Durch ihre millimetergenaue Abstimmung garantieren sie eine dauerhafte und präzise Funktion. Trotz seiner Tragfähigkeit von 400 kg ist das System 25 Prozent leichter als durchschnittliche Lifte seiner Art, was die Nutzlast des Fahrzeugs erhöht. Die Überrollklappe beim Übergang von der Liftplattform ins Wageninnere sowie eine automatische Abrollsicherung gewähren dem Rollstuhlfahrer absolute Sicherheit. Die Plattform des Lifts ist durchsichtig, sodass während der Fahrt eine gute Sicht nach hinten gewährleistet ist.

Statement by the jury
Aluminium, a material commonly employed in the automotive industry, was also used to manufacture the AL1 Solid. This has resulted in its extraordinarily lightweight construction.

Begründung der Jury
Für die Fertigung des AL1 Solid wurde wie in der Automobilindustrie Aluminium eingesetzt, so wurde eine ungewöhnlich leichte Konstruktion erreicht.

Mobility

Dolomite Jazz
Rollator

Manufacturer
Invacare Dolomite AB,
Diö, Sweden
In-house design
Susanne Ek, Tobias Nilsson,
Jos Van Houtem
Web
www.invacare.com
www.dolomite.biz

The Dolomite Jazz rollator has an anthracite-coloured frame fashioned from oval tubes with light grey accents, giving it a lightweight and discreet impression when viewed from the side. The wide frame conveys a feeling of safety and stability to the user. The lateral folding mechanism provides more space for natural walking. The brake system is fully integrated into the rollator and thus prevents the user from getting caught on objects such as furniture.

Der Rollator Dolomite Jazz hat einen anthrazitfarbenen Rahmen mit lichtgrauen Akzenten und ovalem Rohrdurchmesser, sodass er von der Seite einen leichten und diskreten Eindruck macht. Dem Benutzer selbst vermittelt die breite Rahmengestaltung dagegen ein Gefühl von Sicherheit und Stabilität. Der neue seitliche Faltmechanismus stellt mehr Raum für natürliches Gehen zur Verfügung. Das Bremssystem ist voll integriert, sodass der Benutzer mit dem Rollator nicht an Möbeln etc. hängen bleibt.

Wheeldrive
Power Assist Wheels
Elektroantrieb für Rollstühle

Manufacturer
Handicare BV,
Helmond, Netherlands
Design
Indes BV (Wilfred Teunissen),
Enschede, Netherlands
Web
www.handicare.com
www.indes.eu

The small and compact Wheeldrive power assist wheels help wheelchair users increase their mobility and are compatible with most conventional manual wheelchairs with little adjustment. A unique feature of this propulsion system is its dual-rim design. The outer rim offers extra motor support, assisting the user in driving manually. The inner rim provides continuous support while the wheelchair is pushed forwards, thus making driving effortless. Users can decide which kind of propulsion type they wish to use at any given time.

Der kleine und kompakte Elektroantrieb Wheeldrive verhilft Rollstuhlfahrern zu mehr Mobilität und kann bei den meisten handelsüblichen Handrollstühlen ohne großen Aufwand verwendet werden. Das Antriebssystem besteht aus einem äußeren und einem inneren Reifen, was es einzigartig macht. Der äußere Greifreifen wird zusätzlich durch einen Motor unterstützt und erleichtert dem Fahrer den Betrieb per Hand. Der innere Reifen bietet kontinuierliche Unterstützung, wenn der Fahrer den Rollstuhl vorwärts bewegt, und sorgt so für eine mühelose Fortbewegung. Der Fahrer kann selbst entscheiden, welche Antriebsform er wann benutzen möchte.

Statement by the jury
Wheeldrive combines an innovative propulsion system with robust materials and a discreetly sporty look, which suits any manual wheelchair.

Begründung der Jury
Wheeldrive kombiniert ein innovatives Antriebssystem mit robusten Materialien und einem sportlich-dezenten Äußeren, das zu jedem Handrollstuhl passt.

RayoMe
Prosthesis
Prothese

Manufacturer
Teh Lin Prosthetic & Orthopaedic Inc.,
New Taipei, Taiwan
In-house design
Chen-Hsien Chang,
Wei-Kai Wang
Web
www.tehlin.com

The name of this prosthesis, RayoMe, was inspired by the Spanish word "rayo", which means "flash of lightning". This motif has been factored into the sleek contours and the colour scheme. While the colour orange expresses power and energy, the colour black represents the unique personality of the prosthesis. In addition, the design is also boldly inspired by the image of motorcycle sports.

Die Prothese trägt den Namen RayoMe in Anlehnung an das spanische Wort „Rayo", das übersetzt „Blitzstrahl" bedeutet. Das Motiv des Blitzstrahls findet sich in den schnittigen Konturen und dem Farbschema wieder. Während das Orange Kraft und Energie ausdrückt, unterstreicht das Schwarz den einzigartigen Charakter der Prothese. Darüber hinaus ist das Erscheinungsbild bewusst an den Motorsport angelehnt.

1C63 Triton Low Profile
Prosthetic Foot
Prothesenfuß

Manufacturer
Otto Bock HealthCare Ltd.,
Salt Lake City, USA
In-house design
Justin Smith
Web
www.ottobock.com

The 1C63 Triton Low Profile prosthetic foot is suitable for a wide spectrum of applications, ranging from everyday use to recreational sports. It is characterised by a combination of flexible carbon-fibre-composite materials and a base spring made of high-performance polymer. The adapter consists of premium titanium, making the foot extremely robust, stress- and water-resistant. Moreover, its low structural height facilitates flexible use.

Statement by the jury
This prosthetic foot made of high-quality materials expertly imitates the foot's anatomy and enables the wearer to carry out natural movement sequences.

Der Prothesenfuß 1C63 Triton Low Profile ist für viele verschiedenen Anwendungen geeignet, von Alltagsaktivitäten bis zum Freizeitsport. Er zeichnet sich durch eine Kombination aus flexiblen Carbonfaser-Verbundmaterialien und einer Basisfeder aus Hochleistungspolymer aus. Der Adapter besteht aus hochwertigem Titan, sodass der Fuß sehr robust, belastbar und wasserresistent ist. Durch die niedrig aufbauende Konstruktion ist er flexibel anzuwenden.

Begründung der Jury
Dieser Prothesenfuß aus hochwertigen Materialien ahmt die Anatomie des Fußes virtuos nach und ermöglicht dem Träger natürliche Bewegungsabläufe.

Sports Prosthesis
Sportprothese

Manufacturer
Otto Bock HealthCare GmbH,
Duderstadt, Germany
Design
Resolut Design
(Prof. Andreas Mühlenberend),
Dresden, Germany
Web
www.ottobock.com
www.resolutdesign.de

This transfemoral sports prosthesis is suitable for running sports that take place on a wide variety of surfaces. The knee joint 3S80 Sport cushions the flexing and stretching movements in the swing phase, thus enabling a dynamic movement sequence when running, even at a high stride frequency. The robust 1E90 Sprinter carbon blade is ideal for jogging and running due to its exceptionally high energy return. The prosthesis is completed by the 4R204 Sport foot adaptor, which connects the knee joint to the carbon foot.

Statement by the jury
The concept of this sport prosthesis successfully unites high technology, functionality and appealing aesthetics.

Diese Oberschenkel-Sportprothese ist für den Laufsport auf verschiedensten Untergründen geeignet. Das Kniegelenk 3S80 Sport dämpft die Beuge- und Streckbewegungen in der Schwungphase so, dass auch bei hoher Schrittfrequenz während des Laufens ein dynamischer Bewegungsablauf ermöglicht wird. Der widerstandsfähige Carbonfederfuß 1E90 Sprinter ist durch eine besonders hohe Energierückgabe ideal zum Joggen und Laufen. Komplettiert wird die Prothese durch den Fußadapter 4R204 Sport, der das Kniegelenk mit dem Federfuß verbindet.

Begründung der Jury
In dieser Sportprothese vereinen sich auf hohem Niveau Hightech, Funktionalität und eine ansprechende Ästhetik.

Patella Pro
Patella Realignment Orthosis
Knieorthese

Manufacturer
Otto Bock HealthCare GmbH,
Duderstadt, Germany
In-house design
Matthias Vollbrecht
Web
www.ottobock.com

The Patella Pro orthosis for the
treatment of anterior knee pain secures
the patella at all relevant flexion
angles with its dynamic realignment
technology. The guidance of the
patella is specifically adjustable to
each individual patient. The patented
textile combination in conjunction with
the interior vector grip effect ensures
that the orthosis fits perfectly. Due
to a lightweight and slim design, it is
imperceptible under clothing.

Statement by the jury
Patella Pro captivates with a design that
reveals the friction bearing and the rails,
thus providing a visual experience of
this realignment technology.

Die Patella Pro Knieorthese zur Behandlung von Schmerzen im vorderen Knie-
bereich sichert die Kniescheibe mit einer
dynamischen Rezentrierungstechnik in
allen relevanten Beugewinkeln. Die
Führung der Kniescheibe ist auf jeden
Patienten einstellbar. Die patentierte
Textilkombination in Verbindung mit dem
innenseitigen Vektor-Grip-Effekt stellt den
perfekten Sitz der Orthese sicher. Durch
ihre leichte und schlanke Ausgestaltung
fällt sie unter Kleidung nicht auf.

Begründung der Jury
Patella Pro besticht durch die gestal-
terische Offenlegung von Gleitlager und
Schienen, wodurch die innovative
Rezentrierungstechnik auch visuell
erfahrbar wird.

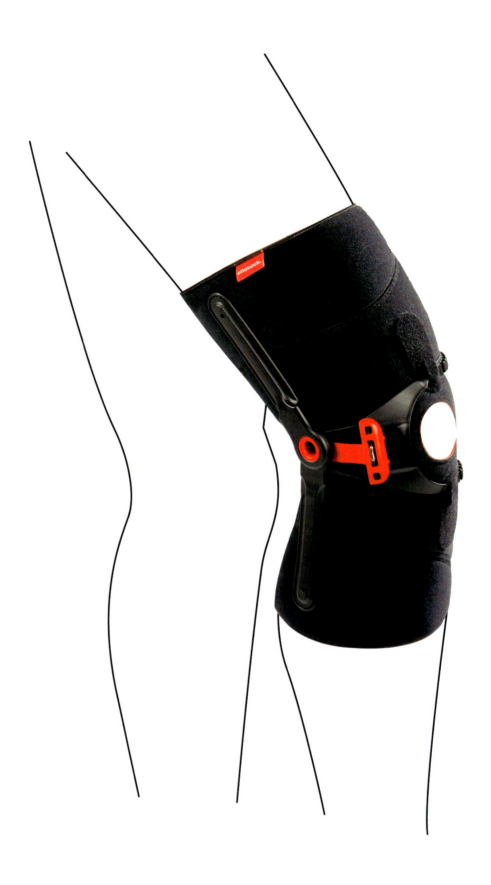

GANYMED Walking Aid
GANYMED Gehhilfen

Manufacturer
Ganymed GmbH,
Berg, Germany
In-house design
Karen Ostertag,
Sigmar Klose
Web
www.ganymed.eu

Walking with conventional crutches requires great physical effort since a so-called "mountain of strength" must be overcome. The Ganymed walking aid, however, avoids such effort expenditure with its S form, thereby facilitating quadrupedal posture and keeping the point of ground contact in view. Moreover, with a hardy bionic construction using the newest of materials, the walking aid is ultralight yet still able to withstand high amounts of stress.

Statement by the jury
The walking aid's asymmetrical grid structure introduces a completely new surface aesthetic, which conveys sportiness, dynamics and elegance.

Das Laufen mit normalen Krücken erfordert einen hohen Kraftaufwand, weil dabei ein sogenannter Kraftberg überwunden werden muss. Bei den Ganymed Gehhilfen ist dieser Aufwand nicht erforderlich, da man durch ihre S-Form einfach im Vierfüßlergang hinter ihnen herläuft und den Aufsetzpunkt stets im Blick hat. Durch die kühne bionische Konstruktion mit neuesten Materialien sind sie zudem ultraleicht und dennoch höher belastbar.

Begründung der Jury
Die asymmetrische Gitterstruktur der Gehhilfen bringt eine ganz eigene Oberflächenästhetik mit sich, die Sportlichkeit, Dynamik und Eleganz ausdrückt.

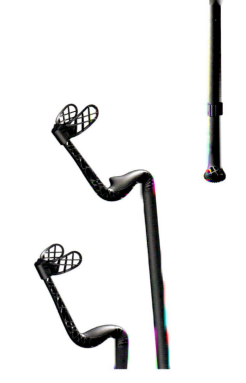

CAMOshoulder
Orthosis after Shoulder Injuries
Orthese nach Schulterverletzungen

Manufacturer
OPED GmbH,
Valley/Oberlaindern, Germany
Design
tomasini formung
(Bernd Tomasini),
Salzburg, Austria
Web
www.oped.de
www.tomasini.com

The CAMOshoulder orthosis provides perfect stability, but also flexibility, after shoulder injuries. The splint can be adjusted on two planes, allowing the patient to rotate the lower arm inwards and outwards. Adduction and abduction is also possible. Thanks to the strapping system, both the healthy arm and the hip relieve the injured shoulder, which prevents tension and back pain.

Statement by the jury
CAMOshoulder impresses with its open and excellently balanced strapping system, which is comfortable to wear even while sleeping.

Optimalen Halt bei gleichzeitiger Flexibilität bietet die Orthese CAMOshoulder nach Schulterverletzungen. Die Schiene ist in zwei Ebenen einstellbar. So kann der Patient den Unterarm nach innen oder außen drehen und auch das Heranziehen oder Abspreizen des Armes ist möglich. Die verletzte Schulter wird dank des Gurtsystems sowohl durch den gesunden Arm als auch durch die Hüfte entlastet. Dadurch werden Verspannungen und Rückenbeschwerden vermieden.

Begründung der Jury
CAMOshoulder überzeugt durch sein offenes, hervorragend ausbalanciertes Gurtsystem, das selbst im Schlaf angenehm zu tragen ist.

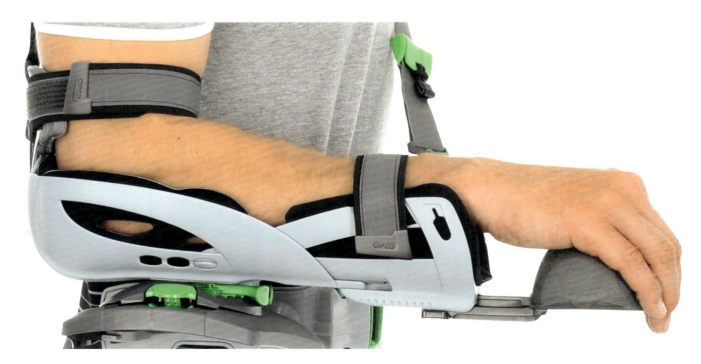

Bimedico
Disposable Splint
Einwegschiene

Manufacturer
Zario,
Moscow, Russia
Design
SmirnovDesign
(Sergey Smirnov, Katerina Grigoryeva),
Moscow, Russia
Web
www.zario.ru
www.smirnovdesign.com

The Bimedico disposable splint has been designed for short-term immobilisation of the cervical spine and also the upper and lower extremities. The movable parts and specific openings make it possible to fixate the extremities, without putting too much strain on them. Informative graphics make the splint easy to assemble, even in stressful situations. It is made mostly of recyclable corrugated cardboard and is thus particularly environmentally friendly and inexpensive. This splint does away with the necessity of sharing reusable splints among ambulance staff and hospital personnel. Moreover, complex cleaning and sterilisation processes become obsolete.

Mit der Einwegschiene Bimedico können die Halswirbelsäule sowie die oberen oder die unteren Extremitäten kurzfristig ruhiggestellt werden. Die beweglichen Teile und speziellen Öffnungen ermöglichen es, die Extremitäten zu fixieren, ohne sie dabei zu stark zu beanspruchen. Durch die informative Grafik gelingt der Zusammenbau sogar in Stresssituationen. Da die Schiene hauptsächlich aus recycelbarer Wellpappe besteht, ist sie umweltfreundlich sowie kostengünstig. Mit ihrem Einsatz ist die Übergabe von wiederverwendbaren Schienen zwischen Rettungshelfer und Krankenhauspersonal nicht mehr notwendig, ebenso entfallen komplexe Reinigungs- oder Sterilisierungsprozesse.

Statement by the jury
Bimedico offers intuitive use and, thanks to its biodegradable material, it is also an excellent way to promote ecological awareness.

Begründung der Jury
Bimedico ist intuitiv anzuwenden und durch das biologisch abbaubare Material darüber hinaus ein Beispiel für umweltbewusstes Handeln.

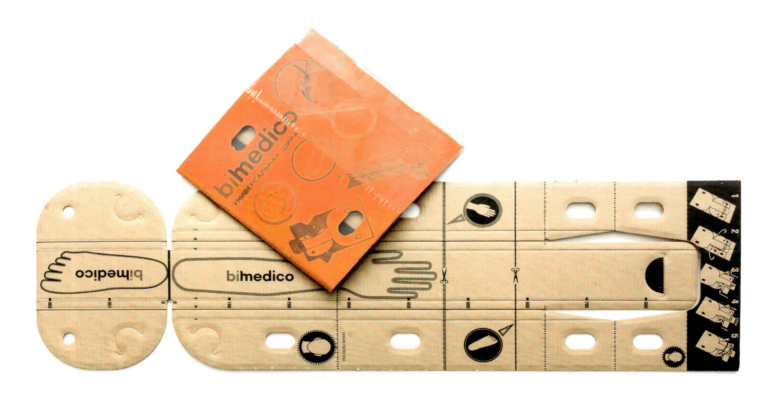

WiTouch Pro
Wireless TENS Unit
Kabellose TENS-Einheit

Manufacturer
Hollywog, LLC,
Chattanooga, USA

Design
Katapult Design Pty Ltd
(Nathan Pollock, Lee Rodezno, Brad Ryan),
Suffolk Park, Australia
Procept Australia, Keysborough, Australia

Web
www.hollywog.com
www.katapultdesign.com.au
www.procept.com.au

reddot design award
best of the best 2013

Mobile relief

Millions of people worldwide suffer from chronic back pain in the lumbar spine. One of the most important treatment methods is the TENS (Transcutaneous Electrical Nerve Stimulation) electrotherapy, which uses small electrical nerve-stimulating pulses to provide non-invasive drug-free pain relief. The WiTouch Pro is a TENS unit that is designed without the cumbersome battery packs and wired electrodes typical of traditional devices of this kind. The innovative wireless unit integrates all necessary elements in a compact and aesthetically appealing housing. A complex dual overmoulded body assembly ensures that the highly flexible electrode wings comfortably hug the body's contours. Thus, the device is easy to transport and can be worn discreetly underneath the clothes for long periods. The WiTouch Pro is intuitively operated, also via remote control, it is pleasant to the touch and highly flexible in use. A combination of low-energy wireless chips and an intelligent software design allows for cost-efficient use even over a long period of time. This design concept presents a paradigm shift for TENS. It provides new treatment qualities for individual pain relief and a high degree of mobility.

Mobile Linderung

Millionen von Menschen leiden weltweit unter chronischen Rückenschmerzen im Bereich der Lendenwirbelsäule. Eine der wichtigsten Behandlungsmöglichkeiten ist die Reizstrombehandlung TENS (transkutane elektrische Nervenstimulation), die durch das Prinzip leicht nervenstimulierender Stromimpulse eine Schmerzlinderung ohne operative Eingriffe oder Medikamente ermöglicht. Die WiTouch Pro ist eine TENS-Einheit, welche ohne die für diese Art von Geräten typischen Batteriepacks und umständlichen Verkabelungen gestaltet ist. Das innovative kabellose Gerät integriert alle nötigen Elemente in einem kompakten und freundlich anmutenden Gehäuse. Eine komplexe Baugruppe aus einem 3-Komponenten-Spritzguss ermöglicht es, dass sich die hochflexiblen Elektrodenflügel dieser TENS-Einheit gut den Körperkonturen des Nutzers anpassen. Die Einheit kann deshalb auf unkomplizierte Weise transportiert und auch diskret für einen langen Zeitraum unter der Kleidung getragen werden. Die WiTouch Pro lässt sich intuitiv auch per Fernbedienung handhaben, sie bietet eine angenehme Haptik und ist flexibel einsetzbar. Eine Kombination kabelloser Niedrigenergie-Chips und ein intelligentes Softwaredesign erlauben einen kostengünstigen Betrieb auch über einen langen Zeitraum hinweg. Das Konzept der Gestaltung dieser TENS-Einheit stellt einen Paradigmenwechsel dar. Sie bietet für eine individuelle Schmerzlinderung neue Qualitäten der Behandlung und ein hohes Maß an Mobilität.

Statement by the jury

Design focusing directly on the patient and a perfect integration of new technologies define the WiTouch Pro wireless TENS unit. This device features a new, elegant design language; it is compact and can be used anytime and anywhere. The possibility of simple, safe and cost-effective pain relief therapy administered by patients themselves, creates a new self-definition for this kind of product.

Begründung der Jury

Design direkt am Patienten und eine perfekte Integration neuer Technologien definieren die kabellose TENS-Einheit WiTouch Pro. Dieses Gerät hat eine neue, elegante Formensprache; es ist kompakt und auch unterwegs jederzeit einsetzbar. Die Möglichkeit der einfachen, sicheren und kostengünstigen Schmerztherapie durch den Patienten selbst schafft ein neues Selbstverständnis eines solchen Produktes.

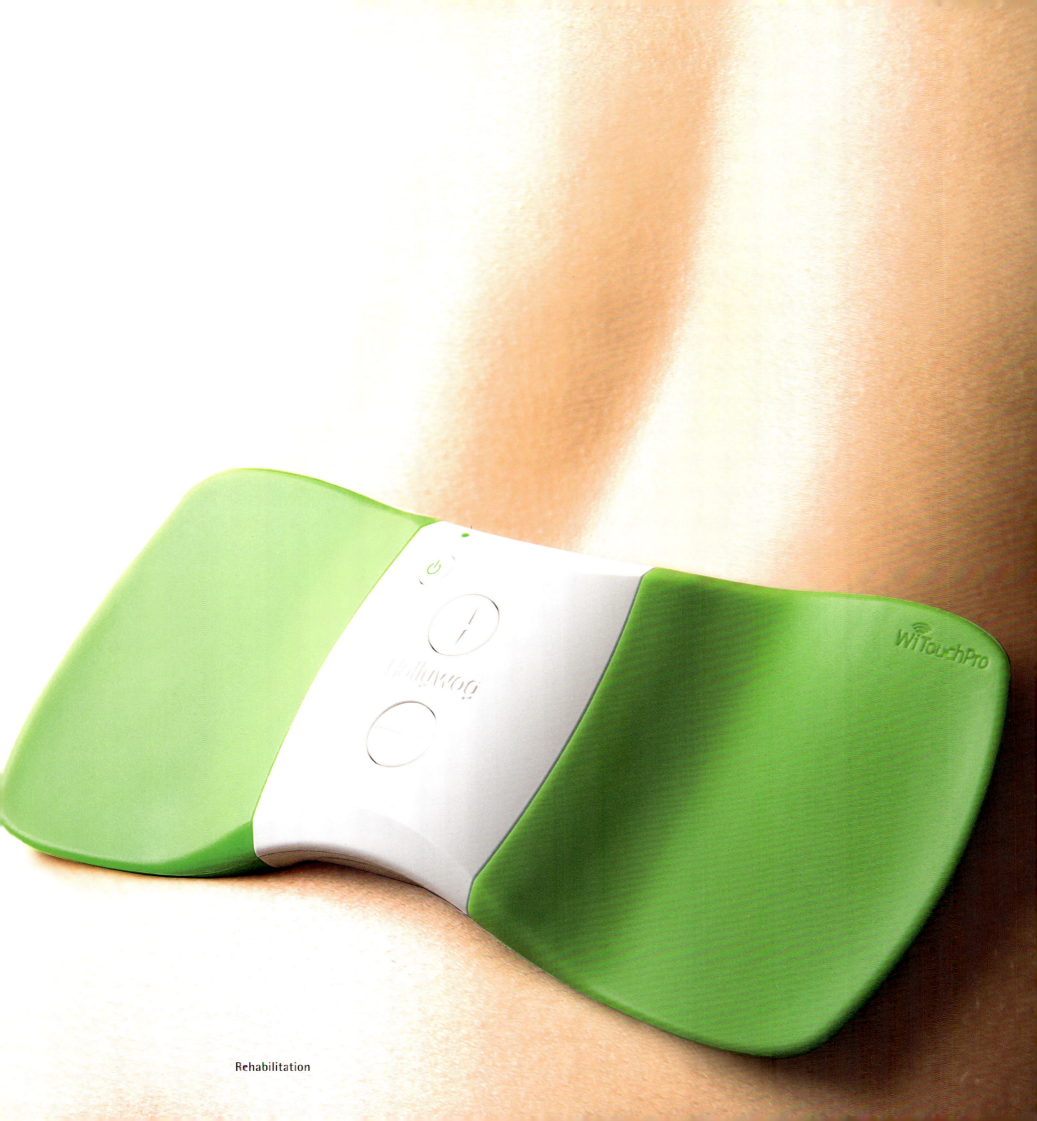

Rehabilitation

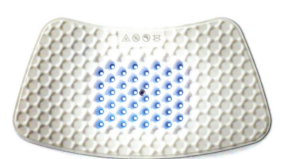

BlueTouch
Pain Relief Patch
Schmerzlinderungspflaster

Manufacturer
Royal Philips Electronics,
Eindhoven, Netherlands
In-house design
Philips Design Lighting Team
Web
www.philips.com

BlueTouch has been developed to ease back pain using blue LED light. The LED diodes are integrated into a flexible back strap, which can be comfortably worn underneath regular clothing. The blue light stimulates the production of nitric oxide in the skin, which activates natural pain relief processes in the body. Sensors in the device detect the wearer's skin type to adjust and optimise the treatment.

Statement by the jury
BlueTouch impresses with its geometric honeycomb look, which promotes air circulation and perfectly aligns with the LED lights.

BlueTouch lindert Rückenschmerzen mithilfe von blauem LED-Licht. Die LED-Dioden sind in ein flexibles Rückenband eingearbeitet, das bequem unter normaler Kleidung getragen werden kann. Das blaue Licht regt die Produktion von Stickstoffmonoxid in der Haut an, das natürliche Schmerzlinderungsprozesse im Körper anregt. Sensoren im Gerät erfassen den Hauttyp des Trägers, um die Behandlung anzupassen und zu optimieren.

Begründung der Jury
BlueTouch überzeugt durch seine geometrische Wabenoptik, die die Luftzirkulation begünstigt und in deren Vertiefungen sich die LED-Dioden perfekt einfügen.

Backpod
Spinal Therapy Device
Rückentherapiegerät

Manufacturer
Plastech Industries,
Christchurch, New Zealand
In-house design
Andrew Wallace,
Nick Laird
Web
www.bodystance.co.nz
Honourable Mention

The Backpod strongly and safely stretches out the upper back which drives most neck problems. The elliptical form of the therapeutic device allows comfortable stretching over the gentler longitudinal curve and greater leverage over the tighter transverse radius. The support cushion is made of TPV material, while the integrated energy compression cones adapt smoothly to the spine to augment the contact surface. A transparent base allows the complexity of the structure to be seen and appreciated.

Statement by the jury
The exciting colour and material effects lend the Backpod a young and agile appearance that underscores its health benefits.

Der Backpod kräftigt und dehnt die Brustwirbelsäule, welche die Ursache für die meisten Nackenschmerzen ist. Dabei erlaubt die elliptische Form mit dem großen Radius in Längsrichtung eine bequeme Dehnung, während der kleine Radius in Querrichtung eine größere Hebelwirkung erzeugt. Das Auflagekissen aus TPV-Kunststoff und die darunter liegenden Massagezapfen passen sich dem Körper an und erhöhen so die Kontaktfläche. Die transparente Basis ermöglicht es, die Komplexität der Struktur zu erkennen.

Begründung der Jury
Die aufregenden Farb- und Materialeffekte verleihen dem Backpod ein junges und agiles Aussehen, das seinen gesundheitlichen Nutzen unterstreicht.

Tarta
Ergonomic Backrest
Ergonomische Rückenlehne

Manufacturer
Tarta Design S.r.l.,
Cividale del Friuli, Italy
In-house design
Marco Galante
Web
www.tartadesign.it

Tarta, an ergonomic backrest, was created by experienced professional technicians who are focused on the study of aiding posture and have translated futuristic materials into an alluring design. Applying a modular principle with several simple yet robust components, this backrest satisfies a whole host of demands, ranging from the need for visually appealing, fashionable seating options to furniture that addresses the serious difficulties faced by the disabled.

Tarta, eine ergonomische Rückenlehne, wurde von erfahrenen Fachleuten entwickelt, die sich mit der Sitzhaltung beschäftigten und futuristische Werkstoffe in ein ansprechendes Design überführt haben. Die im Baukastenprinzip mit wenigen schlichten, aber sehr widerstandsfähigen Komponenten erdachte Rückenlehne wird den unterschiedlichsten Anforderungen gerecht, von der Optik für modische Sitzmöbel bis zu ernsthafteren Problematiken behinderter Mitmenschen.

Statement by the jury
The way in which this backrest combines aesthetic qualities with aspects of ergonomics and functionality is an outstanding creative achievement.

Begründung der Jury
Es ist eine herausragende kreative Leistung, wie diese Rückenlehne ästhetische Qualitäten mit Aspekten der Ergonomie und Funktionalität vereint.

MEDIballs secret
Pelvic Base Exercise Balls
Beckenboden-Trainingskugeln

Manufacturer
redmed GmbH,
Hannover, Germany
In-house design
Jürgen Hantke, Bianca Künnecke,
Ingo Reichmann
Web
www.redmed.de

The pelvic base exercise balls called Mediballs secret consist of a silicone material and are used for medical training purposes. Their ergonomic design simplifies insertion and removal, while the attached return strap is concealed inside the body to ensure particular discretion. Thanks to the balls' innovative structure, they are extremely silent. Located inside each ball is a weight that oscillates back in forth when the body is in motion, thus constantly triggering new stimulation impulses.

Die Mediballs secret für das medizinische Beckenbodentraining bestehen aus Silikomed. Ihre ergonomische Gestaltung erleichtert das Einführen und Entfernen, die im Körper zu tragende Rückholschlaufe ist besonders klein und diskret. Durch die innovative Struktur sind sie flüsterleise. In den Kugeln befindet sich ein Gewicht, das durch die Bewegung des Körpers hin und her schwingt und so permanent neue Stimulationsimpulse auslöst.

Statement by the jury
With a body-friendly material and discreet character, the design of these pelvic base exercise balls strongly takes the needs of women into consideration.

Begründung der Jury
Die Beckenboden-Trainingskugeln gehen durch das körpersympathische Material und ihren diskreten Charakter in hohem Maß auf die Bedürfnisse von Frauen ein.

Revitive LV™
Circulation Booster
Stimulationsgerät
für die Durchblutung

Manufacturer
Actegy Health,
Ascot, GB
Design
Team Consulting Ltd
(David Robinson, Paul Greenhalgh),
Cambridge, GB
Web
www.actegy.com
www.team-consulting.com

Revitive LV alleviates the discomfort in the lower legs and feet that results from poor blood circulation by using electrical impulses to stimulate the muscles. The device was developed particularly for women, which is also communicated through its contrasting colours. It is particularly simple to use and can be understood intuitively. The controls may be comfortably operated with a toe.

Revitive LV lindert Beschwerden in Füßen und Unterschenkeln, die durch eine schlechte Durchblutung verursacht werden. Elektrische Impulse stimulieren dabei die Muskeln. Das Gerät wurde speziell für Frauen entwickelt und kommuniziert dies auch in seinen kontrastierenden Farben. Die Anwendung ist besonders einfach und intuitiv zu verstehen. Die Bedienelemente können bequem mit dem Zeh gesteuert werden.

Statement by the jury
Revitive LV has a captivating circular shape, which is reflected in the controls. The control system is further accentuated by its bright colour scheme.

Begründung der Jury
Revitive LV besticht durch seine Kreisform, die sich in den Bedienelementen widerspiegelt. Zudem wird die Steuerung durch die lebhafte Farbe hervorgehoben.

Communications
Kommunikation

Mobile phones, smartphones, telephones and telephone systems, conference technology, headsets, accessories
Mobiltelefone, Smartphones, Telefone und Telefonanlagen, Konferenztechnik, Headsets, Zubehör

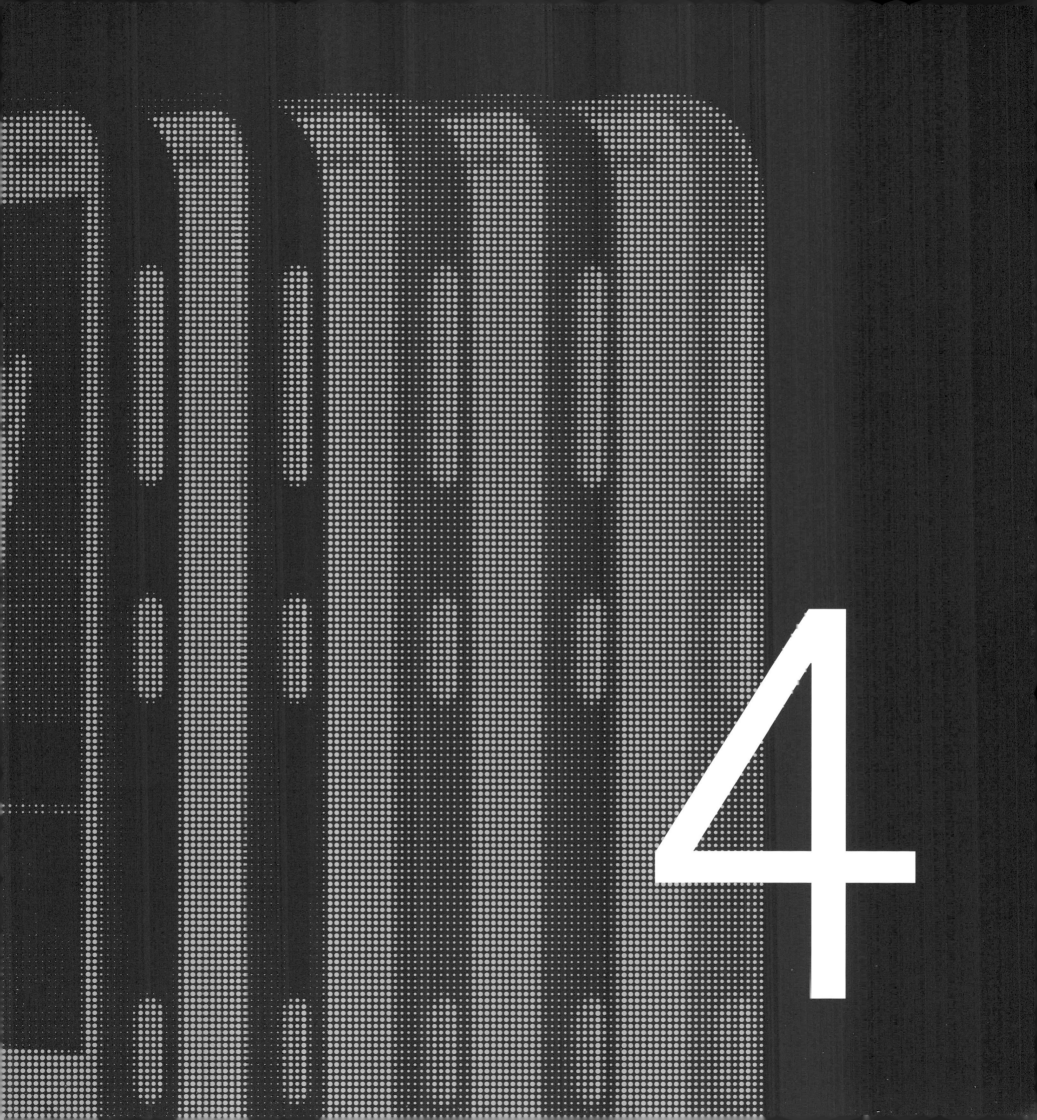

4

Porsche Design P'9981
Smartphone from BlackBerry

Manufacturer
BlackBerry,
Waterloo, Canada

Design
Porsche Design,
Zell am See, Austria

Web
www.blackberry.com
www.porsche-design.com

reddot design award
best of the best 2013

A luxurious tactile experience

The smartphones of the BlackBerry series by Canadian company BlackBerry are highly popular in particular because they allow personal data to be stored, processed and managed in an uncomplicated manner. The design of the Porsche Design P'9981 Black Smartphone presents itself as an impressive reinterpretation of the BlackBerry as lends the smartphone a luxurious tactile quality. The luxurious form factor of this smartphone is based above all on the design of the housing that features high-quality materials all around. It consists of a stainless steel frame that is encased in glossy black plastic and further enhanced by a hand-applied soft genuine leather backing. The high-quality look and feel of the materials is matched by the smartphone offering highly sensitive handling. Additional highlights include a black keyboard with innovatively wide keys that feel intuitive and pleasing to the touch. The keys have been ergonomically optimised for fast typing with less typos. In addition, all displays of the Porsche Design P'9981 smartphone from BlackBerry feature a crystal-clear touch-screen display that is highly responsive and offers smooth high-resolution rendering. The sophisticated design of this smartphone defines it as an object of engineered tactile luxury – it enhances everyday life by delivering a highly appealing touch responsive experience.

Luxuriöser Handschmeichler

Die gleichnamigen Smartphones des kanadischen Unternehmens BlackBerry sind unter anderem deshalb so beliebt, weil sie eine unkomplizierte Möglichkeit der Verwaltung persönlicher Daten bieten. Die Gestaltung des Porsche Design P'9981 Black Smartphone stellt eine beeindruckende Neuinterpretation des BlackBerry dar, denn es verleiht einem Smartphone luxuriös anmutende taktile Qualitäten. Die Anmutung von Luxus basiert bei diesem Smartphone vor allem auf der Gestaltung des Gehäuses, das gänzlich aus hochwertigen Materialien gestaltet ist. Ein mit einem schwarz glänzenden Kunststoff beschichteter Edelstahlrahmen wird auf der Rückseite von Hand mit einem samtweichen Leder bezogen. Die Hochwertigkeit der Materialien verbindet sich bei diesem Smartphone zudem mit einer sensitiven Art seiner Bedienung. Die schwarz gestaltete Tastatur bietet eine innovative Tastengröße, die man intuitiv erfühlen kann. Die Tasten sind ergonomisch gut gestaltet, sodass der Nutzer schnell und fehlerfrei tippen kann. Alle Anzeigen des Porsche Design P'9981 Smartphone von BlackBerry erscheinen zudem in einem kristallklaren Touchdisplay, das leicht anspricht und eine hohe Auflösung bietet. Die feinsinnige Gestaltung dieses Smartphones definiert es als einen luxuriösen Handschmeichler – es bereichert den Alltag auch durch die sinnlich ansprechende Art seiner Nutzung.

Statement by the jury

The Porsche Design P'9981 Black Smartphone from BlackBerry offers users the experience of state-of-the-art technology on the highest level. It is fascinating in that its interface is highly intuitive to operate and makes users feel what is going on inside the phone. The smartphone features a timeless modern appearance and is complemented by a perfectly refined finish.

Begründung der Jury

Das Porsche Design P'9981 Black Smartphone von BlackBerry bietet das Erleben neuester Technologie auf höchstem Niveau. Beeindruckend ist es, wie sich sein Interface intuitiv bedienen lässt und der Nutzer dabei fühlen kann, was im Inneren geschieht. Dieses Smartphone ist zeitlos modern, und sein Finish ist perfekt ausgearbeitet.

BlackBerry Z10
Smartphone

Manufacturer
BlackBerry,
Waterloo, Canada
In-house design
BlackBerry Design Team
Web
www.blackberry.com

The BlackBerry Z10 combines unified communications and mobile computing capabilities into an elegant hardware design. All technical functions are gathered into a solid, clean form that is only 9 mm thick, making it the thinnest BlackBerry so far. Precise engineering, high-grade materials and its colour uniformity in all elements contribute to its classical, timeless impression. Tone-on-tone tactile materials make the BlackBerry Z10 smooth and comfortable to hold, making the hardware a perfect companion for the natural, elegant interaction that the BlackBerry 10 OS delivers.

Das BlackBerry Z10 kombiniert verein-heitlichte Kommunikation und mobile Computerfunktionen in elegantem Hard-waredesign. Alle technischen Funktionen sind in einer soliden, klaren Form mit nur 9 mm Dicke vereint – dem bisher dünnsten BlackBerry-Gerät. Die präzise Technik, hochwertige Materialien und die einheit-liche Farbgebung aller Elemente tragen zu einem zeitlos klassischen Aussehen bei. Die Ton in Ton gestalteten Tastenmateria-lien sorgen für eine angenehme Haptik des BlackBerry Z10 und machen die Hardware zum perfekten Begleiter für die natür-lichen, eleganten Interaktionen, die mit BlackBerry 10 OS möglich sind.

iPhone 5
Smartphone

Manufacturer
Apple, Inc.,
Cupertino, USA
In-house design
Apple Industrial Design Team
Web
www.apple.com

The iPhone 5, at only 7.6 mm thin and 112 grams, is 18 per cent thinner and 20 per cent lighter than the iPhone 4S. It is precisely engineered with a diamond-cut bevelled edge around the display. The 8-megapixel camera is crafted of sapphire crystal glass, and the new 4" Retina display provides a brilliant 1136 x 640 pixel resolution.

Das iPhone 5 ist mit einer Tiefe von nur 7,6 Millimetern und einem Gewicht von 112 Gramm 18 Prozent dünner und 20 Prozent leichter als das iPhone 4S. Es ist präzise gefertigt mit im Diamant-schnitt abgeschrägten Kanten rund um das Display. Die 8-Megapixel-Kamera besteht aus Saphirkristall. Das neue 4"-Retina-Display bietet eine brillante Auflösung von 1136 x 640 Pixeln.

Statement by the jury
Longer, lighter, slimmer – and yet, with its timeless form language and precise engineering, it is undoubtedly the new iPhone.

Begründung der Jury
Länger, leichter, schlanker – und dennoch ist es mit seiner zeitlosen Formensprache und der präzisen Verarbeitung unverkenn-bar das neue iPhone.

Optimus LIFE L-02E
Smartphone

Manufacturer
LG Electronics Inc.,
Seoul, South Korea
In-house design
Youngjoo Cho, Tomoyuki Akutsu
Design
Taku Satoh Design Office Inc.
(Taku Satoh),
Tokyo, Japan
Web
www.lg.com
www.tsdo.jp

The Optimus Life L-02E is based on the consideration that a smartphone today needs not a lot more than a display screen. Furthermore, only the most necessary elements should be "cut out" of the enormous range of possibilities – a design concept which the manufacturer calls "Slice of Life". The form language of the Optimus Life L-02E is accordingly reduced, simple and straightforward.

Dem Optimus Life L-02E liegt der Gedanke zugrunde, dass ein Smartphone heute nicht viel mehr braucht als einen Bildschirm. Darüber hinaus sollten nur die notwendigsten Elemente aus dem riesigen Angebot an Möglichkeiten „herausgeschnitten" werden – ein Gestaltungskonzept, das der Hersteller „Slice of Life" nennt. Dementsprechend reduziert, schlicht und geradlinig ist die Formensprache des Optimus Life L-02E.

Statement by the Jury
The consistent, formal reduction of the Optimus Life L-02E makes a concentration on essentials possible – the information on the display.

Begründung der Jury
Die konsequente formale Reduktion des Optimus Life L-02E ermöglicht eine Konzentration auf das Wesentliche – die Informationen auf dem Display.

MC40
Handheld Mobile Computer

Manufacturer
Motorola Solutions,
Schaumburg, USA
In-house design
Curt Croley, Ian Jenkins,
James Krause, Mark Palmer
Web
www.motorolasolutions.com

The MC40 is built to enhance a retailer's in-store experience. It features applications for inventory look up, mobile payment and checkout, and assisted selling, while doubling as a sales associate communication and productivity tool. It has a rugged aluminium housing built to withstand 4 foot drops and incorporates a credit card reader, 2D barcode scanner and full touch display in a svelte, modern form factor.

Der Motorola MC40 wurde speziell entwickelt, um das Einkaufserlebnis im Einzelhandel zu optimieren. Er bietet zukunftsweisende Funktionen etwa für das Bestandsmanagement oder den mobilen Point of Sale (mPOS) und dient als Kommunikations- und Produktivitätswerkzeug für das Verkaufspersonal. Der MC40 verfügt über ein robustes Aluminiumgehäuse, das Stürze aus bis zu 1,2 Metern übersteht. In seinem schlanken, modernen Gehäuse vereint er zudem einen Kreditkartenleser, einen 2D-Barcode-Scanner und ein Touch-Display.

Statement by the jury
Thanks to its well-considered and compact design, the MC40 is good to operate with its multifunctionality.

Begründung der Jury
Dank seiner durchdachten und kompakten Gestaltung ist der MC40 mit seiner Multifunktionalität gut zu bedienen.

HTC Windows Phone Family
Smartphone

Manufacturer
HTC Corporation,
New Taipei, Taiwan
Design
One & Co,
San Francisco, USA
Web
www.htc.com
www.oneandco.com

The HTC Windows Phone Family is so designed that the form of the smartphone reflects the form of the software interface of the Windows operating system with its characteristic live tiles. Their silhouette is rectangular and contrasts with the softly upholstered exterior. The transitions between the front and the back are flowing; the back is slightly convex. Brilliant colours give the smartphones additional expressiveness.

Die HTC Windows Phone Family ist so gestaltet, dass die Form der Smartphones das Software-Interface des Windows-Betriebssystems mit seinen charakteristischen Live Tiles widerspiegelt. Sie haben eine rechteckige Silhouette, die durch eine weich gepolsterte Oberfläche kontrastiert wird. Die Übergänge zwischen Vorder- und Rückseite sind fließend, die Rückseite ist leicht gewölbt. Leuchtende Farben verleihen den Smartphones zusätzliche Ausdruckskraft.

Statement by the jury
The HTC Windows Phone Family, with their rectangular form, vivid colours and a seamless design are well suited to their Windows operating system.

Begründung der Jury
Die HTC Windows Phone Family passt mit ihrer rechteckigen Form, lebendigen Farben und der nahtlosen Gestaltung gut zu ihrem Windows-Betriebssystem.

Nokia Lumia 820
Smartphone

Manufacturer
Nokia,
Espoo, Finland
In-house design
Tuomas Reivo,
Terence Tan-Han-Yang
Web
www.nokia.com

The components in the interior of the 820 are arranged so that they provide good performance and at the same time a slim, compact design of the housing. Combined with its rounded edges, this means the smartphone lies very well in the hand and can easily be operated with one hand. The shell can be removed and replaced with a wireless charging attachment or with various shells in bright colours.

Die Komponenten im Inneren des Lumia 820 sind so angeordnet, dass sie eine gute Leistung bieten und zugleich eine schlanke, kompakte Gestaltung des Gehäuses erlauben. Im Zusammenspiel mit seinen abgerundeten Kanten führt dies dazu, dass das Smartphone sehr gut in der Hand liegt und leicht mit einer Hand zu bedienen ist. Die Gehäuseschale kann abgenommen und durch einen Aufsatz zum kabellosen Laden oder Schalen in leuchtenden Farben ersetzt werden.

Statement by the jury
The Lumia 820 combines pleasant haptics and high usability in a slim, customisable housing.

Begründung der Jury
Das Lumia 820 vereint eine angenehme Haptik und hohe Gebrauchstauglichkeit in einem schlanken, individualisierbaren Gehäuse.

Nokia Lumia 620
Smartphone

Manufacturer
Nokia,
Espoo, Finland
In-house design
Daniel Dhondt,
Sawa Tanaka
Web
www.nokia.com

The Nokia Lumia 620 is a compact, elegant and affordable smartphone. The Lumia 620 delivers a full Windows Phone 8 experience through a 3.8" display. It also boasts a fast 5 megapixel camera, NFC and a front facing camera. The Lumia 620 has a striking look through the use of an innovative, dual-layer polycarbonate. This special feature generates deep, translucent tones, and high colour interiors encased in transparent top-coats.

Das Nokia Lumia 620 ist ein kompaktes, elegantes und erschwingliches Smartphone. Sein 3,8"-Display ermöglicht ein umfassendes Windows Phone 8-Erlebnis. Außerdem bietet es eine schnelle 5-Megapixel-Kamera, NFC und eine Frontkamera. Sein markantes Erscheinungsbild erhält das Lumia 620 durch die Verwendung von innovativem, zweilagigem Polycarbonat. Dadurch werden satte, durchscheinende Farbtöne erzeugt, die in transparente Decklacke eingebettet sind.

Statement by the jury
The compact, convenient smartphone Lumia 620 draws attention due to its brilliant colouring and delights with its easy operation.

Begründung der Jury
Das kompakte, handliche Smartphone Lumia 620 erregt mit seiner leuchtenden Farbgestaltung Aufmerksamkeit und begeistert mit seinem Bedienkomfort.

Optimus G Pro (F240)
Smartphone

Manufacturer
LG Electronics Inc.,
Seoul, South Korea
In-house design
Young-Ho Kim,
Hyung-Gon Ryu,
Yoo-Shin Ahn
Web
www.lg.com

The smartphone F240 is a premium device with a 5.5" full HD screen. The case is of minimalist design, highlighting the screen further and providing from a haptic viewpoint a secure and pleasing feeling. A discreet metal ring unites the front and the back, giving it an elegant appearance. The MMR-inspired, glare-free structure pattern of the battery cover gives the F240 a special impression.

Das Smartphone F240 ist ein Premium-Gerät mit einem 5,5"-Full HD-Bildschirm. Das Gehäuse unterstreicht mit seiner minimalistischen Gestaltung den Bildschirm noch und vermittelt unter haptischen Aspekten ein sicheres und angenehmes Gefühl. Ein dezenter Metallring vereint Vorder- und Rückseite und sorgt für eine elegante Erscheinung. Das MMR-inspirierte, blendfreie Strukturmuster der Batterieklappe verleiht dem F240 eine besondere Anmutung.

Statement by the jury
The upmarket appearance of the F240 is a result of its large, full HD display and is further highlighted by the detailed design of the housing.

Begründung der Jury
Die hochwertige Erscheinung des F240 wird durch sein großes Full HD-Display geprägt und durch die detaillierte Gestaltung des Gehäuses noch unterstrichen.

Optimus L3II (E430)
Smartphone

Manufacturer
LG Electronics Inc.,
Seoul, South Korea
In-house design
Jin-Ki Min,
Dong-Soon Kim
Web
www.lg.com

Optimus L3II combines the advantages of a Smartphone with simple operation. It has a powerful 1.0 gigahertz processor, a long battery life and a user-friendly 3 megapixel camera, speech activated. It takes on new elements such as laser-cut contours, and intelligent LED lighting by means of which the user can set the light colour of the home button individually. With regard to ergonomic handling, the device is compact and perfectly balanced.

Statement by the jury
High user-friendliness and sophisticated technology combine in the Optimus L3II to produce a compact form with reduced and yet elegant lines.

Optimus L3II vereint die Vorzüge eines Smartphones mit einer einfachen Bedienbarkeit. Es hat einen leistungsstarken 1,0-Gigahertz-Prozessor, eine lange Akku-Laufzeit und eine 3-Megapixel-Kamera mit Sprachauslöser-Funktion. Neue Gestaltungselemente sind Laser-Cut-Konturen und eine intelligente LED-Beleuchtung, dank derer die Lichtfarbe des Home Buttons individuell eingestellt werden kann. Das Gerät ist mit Blick auf eine ergonomische Handhabung kompakt und perfekt ausbalanciert.

Begründung der Jury
Hohe Nutzerfreundlichkeit und ausgereifte Technik treffen beim Optimus L3II auf eine kompakte Form mit reduzierter und zugleich eleganter Linienführung.

Lucid 4G (VS840)
Smartphone

Manufacturer
LG Electronics Inc.,
Seoul, South Korea
In-house design
Do-Young Park,
Hong-Kyu Park,
Sung-Hee Han,
Hyun Lee
Web
www.lg.com

The LTE smartphone Lucid 4G with its side lines emphasised by a metal strip appears very filigree. Stability is provided in the area of the power button with this side line which runs right round the case. The slim impression is further emphasised by the ergonomic recesses on the outside of the back panel. The black-red, digital print pattern on the battery housing gives the smartphone an additional conciseness.

Statement by the jury
Regarding its form, Lucid is characterised by an elegant, all round metal line which provides the smartphone with a high-grade and slim appearance.

Das LTE-Smartphone Lucid 4G wirkt durch seine durch einen Metallstreifen betonte Seitenlinie sehr filigran. Die Linie sorgt im Bereich des Power-Knopfes für Stabilität und zieht sich einmal längs um das Gehäuse herum. Der schlanke Eindruck wird durch die ergonomischen Vertiefungen an der Außenseiten der Rückseite noch verstärkt. Schwarz-rote, digitale Druckmuster auf der Batterieabdeckung verleihen dem Smartphone zusätzliche Prägnanz.

Begründung der Jury
Lucid wird in formaler Hinsicht durch eine elegante, umlaufende Metalllinie geprägt, die dem Smartphone eine hochwertige und schlanke Anmutung verleiht.

MEDIAS W N-05E
Smartphone

Manufacturer
NEC CASIO Mobile Communications, Ltd.,
Kawasaki, Japan
Design
NEC Design & Promotion, Ltd.
(Kazuya Matsumoto)
NEC CASIO Mobile Communications, Ltd.,
Creative Studio (Toshiaki Sato)
Web
www.n-keitai.com

Medias W N-05E is a smartphone which is small and therefore easy to carry around; at the same time, however, it has a large, folding screen. At the front and the back of the device is a 4.3" LCD panel for normal use. Both these screens can be folded out to make one 5.6" screen. The user can thus choose either the smaller or the larger display according to requirements.
Thanks to its thin material, the separating line between the two displays of the large screen is reduced to a minimum.

Medias W N-05E ist ein Smartphone, das klein und damit ideal zum Mitnehmen ist, gleichzeitig aber auch einen großen aufklappbaren Bildschirm hat. Auf der Vorder- und Rückseite des Gerätes befindet sich jeweils ein 4,3"-Flüssigkristallbildschirm für den normalen Gebrauch. Diese beiden Displays verwandeln sich zusammen in einen 5,6"-Bildschirm, wenn sie nach außen aufgeklappt werden. Je nach Anwendung kann der Nutzer so entweder eines der kleineren oder das große Display nutzen. Dank des dünnen Materials wird die Trennlinie zwischen den beiden Displays beim großen Bildschirm auf ein Minimum reduziert.

Statement by the jury
The Medias W N-05E convinces with its intelligent and high-grade design which allows it to merge two small displays into one big one.

Begründung der Jury
Das Medias W N-05E überzeugt mit seiner intelligenten und hochwertigen Gestaltung, die es erlaubt, zwei kleine Displays zu einem großen zusammenzuführen.

Xperia sola
Smartphone

Manufacturer
Sony Mobile Communications
International AB,
Lund, Sweden
In-house design
Sony Mobile UX Creative Design Team
Web
www.sonymobile.com

Xperia sola is a slim and compact smartphone with many functions for daily use. Due to the simple structure of the Xperia sola, separation lines between the various body parts are hardly visible, and reinforce the impression of a solid product of high quality. An innovative, floating touch display allows operation without direct contact with the touchscreen.

Statement by the jury
The distinctive and simultaneously reduced form language of the Xperia sola attracts attention to the innovative display and provides ease of operation.

Xperia sola ist ein schlankes und kompaktes Smartphone mit vielen Funktionen für den täglichen Gebrauch. Durch die einfache Struktur des Xperia sola konnten die Trennfugen zwischen den verschiedenen Gehäuseteilen so gestaltet werden, dass sie kaum sichtbar sind und den Eindruck eines soliden, qualitativ hochwertigen Produktes verstärken. Ein innovatives Floating Touch-Display erlaubt eine Bedienung ohne Berührung des Bildschirms.

Begründung der Jury
Die markante und gleichzeitig reduzierte Formensprache des Xperia sola lenkt das Augenmerk auf das innovative Display und unterstützt eine komfortable Bedienung.

Xperia U
Smartphone

Manufacturer
Sony Mobile Communications
International AB,
Lund, Sweden
In-house design
Sony Mobile UX Creative Design Team
Web
www.sonymobile.com

The Xperia U is a compact, clearly designed smartphone, whereby the bottom caps of the housing are interchangeable. The black or white main housing can be combined with a yellow, pink, black or white bottom cap. A transparent, illuminating element under the display provides additional liveliness by changing colour to match each image shown on the album cover or background.

Statement by the jury
The clear, practical form language of the Xperia U contrasts pleasantly with the illuminating band which keeps changing colour.

Das Xperia U ist ein kompaktes, klar gestaltetes Smartphone, dessen untere Gehäusekante austauschbar ist. Je nach persönlicher Vorliebe kann das schwarze oder weiße Basisgehäuse mit einem gelben, pinken, schwarzen oder weißen Bodenstück kombiniert werden. Eine transparente, leuchtende Leiste unterhalb des Displays sorgt für zusätzliche Lebendigkeit, indem sie sich farblich an das jeweils angezeigte Foto, Album-Cover oder Hintergrundbild anpasst.

Begründung der Jury
Die klare, sachliche Formensprache des Xperia U wird durch die sich farblich immer wieder verändernde Leuchtleiste angenehm kontrastiert.

Xperia S
Smartphone

Manufacturer
Sony Mobile Communications
International AB,
Lund, Sweden
In-house design
Sony Mobile UX Creative Design Team
Web
www.sonymobile.com

The Xperia S attracts attention with its HD touchscreen which appears to float in the air. This impression is created by a transparent element beneath the display which seems to isolate it from the rest of the housing. This component is not only a design element but also, together with the bottom foundation, serves as main antenna. Furthermore it indicates incoming calls or other communications by means of a light signal, visible from all angles.

Statement by the jury
The Xperia S convinces due to a technically sophisticated display, whose design is emphasised by the multifunctional, transparent light strip.

Das Xperia S erregt Aufmerksamkeit mit einem HD-Touchscreen, der zu schweben scheint. Dieser Eindruck wird durch ein transparentes Element unterhalb des Displays erreicht, das es vom Rest des Gehäuses zu isolieren scheint. Dieses Bauteil ist nicht nur Designelement, sondern dient gemeinsam mit dem Bodenteil auch als Hauptantenne. Zudem kommuniziert es eingehende Anrufe oder Ähnliches per Lichtsignal, das von allen Seiten aus sichtbar ist.

Begründung der Jury
Das Xperia S überzeugt durch ein technisch hochwertiges Display, das gestalterisch durch die multifunktionale transparente Lichtleiste betont wird.

Xperia SX
Smartphone

Manufacturer
Sony Mobile Communications
International AB,
Lund, Sweden
In-house design
Sony Mobile UX Creative Design Team
Web
www.sonymobile.com

Xperia SX is a slim, minimalist smartphone. Its design is compact and ergonomic, and therefore lies pleasantly in the hand. The surfaces of the four main blocks are simultaneously separated and visually held together via two thin metal lines which cross at the sides. The lines elegantly emphasise the slim silhouette of the smartphone. The Xperia SX is available in black, white, pink or orange.

Statement by the jury
Xperia SX delights with its precise workmanship and its overall very high quality and elegant impression.

Xperia SX ist ein schlankes, minimalistisches Smartphone. Seine Gestaltung ist kompakt und ergonomisch, wodurch es gut in der Hand liegt. Die Oberflächen der vier Haupt-Gehäuseteile werden durch zwei dünne Metall-Linien, die sich an den Seiten kreuzen, gleichzeitig getrennt und optisch zusammengehalten. Die Linien betonen die schlanke Silhouette des Smartphones auf elegante Weise. Das Xperia SX ist in Schwarz, Weiß, Pink oder Orange erhältlich.

Begründung der Jury
Xperia SX begeistert mit seiner präzisen Verarbeitung und seiner insgesamt sehr hochwertigen und eleganten Anmutung.

Xperia go
Smartphone

Manufacturer
Sony Mobile Communications
International AB,
Lund, Sweden
In-house design
Sony Mobile UX Creative Design Team
Web
www.sonymobile.com

Xperia go is a highly resistant, dust- and waterproof smartphone. Its design is sleek, the screen emphasised by being slightly raised from the body. This makes it on the one hand easy to clean, on the other hand simple to recognise in the pocket by feel. It is available in matt white, textured black and striking yellow. This allows the user to choose the appropriate smartphone according to style – elegant, functional or sporty.

Statement by the jury
Xperia go, with its sleek, functional design proves that a robust and watertight smartphone can also be very elegant.

Xperia go ist ein widerstandsfähiges, staub- und wasserdichtes Smartphone. Seine Gestaltung ist schlicht, der Bildschirm durch eine leichte Anhebung betont. Das erleichtert zum einen die Reinigung, zum anderen die Orientierung beim Griff in die Tasche. Es ist in mattem Weiß, strukturiertem Schwarz und einem Signalgelb erhältlich. Das erlaubt dem Nutzer, das passende Smartphone zum eigenen Stil zu wählen – elegant, funktional oder sportlich.

Begründung der Jury
Xperia go beweist mit seiner schlichten, funktionellen Gestaltung, dass ein strapazierfähiges und wasserdichtes Smartphone auch sehr elegant sein kann.

Xenium mobile phone W737
Smartphone

Manufacturer
Royal Philips Electronics,
Eindhoven, Netherlands
In-house design
Philips Design Consumer Lifestyle Team
Design
Sang Fei Design Team,
Shenzhen, China
Web
www.philips.com
www.sangfei.com

This Xenium smartphone catches the eye due to its slim, slightly wedge-shaped design, its elegant housing and its well-considered user interface. Its energy is provided by a long-life Xenium battery. A particular feature of the smartphone is a special energy-saving switch which is located at the side of the housing. Thus the user does not need to first dive into the depths of the user menu in order to switch into energy saving mode.

Dieses Xenium-Smartphone fällt durch seine schlanke, leicht keilförmige Gestaltung, sein elegantes Gehäuse und seine durchdachte Benutzeroberfläche auf. Seine Energie bezieht es aus einer langlebigen Xenium-Batterie. Besonderes Merkmal des Smartphones ist ein spezieller Energiesparschalter, der seitlich am Gehäuse angeordnet ist. So muss der Nutzer nicht erst in die Tiefen des Benutzermenüs eintauchen, wenn er in den Energiesparmodus wechseln will.

Statement by the jury
This technically and creatively high-end smartphone impresses with a small but convincing feature: its physical energy saving button.

Begründung der Jury
Dieses technisch und gestalterisch hochwertige Smartphone besticht mit einer kleinen, aber überzeugenden Besonderheit: seinem physischen Energiesparknopf.

Kid's Phone HW-01D
Mobile Phone
Mobiltelefon

Manufacturer
Huawei Device Co., Ltd.,
Shenzhen, China
In-house design
Xudong Wang
Web
www.huaweidevice.com
Honourable Mention

Kid's Phone HW-01D is a mobile phone
with many properties particularly
suitable for children: with its compact
form it fits nicely in a child's hands and
is easy to operate. Texting is made easy
by text templates. The screen is made of
scratch-resistant material; the housing
is watertight and impact-resistant. Kid's
Phone also has a safety alarm which is
activated by pulling a ring, automatically
sending an SOS signal with details of the
location.

Kid's Phone HW-01D ist ein Mobiltelefon
mit vielen kindgerechten Eigenschaften:
Mit seiner kompakten Form passt es gut in
Kinderhände und lässt sich leicht bedienen.
Das Texten wird durch Vorlagen erleichtert.
Der Bildschirm besteht aus kratzfestem
Material, das Gehäuse ist wasserdicht und
stoßfest. Kid's Phone hat zudem einen
Sicherheitsalarm, der über das Ziehen
eines Rings aktiviert wird und automatisch
eine SOS-Nachricht mit Standortangabe
versendet.

Statement by the jury
The sturdy Kid's Phone is optimally
suited to the special needs of children
and is very easy to operate

Begründung der Jury
Das robuste Kid's Phone ist optimal auf
die besonderen Bedürfnisse von Kindern
abgestimmt und kinderleicht zu bedienen.

SmartWatch
Bluetooth Device
Bluetooth-Gerät

Manufacturer
Sony Mobile Communications
International AB,
Lund, Sweden
In-house design
Sony Mobile UX Creative Design Team
Web
www.sonymobile.com

SmartWatch is a Bluetooth multifunctional device which can be worn on the wrist like a watch. It is connected to a smartphone by Bluetooth technology and displays messages or incoming calls on its screen. The minimalist housing format with a polished aluminium frame is discreet; the rugged, multi-touch display is the centrepiece. Thanks to an adapter, the SmartWatch can be worn with almost any watchband.

SmartWatch ist ein Bluetooth-Multifunktionsgerät, das wie eine Uhr am Handgelenk getragen werden kann. Via Bluetooth wird es mit einem Smartphone verbunden und zeigt Nachrichten oder eingehende Anrufe auf seinem Bildschirm an. Die minimalistische Gehäuseform mit einem Rahmen aus gebürstetem Aluminium ist dezent, das robuste Multi-Touch-Display steht im Mittelpunkt. Dank eines Adapters kann die SmartWatch mit fast jedem Uhrenband getragen werden.

Statement by the jury
Its discreet design and its intelligent multifunctionality make the Smart-Watch a useful and very wearable smartphone accessory.

Begründung der Jury
Ihre dezente Gestaltung und ihre intelligente Multifunktionalität machen die SmartWatch zu einem sinnvollen und sehr tragbaren Smartphone-Accessoire.

Limmex Emergency Watch
Limmex Notruf-Uhr
Emergency Call System
Notrufsystem

Manufacturer
Limmex,
Zürich, Switzerland
In-house design
Design
Sandra Kaufmann,
Zürich, Switzerland
Brunner Mettler Ltd.
(Thilo Alex Brunner, Jörg Mettler),
Zürich, Switzerland
Web
www.limmex.com
Honourable Mention

This Emergency Watch looks like a classical watch and thus counteracts stigmatisation of the wearer. The emergency call button is located on the right side of the housing to avoid false alarms. When the emergency call is made, thanks to an integrated GSM module, loudspeaker and microphone, it can set up a telephone call to people whose numbers are stored. Despite complex technology, which is located in the smallest space, the watch is very light weighing only 46 grams.

Statement by the jury
The emergency watch convinces with a well-conceived style in classical watch design which counters the stigmatisation of people in need of help.

Diese Notruf-Uhr sieht wie eine klassische Uhr aus und wirkt so einer Stigmatisierung des Trägers entgegen. Der Notrufknopf ist an der rechten Gehäuseseite angebracht, um Fehlalarme zu vermeiden. Wird der Notruf ausgelöst, kann die Uhr dank eingebautem GSM-Modul, Lautsprecher und Mikrofon eine Telefonverbindung zu gespeicherten Personen aufbauen. Trotz komplexer Technik, die auf kleinstem Raum untergebracht ist, ist die Uhr mit 46 Gramm sehr leicht.

Begründung der Jury
Die Notruf-Uhr überzeugt mit einer durchdachten Gestaltung im klassischen Uhren-Design, die der Stigmatisierung hilfsbedürftiger Personen entgegenwirkt.

eNest
Mobile Safety System
Mobiles Sicherheitssystem

Manufacturer
The Nest Network,
Madrid, Spain
Design
Mormedi
(Jaime Moreno),
Madrid, Spain
Web
www.nestwork.eu
www.mormedi.com

eNest is a security system consisting of a main device and a watchstrap, both of which are equipped with GPS technology. In an emergency, an SOS is sent to a security service, activated by pressing a panic button, or is self-activated, for example, when a child leaves a defined safety area. Information on the location is given in real time and can be obtained online.

Statement by the jury
eNest is an intelligent security system which convinces with its discrete, friendly design and high degree of user-friendliness.

eNest ist ein Sicherheitssystem, das aus einem Hauptgerät und einem Armband besteht, die beide mit GPS-Technologie ausgestattet sind. Im Falle einer Notsituation wird eine SOS-Nachricht an einen Sicherheitsdienst gesendet, die durch das Drücken des Panikknopfs aktiviert wird oder durch Selbstauslösung, etwa wenn ein Kind ein definiertes Sicherheitsareal verlässt. Die Standortinformationen werden in Echtzeit aktualisiert und können online abgerufen werden.

Begründung der Jury
eNest ist ein intelligentes Sicherheitssystem, das mit seiner diskreten, freundlichen Gestaltung und einer hohen Benutzerfreundlichkeit überzeugt.

Pocket Geiger Type 4
**Radiation Detector
for Smartphones**
Geigerzähler
für Smartphones

Manufacturer
Radiation-watch.org,
Yaguchi Electric Corporation,
Ishinomaki City, Miyagi Prefecture, Japan
In-house design
Yang Ishigaki,
Yoshihisa Tanaka
Design
Masashi Ogasawara,
Kouno City, Osaka Prefecture, Japan
Web
www.radiation-watch.org

Pocket Geiger is a compact radiation detector, used in connection with smartphones. It was developed after the reactor accident in Japan by engineers and scientists of the organisation Radiation-Watch.org in order to ease radiation measurement for the population. Within two minutes the device measures the radiation contamination in ambient air by means of photodiodes – the values are displayed on the smartphone and collected centrally.

Pocket-Geiger ist ein kompakter Strahlungsdetektor, der in Verbindung mit Smartphones genutzt wird. Er wurde nach dem Reaktorunfall in Japan von Ingenieuren und Wissenschaftlern der Organisation Radiation-Watch.org entwickelt, um der Bevölkerung die Strahlenmessung zu erleichtern. Binnen zwei Minuten misst das Gerät mithilfe von Fotodioden die Strahlenbelastung der Umgebungsluft – die Werte werden auf dem Smartphone angezeigt und zentral gesammelt.

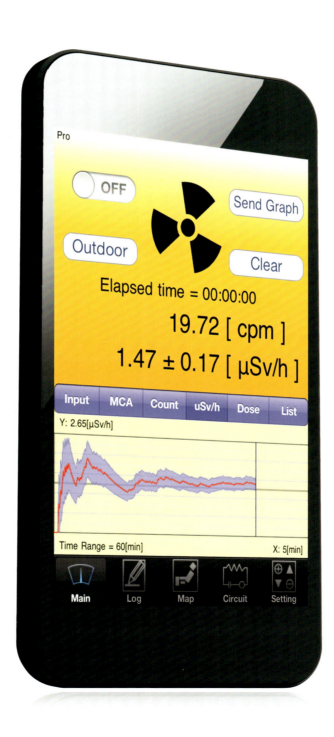

Smartphone Home
Smartphone Holder
Smartphone-Halter

Manufacturer
Simon Schreinerwerkstätte GmbH,
Hupperath, Germany
In-house design
Gerhard Simon
Web
www.jg-simon.de
Honourable Mention

This smartphone holder catches the eye at first glance due to its unusual material: it is made of exotic wood which means that each holder is unique. It has a suction pad on the resting surface which only with microporous suction pad technology holds every smartphone securely – vertically as well as horizontally. The underside of the holder is rubberized; connection of charger or PC connector cables is possible from all sides.

Statement by the jury
This smartphone holder combines the natural material wood with an innovative holding technology which provides enhanced functionality.

Dieser Smartphone-Halter fällt schon auf den ersten Blick durch sein ungewöhnliches Material auf: Er wird aus Edelhölzern gefertigt, die jeden Halter zum Unikat machen. Auf der Auflagefläche hat er ein Haft-Pad, das allein mit mikroporöser Saugnapftechnik jedes Smartphone sicher hält – vertikal oder horizontal. Die Unterseite des Halters ist gummiert, der Anschluss von Lade- oder PC-Verbindungskabeln von allen Seiten möglich.

Begründung der Jury
Dieser Smartphone-Halter kombiniert das natürliche Material Holz mit einer innovativen Haft-Technologie, die zu einer erhöhten Funktionalität führt.

Milo Aluminium
Phone Stand
Mobiltelefon-Ständer

Manufacturer
Bluelounge,
Singapore
In-house design
Dominic Symons (Creative Director)
Max Wijoyoseno (Product Designer)
Kevin Keller (Engineer)
Rikky Risdiansah (Art Director)
Web
www.bluelounge.com

Milo is an elegant, minimalist accessory for the desktop – a place to hold smartphones or other mobile devices. The sleek aluminium stand uses high-tech, Japanese, micro-suction technology to hold devices with smooth surfaces at a desired angle without leaving a residue. Milo is perfect for making phone calls, watching videos or simply for docking.

Statement by the jury
Milo Aluminium convinces with its ease of handling and plain form language which creates an impression of tidiness at the workplace.

Milo ist ein elegantes minimalistisches Accessoire für den Schreibtisch – ein Platz zum Ablegen von Smartphones oder anderen mobilen Geräten. Der schlanke Aluminiumständer hat eine Auflagefläche, die japanische High-Tech Mikro-Hafttechnologie nutzt, um Geräte mit glatter Oberfläche in jedem gewünschten Winkel zu halten – je nachdem, ob sie für Gespräche genutzt werden, um Videos anzuschauen oder einfach als Dockingstation.

Begründung der Jury
Milo Aluminium überzeugt mit einer einfachen Handhabung und einer schlichten Formensprache, die zu einem aufgeräumten Eindruck des Arbeitsplatzes beiträgt.

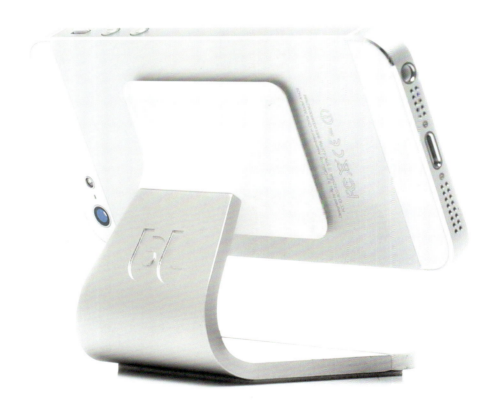

iXY
Microphone
Mikrofon

Manufacturer
RØDE Microphones,
Silverwater, Australia
In-house design
Peter Cooper, Sam Trewartha
Web
www.rodemic.com

The iXY is an up-market stereo microphone for mobile Apple devices whose design was based on functional considerations: The distinctive configuration of the cylindrical capsule housing follows the exact definition of "XY microphone alignment", a professional recording process for an accurate stereo image. The capsule housing and base have been developed with their acoustic properties in mind, to avoid interference from reflections and thus assure optimal directional and frequency characteristics. The iXY is made completely of metal and provides effective shielding from the electronics. The design of the outside and other elements is adapted to a harmonious interplay with Apple devices which use the microphone.

iXY ist ein hochwertiges Stereomikrofon für mobile Apple-Geräte, das insbesondere unter funktionellen Gesichtspunkten gestaltet wurde: Die markante Anordnung der zylindrischen Kapselgehäuse folgt exakt den Vorgaben der „XY-Mikrofonie" – ein professionelles Aufnahmeverfahren für eine akkurate Stereo-Abbildung. Auch Kapselgehäuse und Sockel wurden im Hinblick auf ihre akustischen Eigenschaften entwickelt, um störende Reflexionen zu vermeiden und so eine optimale Richt- und Frequenzcharakteristik zu gewährleisten. Das iXY ist komplett aus Metall gefertigt und bietet eine effektive Abschirmung der Elektronik. Die Gestaltung der Oberflächen und anderer Elemente ist auf ein harmonisches Zusammenspiel mit den Apple-Geräten, mit denen das Mikrofon verwendet wird, abgestimmt.

Statement by the jury
The iXY is designed in an upmarket and functional manner and, thanks to its compactness, is a practical alternative to conventional sound recorders. It converts iPhone or iPad into a recorder of professional recording quality in no time.

Begründung der Jury
Das iXY ist hochwertig und funktionell gestaltet und dank seiner Kompaktheit eine praktische Alternative zu herkömmlichen Tonrekordern. Es verwandelt iPhone oder iPad im Handumdrehen in einen Rekorder mit professioneller Aufnahmequalität.

LINKASE

Signal Enhancing Case for the iPhone 5
Signalverstärkende Hülle für das iPhone 5

Manufacturer
Absolute Technology Co., Ltd.,
Taipei, Taiwan
In-house design
Angel Sanz Correa,
Jenny Chih-Chieh Teng
Web
www.absolute.com.tw
Honourable Mention

Linkase is a protective case for the iPhone 5 which is at the same time able to enhance Wi-Fi signals. For this it is equipped with a special technology (EMW) which strengthens the signal from the antenna considerably. The EMW element is incorporated in the back of the case; for a better signal it is pushed upwards. The case and EMW element are available in various colours and can be combined freely.

Linkase ist eine Schutzhülle für das iPhone 5, die gleichzeitig in der Lage ist, Wi-Fi-Signale zu verbessern. Dafür wurde sie mit einer speziellen Technologie (EMW) ausgestattet, die das Signal der Antenne deutlich verstärkt. Das EMW-Element verbirgt sich in einem Schiebemechanismus an der Rückseite, für ein besseres Signal wird es nach oben herausgeschoben. Hülle und EMW-Element sind in verschiedenen Farben erhältlich und frei kombinierbar.

Statement by the jury
Linkase combines all advantages of a good case with special technology and thus extends the functionality of the smartphone.

Begründung der Jury
Linkase verbindet alle Eigenschaften einer guten Schutzhülle mit einer besonderen Technologie und erweitert damit die Funktionalität des Smartphones.

hipKey
Wireless Smartphone Accessory
Kabelloses Smartphone-Zubehör

Manufacturer
hippih ApS,
Kongens Lyngby, Denmark
In-house design
Martin Staunsager Larsen
Web
www.hippih.com
Honourable Mention

hipKey is an innovative wireless smartphone accessory that helps you keep an eye on your valuables. hipKey can be easily configured via your smartphone or tablet and can be attached to e.g. keys or a bag. It sends an alert when the devices are separated by distance. hipKey consists of a ring of anodised aluminium, which is enclosed by two robust, curved, plastic components. Its keypad is well protected in a concave recess.

Statement by the jury
This intelligent accessory is eye-catching and versatile, yet also ergonomically designed, so that it proves to be a true hand charmer.

hipKey ist ein innovatives drahtloses Smartphone-Accessoire, das dabei hilft, Wertsachen im Auge zu behalten. Es lässt sich einfach via Smartphone oder Tablet-PC konfigurieren und löst einen Alarm aus, wenn der Gegenstand, an dem es befestigt ist, außer Reichweite gerät. hipKey besteht aus einem Ring aus eloxiertem Aluminium, der von zwei robusten, gewölbten Kunststoffteilen umschlossen ist. Sein Bedienfeld liegt gut geschützt in einer konkaven Aussparung.

Begründung der Jury
Dieses intelligente Accessoire ist prägnant und vielseitig, dabei jedoch so ergonomisch gestaltet, dass es sich als wahrer Handschmeichler erweist.

Lightning Cap
Anschluss-Abdeckung

Manufacturer
Fruitshop International Co., Ltd.,
New Taipei, Taiwan
In-house design
Reads Lin
Web
www.bonecollection.com
Honourable Mention

The connector is for many devices a weak-point with regard to penetration of damp or dust. The Lightning Cap protects iPhone 5, iPad, iPad mini and iPod touch from this in a charming, simple manner: A small cap in the form of an animal can be simply pushed onto the device concerned and lies protectively with its silicone body over the connection terminal and home button.

Statement by the jury
The Lightning Caps are not only a sensible protection for the lightning connection, these nice little companions provide the devices with an individual touch.

Ihre Schnittstelle ist bei vielen Geräten ein Schwachpunkt, wenn es um das Eindringen von Feuchtigkeit oder Staub geht. Die Lightning Cap schützt iPhone 5, iPad, iPad mini und iPod touch auf eine charmante, einfache Weise davor: Eine kleine Kappe in Tierform kann einfach auf das jeweilige Gerät gesteckt werden und legt sich mit seinem Silikonkörper schützend über Anschlussstelle und Home-Button.

Begründung der Jury
Die Lightning Caps sind nicht nur ein sinnvoller Schutz für den Lightning-Anschluss, diese netten kleinen Gesellen verleihen den Geräten auch eine individuelle Note.

Steffany
Stylus for Smartphones and Tablets
Eingabestift für Smartphones und Tablets

Manufacturer
Fadtronics Innovation Limited,
Hong Kong
In-house design
Sui Wang,
Chris Ng
Web
www.oic-concept.com
www.fadtronics.com
Honourable Mention

Steffany is simultaneously piece of jewellery and a useful accessory for smartphones and tablets. The aluminium ring is light, has a shiny surface and fits pleasantly onto the finger. In place of a jewel stone, a rubber ball sits in its socket, turning the ring into an elegant stylus for capacitive displays.

Statement by the jury
At first sight a decorative piece of jewellery, this ring amazes with its extended functionality as a compact stylus for devices with a touchscreen.

Steffany ist gleichzeitig Schmuckstück und nützliches Accessoire für Smartphones und Tablets. Der Ring aus Aluminium ist leicht, hat eine glänzende Oberfläche und ist angenehm am Finger zu tragen. Statt eines Schmucksteins ruht in seiner Fassung jedoch eine Gummikugel, die den Ring zu einem eleganten Stylus für kapazitive Displays macht.

Begründung der Jury
Dem ersten Anschein nach ein dekoratives Schmuckstück, verblüfft dieser Ring mit seiner erweiterten Funktionalität als kompakter Stylus für Geräte mit Touchscreen.

iPhone 5 Rechargeable Battery with Kickstand
iPhone 5-Akku mit Ständer

Manufacturer
ODOYO International Limited,
California, USA
In-house design
Kevin Leung,
Jae Au
Web
www.odoyo.com
Honourable Mention

The Power+Shell rechargeable battery with kickstand is a portable energy source in a slim-lined, protective frame for the iPhone 5. It provides a high battery power of 2200 mAh, which is sufficient to fully charge the iPhone one time. The quick-charge technology supplies this power at a very fast rate. The back of the stand can be folded out and supports the iPhone 5 at an angle which gives an optimal view of the display.

Der Power+Shell-Akku mit Ständer ist eine tragbare Energiequelle in einem schlanken, schützenden Rahmen für das iPhone 5. Er bietet eine hohe Akkuleistung von 2200 mAh, die ausreicht, um das iPhone einmal komplett aufzuladen. Die Schnell-lade-Technologie liefert diese Energie sehr rasch. Der Ständer lässt sich an der Rückseite ausklappen und stützt das iPhone 5 in einem Winkel ab, der einen optimalen Blick auf das Display erlaubt.

Statement by the jury
With the Power+Shell rechargeable battery, the look of the iPhone is preserved; at the same time it convinces with its power and the aid of a useful stand.

Begründung der Jury
Mit dem Power+Shell-Akku wird die Optik des iPhone 5 beibehalten, gleichzeitig überzeugt er mit seiner Leistung und einer nützlichen Standhilfe.

Leather Wrap Case for iPhone 5
Mobile Phone Case
Mobiltelefon-Hülle

Manufacturer
Adopted, Inc.,
Shanghai, China
In-house design
David Watkins,
Pablo Spagnoletti
Web
www.getadopted.com

The Leather Wrap Case consists of leather and a stable metal frame. With its leather appearance it reminds one a little of vintage binoculars. The flowing lines of the case and the elegantly contrasting materials assure good haptics. With its slim-line silhouette, this case fits the iPhone 5 perfectly, at the same time protecting it reliably. The Leather Wrap Case is available in eight different versions.

Statement by the jury
The Leather Wrap Case pays homage to a classical optical device and provides an exciting contrast to the technology of the iPhone.

Die Leather Wrap-Hülle besteht aus Leder und einem stabilen Metallrahmen. Mit ihrer Leder-Optik erinnert die Hülle ein wenig an Vintage-Ferngläser. Die fließenden Linien der Hülle und die sich elegant kontrastierenden Materialien sorgen für eine gute Haptik. Mit ihrer schmalen Silhouette passt diese Hülle perfekt zum iPhone 5, gleichzeitig schützt sie es zuverlässig. Die Leather Wrap-Hülle ist in acht verschiedenen Ausführungen erhältlich.

Begründung der Jury
Die Leather Wrap-Hülle ist eine Hommage an klassische optische Geräte und bildet einen spannenden Kontrast zur Technologie des iPhones.

gosh! Parallel
Rechargeable Battery for the iPhone 5
iPhone 5-Akku

Manufacturer
Appcessory,
Singapore
In-house design
Ryan Xie
Web
www.whygosh.com

Parallel is a protective case with detachable, rechargeable battery which was developed especially for the iPhone 5 and dimensioned as exactly for it. For this reason the iPhone remains the same length and is only a little thicker when the battery is pushed onto the case for charging. Thus the iPhone 5 lies compactly in the hand, also while charging. The battery has a storage capacity of 2500 mAh.

Statement by the jury
Parallel is designed so that it fits the iPhone 5 perfectly und thus does not restrict the use of the iPhone while it is charging.

Parallel ist ein Schutzgehäuse mit abnehmbarem Akku, das speziell für das iPhone 5 entwickelt und hinsichtlich seiner Größe exakt auf dieses abgestimmt wurde. Damit wird das iPhone bei gleichbleibender Länge nur ein wenig dicker, wenn der Akku zum Laden auf das Gehäuse geschoben wird. So liegt das iPhone 5 auch während des Ladevorgangs kompakt in der Hand. Der Akku hat eine Speicherkapazität von 2500 mAh.

Begründung der Jury
Parallel ist so gestaltet, dass es perfekt auf das iPhone 5 abgestimmt ist und dadurch die Benutzung des iPhones während des Ladevorgangs nicht einschränkt.

O!coat FaaGaa
Protective Case
for iPhone 5
iPhone 5-Hülle

Manufacturer
Ozaki International Co., Ltd.,
Taipei, Taiwan
In-house design
Freeman Liu, Henry Mao
Web
http://ozakiverse.com

This protective case for the iPhone 5 is eye-catching due to its colourful design with animal faces. The nose-mouth section of every animal face can be folded out. It is designed on the back like a tongue and is used as a standing aid. The case is made of sturdy polycarbonate and has recesses for connections and camera, so that they are accessible at all times.

Diese Schutzhülle für das iPhone 5 fällt durch seine farbenfrohe Gestaltung mit Tiergesichtern auf. Die Nase-Mund-Partie kann bei allen Gesichtern herausgebogen werden. Sie ist auf der Rückseite wie eine Zunge gestaltet und dient als Standhilfe. Die Hülle ist aus widerstandsfähigem Polycarbonat gefertigt und hat Aussparungen für Anschlüsse und Kamera, sodass diese jederzeit frei zugänglich sind.

Statement by the jury
O!coat FaaGaa is a colourful protective case which, due to its standing aid, acquires playful components.

Begründung der Jury
O!coat FaaGaa ist eine farbenfrohe Schutzhülle, die durch ihre Standhilfe eine spielerische Komponente bekommt.

Bubblepack Playcase
Protective Case for iPhone 5
iPhone 5-Hülle

Manufacturer
SSONGSDESIGN,
Anyang, South Korea
In-house design
Yoonjae Jung,
Sungho Bai,
Jungwook Choi
Web
www.bubblepack.co.kr

Bubblepack Playcase is a protective case for the iPhone 5, which can easily be attached. It is available in brilliant colours and designed so that it fits exactly onto the iPhone. Thanks to its lightly padded surface it lies nicely in the hand. Furthermore, the case serves as pouch for credit cards or tickets which are simply hidden between iPhone and case and are secured safely by an additional element, so that they can be easily carried along.

Statement by the jury
Bubblepack Playcase convinces with its extended functionality as repository for credit cards or RFID chip cards.

Bubblepack Playcase ist eine Schutzhülle für das iPhone 5, die einfach aufgesteckt werden kann. Sie ist in leuchtenden Farben erhältlich und so gestaltet, dass sie genau auf das iPhone passt. Dank der leicht gepolsterten Oberfläche liegt sie angenehm in der Hand. Darüber hinaus dient die Hülle als Fach für Kredit- oder Fahrkarten, die einfach zwischen iPhone und Hülle versteckt und, durch ein zusätzliches Element sicher geschützt, mitgeführt werden können.

Begründung der Jury
Bubblepack Playcase überzeugt mit seiner erweiterten Funktionalität als Aufbewahrungsort für Kredit- oder RFID-Chipkarten.

Mobile Phone Common Water-Proof Case
Wasserdichte Mobiltelefon-Hülle

Manufacturer
Ningbo Cooskin Stationery Co., Ltd.,
Ningbo, China
In-house design
Xingxing Li
Web
www.cooskin.com
Honourable Mention

This water-proof mobile phone case can be used for all devices up to a display size of 4.7". It is particularly suitable for protection during activity holidays, for instance diving, swimming or surfing. The case is made of sturdy nylon and uses silica gel sealing technology, making it dust and impact resistant as well as water-tight in accordance with the IPX-8 standard. With this case, the smartphone can still be easily operated.

Statement by the jury
This slim and light protection for mobile telephones is highly functional and with its vivid colours suitable for an active lifestyle.

Diese wasserdichte Hülle für Mobiltelefone kann für alle Geräte bis zu einer Displaygröße von 4,7" genutzt werden. Sie eignet sich besonders zum Schutz in Aktivurlauben, etwa beim Tauchen, Schwimmen oder Surfen. Die Hülle ist durch widerstandfähiges Nylon und eine Versiegelungstechnologie unter Verwendung von Kieselgel staub- und stoßfest sowie wasserdicht gemäß IPX-8-Standard. Auch mit der Hülle lässt sich das Smartphone noch gut bedienen.

Begründung der Jury
Dieser dünne und leichte Schutz für Mobiltelefone ist höchst funktionell und passt mit seinen leuchtenden Farben zu einem aktiven Lebensstil.

i-Pooding Grip Cover
iPhone Case
iPhone-Hülle

Manufacturer
sumneeds,
Seoul, South Korea
In-house design
Seol Hee Son, Ji Seung Song,
Seung Reol Oh, Won Ho Son
Web
www.sumneeds.com

i-Pooding is a smartphone case which is filled with paraffin oil and is manufactured by means of high frequency technology. Due, above all, to the use of liquid in its interior, which reacts to the user's every movement, the case has an innovative and unusual effect and provides a completely new haptic experience. It also lies pleasantly in the hand and protects the smartphone reliably from impact and vibration.

Statement by the jury
With its soft cover filled with liquid, i-Pooding provides smartphones with a charming, sensual component.

i-Pooding ist eine Smartphone-Hülle, die mit flüssigem Paraffinöl gefüllt ist und unter Einsatz von Hochfrequenztechnik hergestellt wird. Vor allem durch die Flüssigkeit in ihrem Inneren, die auf jede Bewegung des Nutzers reagiert, wirkt die Hülle innovativ und ungewöhnlich und bietet eine völlig neue haptische Erfahrung. Darüber hinaus liegt sie gut in der Hand und schützt das Smartphone zuverlässig vor Stößen und Erschütterungen.

Begründung der Jury
Mit ihrem weichen, mit Flüssigkeit gefüllten Cover verleiht i-Pooding Smartphones eine charmante, sinnliche Komponente.

ULTRA'GO mini
Portable Power Station
with Vibration Signalling
Mobiler Akku mit Vibrationsalarm

Manufacturer
Calibre Style Ltd.,
Taipei, Taiwan
In-house design
Johnny Tsai
Web
www.calibre-style.com

Ultra'Go mini is a portable charger battery (power bank) for iPhone, iPad, iPod and other USB-powered devices. This gadget fits easily into any pocket with its ultra-slim, lightweight and easy-to-hold design offering ultimate portability, whilst the screwless aluminium alloy enclosure provides outstanding protection. The "Vibration Signalling" feature indicates battery power status by varying vibration patterns.

Ultra'Go mini ist ein tragbarer Akku für iPhone, iPad und iPod sowie andere mobile Geräte mit USB-Anschluss. Sein extrem leichtes und kompaktes Aluminium-Gehäuse kommt ohne Schrauben aus. Es ist ergonomisch gestaltet, passt in jede Tasche und schützt den Akku. Ultra'Go mini kommuniziert seinen Energiezustand über verschiedene Vibrationssignale.

Gum

Power Pack
Back-up-Akku

Manufacturer
Just Mobile Ltd.,
Taichung, Taiwan
In-house design
Erich Huang,
Nils Gustafsson
Web
www.just-mobile.com

Gum is a very compact and elegant back-up power pack for smartphones and other USB devices. Almost thumb-size, Gum surprises with sufficient capacity to fully load a smartphone. Gum itself is charged via a micro-USB connection. Three green LEDs indicate its charge status. Also belonging to the power pack with top-quality aluminium housing is a micro-USB cable and a cable with docking connector for iPhone or iPod touch.

Gum ist ein sehr kompakter und eleganter Back-up-Akku für Smartphones und andere USB-Geräte. Kaum mehr als daumengroß, überrascht Gum mit ausreichender Kapazität, um ein Smartphone komplett aufzuladen. Gum selbst wird über einen Mikro-USB-Anschluss geladen; drei grüne LEDs zeigen seinen Ladezustand an. Zu dem Akku mit edlem Aluminiumgehäuse gehören außerdem ein Mikro-USB-Kabel und ein Kabel mit Dock-Anschluss für iPhone oder iPod touch.

Statement by the jury
The design of this power pack is so extremely compact that it can really be carried in every trouser pocket in order to then reliably fulfil its purpose.

Begründung der Jury
Dieser Akku ist so extrem kompakt gestaltet, dass er sich wirklich in jeder Hosentasche mitnehmen lässt, um dann verlässlich seinen Dienst zu versehen.

Chocolate
Modular External Battery Pack
Modularer Akku

Manufacturer
Hong Kong Mricetechnology Co., Ltd.,
Hong Kong
Design
LKK Design Co., Ltd.
(Yichao Li, Jiuzhou Zhang),
Shenzhen, China
Web
www.mrice.cn
www.lkkdesign.com
Honourable Mention

This product is inspired from a bar of chocolate and promises to supply exactly as much energy as is required at the time. One piece of the modular battery pack is sufficient, for instance, to enable continued use for a while yet of a device whose battery is discharged. Two or more pieces of Chocolate will provide the energy requirement of the device, for instance, during a business trip. The design of the modular chocolate bar makes Chocolate very flexible in its use.

Statement by the jury
The design of the modular Chocolate battery pack is based on a charming and easily understood idea which leads to its intuitive use.

Dieses Produkt ist von einer Schokoladentafel inspiriert und verspricht, genau so viel Energie zu liefern, wie gerade benötigt wird. Ein Stück des modularen Akku-Baukastens reicht beispielsweise, um ein Gerät, dessen Akku leer ist, noch eine Weile nutzen zu können. Zwei oder mehr Chocolate-Stücke stillen den Energiebedarf des Geräts etwa während einer Geschäftsreise. Die Gestaltung als modulare Schokoladentafel macht Chocolate sehr flexibel in der Anwendung.

Begründung der Jury
Der Gestaltung des modularen Akkus Chocolate liegt eine charmante und leicht verständliche Idee zugrunde, die zu einer intuitiven Benutzung führt.

Xtorm Power Bank 7300
External Rechargeable Battery
Externer Akku

Manufacturer
A-solar BV,
Houten, Netherlands
In-house design
Ralph Both
Web
www.a-solar.eu

The Xtorm Power Bank 7300 is constructed of brushed aluminium and has an internal energy storage capacity of 7300 mAh – sufficient to charge a smartphone up to five times. The device has two integrated cables: a standard USB cable for charging the power bank and a micro-USB cable for charging mobile devices. In addition it has a standard USB interface which means that it is truly compatible with all mobile devices.

Statement by the jury
This external rechargeable battery is compact and of high design quality and impresses with its high storage capacity, making it a trustworthy companion.

Die Xtorm Power Bank 7300 ist aus gebürstetem Aluminium gefertigt und hat eine interne Energiespeicherkapazität von 7300 mAh – ausreichend, um ein Smartphone bis zu fünfmal aufzuladen. Das Gerät hat zwei integrierte Kabel: ein Standard-USB-Kabel zum Aufladen der Power Bank und ein Micro-USB-Kabel für das Laden mobiler Geräte. Sie verfügt außerdem über eine Standard-USB-Schnittstelle, sodass sie auch wirklich mit allen mobilen Geräten kompatibel ist.

Begründung der Jury
Dieser externe Akku ist kompakt und hochwertig gestaltet und beeindruckt mit seiner hohen Speicherkapazität, die ihn zu einem verlässlichen Begleiter macht.

Power Solution
Digital Charger
Digitales Ladegerät

Manufacturer
Lifetrons Switzerland AG,
Dicken, Switzerland
In-house design
Alynn Lutz-Ramoie
Web
www.lifetrons.com

This charger with its elegant metallic finish has a capacity of 6000 mAh and by means of a sophisticated power-saving function, provides many additional hours of battery power. A patent LED digital display indicates how much energy is still available. The Power Solution Digital Charger has two USB ports, allowing two devices to be charged simultaneously, five integrated safety levels and an automatic shut-off function.

Statement by the jury
This upmarket charger delights with its energy efficiency which can be followed at any time due to its clear LED digital display.

Dieses Ladegerät mit elegantem Metallic-Finish hat eine Kapazität von 6000 mAh und bietet durch eine raffinierte Stromspar-Funktion viele Stunden zusätzliche Akkulaufzeit. Eine patentierte LED-Digitalanzeige zeigt an, wie viel Energie noch verfügbar ist. Der Power Solution Digital Charger hat zwei USB-Anschlüsse, an denen sich zwei Geräte gleichzeitig laden lassen, fünf eingebaute Sicherheitsebenen und eine automatische Abschaltfunktion.

Begründung der Jury
Dieses hochwertige Ladegerät begeistert mit seiner Energieeffizienz, die anhand der klaren LED-Digitalanzeige jederzeit nachvollzogen werden kann.

U-STONE
Power Bank Series
Mobiler Akku

Manufacturer
Shenzhen LEPOW
Creative Technology Co., Ltd.,
Shenzhen, China
In-house design
Gang Liu
Web
www.lepow.hk

The U-Stone is a high-performance battery which catches the eye due to a seamless, U-form design which reminds one of the form of a classical key case. The battery displays the energy status by means of LEDs which are activated by shaking the battery. The uncomplicated retro appearance of the U-Stone is emphasised by the matt surface with its natural haptics. The USB interface is concealed on a cable in the battery opening.

Der U-Stone ist ein Hochleistungs-Akku, der durch eine nahtlose U-förmige Gestaltung auffällt, die an die Formgebung eines klassischen Schlüsseletuis erinnert. Der Akku zeigt seinen Energiezustand mithilfe von Leuchtdioden an, die durch Schütteln des Akkus aktiviert werden. Die schlichte Retro-Anmutung des U-Stone wird durch die mattierte Oberfläche mit ihrer natürlichen Haptik betont. Die USB-Schnittstelle verbirgt sich an einem Kabel in der Akku-Öffnung.

Qimini Pocket
Wireless Charging Plate
Kabellose Ladestation

Manufacturer
Tektos Limited,
Hong Kong
In-house design
François Hurtaud
Web
www.tektosworld.com

Qimini is a wireless charging station which is compatible with all Qi-enabled devices. It starts to charge as soon as a device is placed on it. With a diameter of 9 mm it is extremely thin and convenient. All components have been designed so that they fit in the compact housing at maximal power. Qimini can be charged via an integrated USB cable using any USB power source. The cable disappears into the housing when not in use.

Qimini ist eine kabellose Ladestation, die mit allen Qi-fähigen Geräten kompatibel ist. Der Ladevorgang startet, sobald ein Gerät aufgelegt wird. Mit einem Durchmesser von 9 mm ist sie extrem flach und handlich. Alle Komponenten wurden so gestaltet, dass sie bei maximaler Leistung in das kompakte Gehäuse passen. Qimini kann über ein integriertes USB-Kabel mit jeder USB-Stromquelle geladen werden. Nicht genutzt verschwindet das Kabel im Gehäuse.

Statement by the jury
Qimini makes an uncomplicated impression which is confirmed in use – a reliable, pleasant charging station in pocket format.

Begründung der Jury
Qimini macht einen unkomplizierten Eindruck, der sich im Gebrauch bestätigt – eine verlässliche, sympathische Ladestation im Hosentaschenformat.

WCP-400
Wireless Charging Orb
for Nexus 4
Kabellose Ladestation
für Nexus 4

Manufacturer
LG Electronics Inc.,
Seoul, South Korea
In-house design
Sea-La Park,
Cheol-Woong Shin,
Hyun-Woo Yoo,
Byung-Hyun Yi
Web
www.lg.com

The Charging Orb was specially developed to provide wireless charging of the Google smartphone Nexus 4; it is, however, also compatible with other Qi-enabled devices. The form is that of an angular semi-sphere where the smartphone lies, so that it can easily be used while it charges. Charging Orb is well engineered and available in matt black or high gloss.

Statement by the jury
A geometric form language, matt-high-gloss contrast and high functionality combine in Charging Orb to produce an elegant charging station.

Die Ladestation Charging Orb wurde speziell im Hinblick auf das kabellose Laden des Google-Smartphones Nexus 4 entwickelt, ist jedoch auch mit anderen Qi-fähigen Geräten kompatibel. Sie hat die Form einer abgeschrägten Halbkugel, auf der das Smartphone so liegt, dass es auch während des Ladens bequem benutzt werden kann. Charging Orb ist hochwertig verarbeitet und in Mattschwarz oder hochglänzend erhältlich.

Begründung der Jury
Eine geometrische Formensprache, Matt-Hochglanz-Kontraste und eine hohe Funktionalität vereinen sich in Charging Orb zu einer eleganten Ladestation.

WCP-300
Wireless Charging Cradle
Kabellose Ladestation

Manufacturer
LG Electronics Inc.,
Seoul, South Korea
In-house design
Bo-Ra Choi,
Seung-Beom Park
Web
www.lg.com

This Qi compatible charging station makes it possible to charge Qi-enabled devices by induction simply by putting it in place. The housing, with a diameter of a little less than 7 cm is extremely compact, round and very thin and can simply be carried in any pocket. The charging station is monochrome, its form language minimalist and clear. The charging station thus appears timeless and fits into any environment.

Statement by the jury
This charging station with its reduced, geometric form language and the colouring in black or white has very classical aesthetics which contribute to its long life.

Dieses Qi-kompatible Ladegerät ermöglicht das Laden von Qi-fähigen Geräten mittels Induktion durch einfaches Auflegen. Das mit einem Durchmesser von nur knapp 7 cm extrem kompakte und sehr flache runde Gehäuse lässt sich einfach in jeder Tasche transportieren. Die Ladestation ist einfarbig, ihre Formensprache minimalistisch und klar. So wirkt das Ladegerät zeitlos und fügt sich in jede Umgebung ein.

Begründung der Jury
Diese Ladestation hat mit ihrer reduzierten geometrischen Formensprache und der Farbgebung in Schwarz oder Weiß eine sehr klassische Ästhetik, die zu ihrer Langlebigkeit beiträgt.

WCD-800
Wireless Charging Cradle
Kabellose Ladestation

Manufacturer
LG Electronics Inc.,
Seoul, South Korea
In-house design
Seung-Hwan Song,
Seung-Bum Park
Web
www.lg.com

The WCD-800 is a charging cradle with "Free Position" technology which allows it to charge a mobile telephone, whether it is lying in the cradle vertically or standing horizontally. The contact surface is softly coated; the charging area is double that of previous models, so that the cradle can be used for various mobile phones. The purist design with metal-black contrast indicates a high quality rating.

WCD-800 ist eine Ladeschale mit „Free Position"-Technologie, die es erlaubt, ein Mobiltelefon zu laden, egal ob es horizontal oder vertikal in der Ladeschale liegt. Die Kontaktfläche ist weich beschichtet, der Ladebereich wurde im Vergleich zu früheren Modellen verdoppelt, sodass die Schale mit verschiedenen Mobiltelefonen verwendet werden kann. Die puristische Gestaltung mit Metall-Schwarz-Kontrast kommuniziert eine hohe Wertigkeit.

Statement by the jury
The WCD-800 charging cradle provides a high degree of functionality with well-considered details.

Begründung der Jury
Die Ladeschale WCD-800 bietet eine hohe Funktionalität und flexible Nutzung mit durchdachten Details.

Topp Wall Charger SPAC03
Wall Charger
Wand-Ladegerät

Manufacturer
MiPow Ltd.,
Shenzhen, China
In-house design
Stanley Wai Yung Yeung
Web
www.mipow.com
Honourable Mention

The Topp Wall Charger SPAC03 is a USB port adaptor with a retractable micro-USB charging cable with which smartphones and other micro-USB devices can be charged. Via an additional USB port, for example, Apple devices can be charged using their own cable. The Wall Charger has a foldable plug, making it the ideal travel companion. An automatic shut-down function protects it from overcharge.

Statement by the jury
The Topp Wall Charger SPAC03 is a compact, handy charger which is convincing due to its clear form language and high degree of functionality.

Der Topp Wall Charger SPAC03 ist ein USB-Steckdosenadapter mit einem ausziehbaren Micro-USB-Ladekabel, mit dem Smartphones sowie andere Micro-USB-Geräte aufgeladen werden können. Über die zusätzliche USB-Buchse können beispielsweise auch Apple-Geräte mit ihrem eigenen Ladekabel aufgeladen werden. Der Wall Charger hat einklappbare Stecker, was ihn zum idealen Reisebegleiter macht. Eine automatische Abschaltfunktion schützt vor Überladung.

Begründung der Jury
Der Topp Wall Charger SPAC03 ist ein kompaktes, handliches Ladegerät, das mit einer klaren Formensprache und hoher Funktionalität überzeugt.

Car Charger CCU2000
Retractable Car Charger
Ladegerät fürs Auto

Manufacturer
Unplug Ltd,
Hong Kong
Design
Christoph Behling Design
(Christoph Behling),
London, GB
Web
www.unplug.com.hk
www.christophbehlingdesign.com

CCU2000 is a charger for the car which allows two mobile devices to be charged at the same time almost twice as quickly as comparable products. Integrated sensors cut off the charging process automatically if the heating level gets too high. The charging elements have been developed with high energy efficiency in mind and have automatic fuses. CCU2000 is available with Apple or Micro-USB charger cable and features a retractable cable.

Statement by the jury
The charger CCU2000 has a clear form language and gives a robust effect; at the same time it is technically sophisticated and has a high degree of plausibility.

CCU2000 ist ein Ladegerät für das Auto, an dem zwei mobile Geräte gleichzeitig fast doppelt so schnell geladen werden können wie bei vergleichbaren Produkten. Eingebaute Sensoren beenden den Ladeprozess automatisch, wenn die Hitzeentwicklung zu hoch wird. Die Ladeelemente wurden mit Blick auf hohe Energieeffizienz entwickelt und haben automatische Sicherungen. CCU2000 ist mit Apple- oder Micro-USB-Ladekabeln erhältlich und hat eine Kabelaufrollautomatik.

Begründung der Jury
Das Ladegerät CCU2000 hat eine klare Formensprache und wirkt robust, zugleich ist es technisch ausgereift und hat eine hohe Selbsterklärungsqualität.

Gigaset E630/E630A
Cordless Phone
Schnurlostelefon

Manufacturer
Gigaset Communications GmbH,
Munich, Germany
In-house design
Hans-Henning Brabänder,
Peter Kolin
Design
platinumdesign
(Andreas Dimitriadis, Pablo Bernal),
Stuttgart, Germany
Web
www.gigaset.com
www.platinumdesign.com

The Gigaset E630, thanks to a coherent material concept which combines structured polycarbonate with soft components which lie securely in the hand, gives an upmarket and functional effect. This functionality is also shown in its high grade of ergonomics in operation and its insensitivity to dust, shock and water. Incoming calls are indicated by an LED which also acts as a torch. The E630 is also available with integrated answering service (E630A).

Das Gigaset E630 wirkt dank eines schlüssigen Materialkonzepts, das strukturiertes Polycarbonat mit griffigen Weichkomponenten zusammenbringt, hochwertig und funktionell. Diese Funktionalität zeigt sich auch in seiner hohen Bedienergonomie und seiner Unempfindlichkeit gegen Staub, Erschütterungen und Wasser. Eine LED signalisiert eingehende Anrufe und dient als Taschenlampe. Das E630 ist auch mit integriertem Anrufbeantworter (E630A) verfügbar.

Statement by the jury
The Gigaset E630 convinces not least because of its rational material combination, which means the telephone lies securely in the hand and is robust.

Begründung der Jury
Das Gigaset E630 überzeugt nicht zuletzt durch eine sinnvolle Materialkombination, die dazu führt, dass das Telefon besonders griffig und robust ist.

M550 Mira
DECT Phone
DECT-Telefon

Manufacturer
Royal Philips Electronics,
Eindhoven, Netherlands
In-house design
Philips Design Consumer Lifestyle Team
Web
www.philips.com

The Mira cordless telephone is easy to operate, thanks to its special form and configuration. The softly curved handset can be placed on the base station in both directions. A modern sound monitor provides excellent voice quality; the ECO+Modus assures low energy consumption. The Mira styling is iconic and of minimalist design, allowing it to fit into any environment. It is available in white and dark grey.

Das Schnurlostelefon Mira ist dank seiner besonderen Form und Konfiguration einfach zu bedienen. Das sanft geschwungene Mobilteil kann in beide Richtungen auf der Basisstation platziert werden. Eine moderne Klangüberprüfung sorgt für hervorragende Sprachqualität, der ECO+Modus für niedrigen Stromverbrauch. Mira fügt sich mit seiner ikonenhaften, minimalistischen Gestaltung in jede Umgebung ein und ist in Weiß und Dunkelgrau verfügbar.

Statement by the jury
The Mira convinces with a fresh, innovative form language for a telephone, making it a true eye-catcher.

Begründung der Jury
Das Mira überzeugt mit einer für ein Telefon frischen, innovativen Formensprache, die es zu einem echten Blickfang macht.

DCN-D

Discussion Unit
Diskussionseinheit

Manufacturer
Bosch Sicherheitssysteme GmbH,
Grasbrunn, Germany
Design
Teams Design Consulting
(Shanghai) Co., Ltd. (An Luo),
Shanghai, China
Web
www.boschsecurity.com
www.teamsdesign.com

The DCN-D, with its rounded form and the combination of timeless silver and matt black, integrates easily in every conference room. Each discussion unit has an integrated loudspeaker, microphone and headset socket and can be configured as chairman unit. Furthermore, the unit offers voting facilities and interpreter channels for up to 31 languages. Thanks to its ergonomic keys it is simple to operate.

Das DCN-D integriert sich mit seiner geschwungenen Form und der Kombination von zeitlosem Silber und mattem Schwarz leicht in jeden Konferenzraum. Jede Diskussionseinheit hat einen integrierten Lautsprecher, Mikrofon und Kopfhöreranschluss und lässt sich als Vorsitzenden-Einheit konfigurieren. Zudem ermöglicht die Einheit elektronische Abstimmungen und das Dolmetschen in bis zu 31 Sprachen. Dank großer ergonomischer Tasten ist sie einfach zu bedienen.

SX20 Quick Set
Video Conferencing System
Videokonferenz-System

Manufacturer
Cisco Systems,
Lysaker, Norway
In-house design
Glenn Robert Grimsrud Aarrestad,
Knut Helge Teppan
Design
Designit (Bjørn Saunes),
Oslo, Norway
Web
www.cisco.com
www.designit.com

The SX20 Quick Set makes the conversion of a conference room into a video conference room easy. All components of this system have been developed bearing in mind wall mounting, together with flat screens and designed in a way that the system integrates well into various interiors. The system is easily installed and operated; it contains a wide-angle camera and high-end audiovisual components.

Statement by the jury
This video conferencing system convinces with its purist design by which it integrates discreetly and harmoniously in every environment.

Das SX20 Quick Set ermöglicht auf einfache Weise die Umrüstung eines Tagungsraums zu einem Videokonferenzraum. Alle Bestandteile dieses Systems wurden im Hinblick auf eine Wandmontage zusammen mit Flachbildschirmen entwickelt und so gestaltet, dass sie sich gut in verschiedene Interieurs integrieren. Das Set kann leicht installiert und bedient werden, verfügt über eine Weitwinkelkamera und hochwertige audiovisuelle Komponenten.

Begründung der Jury
Dieses Videokonferenz-System überzeugt mit einer puristischen Gestaltung, durch die es sich diskret und harmonisch in jedes Umfeld integriert.

PT-3211
Smart IP Camera
IP-Kamera

Manufacturer
Alpha Networks Inc.,
Hsinchu, Taiwan
In-house design
Pei-Li Hu
Web
www.alphanetworks.com

The PT-3211 is a surveillance device which is controlled via apps and provides a high degree of security. The PT-3211 features a two megapixel camera with 10x optical zoom lens which supports the H.264 format. It also has an intercom and a night vision function, it can monitor room temperature and dampness or start automatically recording and send messages as soon as noises or motion are detected.

Statement by the jury
PT-3211 seems to stick out its neck to be able to see everything. That makes this functional camera likable, at the same time presenting a realistic form and colouring.

Die PT-3211 ist ein Überwachungsgerät, das sich einfach via App steuern lässt und hohe Sicherheit bietet. Die PT-3211 ist eine Zwei-Megapixel-Kamera mit zehnfachem optischem Zoom, die das H.264-Format unterstützt. Sie hat außerdem eine Gegensprech- und eine Nachtsichtfunktion, kann Raumtemperatur und -feuchtigkeit überwachen oder automatisch Aufzeichnungen starten und Benachrichtigungen versenden, sobald Geräusche oder Bewegungen erkannt werden.

Begründung der Jury
PT-3211 scheint den Hals zu recken, um alles sehen zu können. Das macht diese hoch funktionelle Kamera sympathisch bei gleichzeitig sachlicher Form- und Farbgebung.

Jabra Speak 510
Mobile Speakerphone for UC Conference Calls
Mobile Freisprechlösung für UC-Konferenzen

Manufacturer
GN Netcom A/S,
Ballerup, Denmark
Design
Klaus Rath Design
(Klaus Rath),
Copenhagen, Denmark
Web
www.jabra.com
www.rathdesign.dk

Jabra Speak 510 is a compact speakerphone for conference discussions, designed with special relevance to the sound quality and simple operation. The slim device houses a Bluetooth supported plug-and-play solution and can be connected to various terminal devices. Microphone and speaker are concealed under the cover which contains punched holes. The operating buttons at the outer ring are operated intuitively.

Statement by the jury
Jabra Speak 510 is a speakerphone which, due to its compact, round form appears to involve all conference members in the discussion with equal rights.

Jabra Speak 510 ist eine kompakte Freisprecheinrichtung für Konferenzgespräche, die mit Blick auf die Klangqualität und eine einfache Handhabung gestaltet wurde. Das schlanke Gerät beherbergt eine Bluetooth-fähige Plug-and-Play-Lösung und lässt sich mit verschiedenen Endgeräten verbinden. Mikrofon und Lautsprecher verbergen sich unter der mit gestanzten Löchern versehenen Abdeckung. Die Funktionstasten am Außenring sind intuitiv zu bedienen.

Begründung der Jury
Jabra Speak 510 ist eine Freisprechlösung, die mit ihrer kompakten runden Form alle Konferenzteilnehmer gleichberechtigt mit in das Gespräch einzubeziehen scheint.

xqPRO
Bluetooth Speaker and Hands-Free Set
Bluetooth-Lautsprecher und -Freisprecheinrichtung

Manufacturer
Strax GmbH,
Troisdorf, Germany
In-house design
Timo Treudt
Design
iui design
(Michael Tse),
Hong Kong
Web
www.strax.com
www.xqisit.com
www.iuidesign.com

xqPRO makes it possible to listen to music and make calls wirelessly via hands-free technology. The high-gloss device has a dynamic form language and high-grade technical attributes. By means of APTX and AAC Bluetooth protocols it provides a high rate and good quality transmission. Two speakers and a passive subwoofer provide good audio quality. Operation is by means of touch panel of the xqPRO or the smartphone.

xqPRO ermöglicht kabelloses Musikhören und Telefonieren per Freisprechanlage. Das hochglänzende Gerät hat eine dynamische Formensprache und eine hochwertige technische Ausstattung. Mithilfe von APTX- und AAC-Bluetooth-Protokollen liefert es eine hohe Übertragungsgeschwindigkeit und -qualität. Zwei Lautsprecher und ein passiver Subwoofer sorgen für einen guten Klang. Die Bedienung erfolgt über die Touch-Oberfläche des xqPRO oder das Smartphone.

BT-V37
Bluetooth Headset

Manufacturer
emporia Telecom,
Linz, Austria
Design
Mango Design
(Markus Anlauff),
Braunschweig, Germany
Web
www.emporia-zubehoer.at
www.mango-design.de

The Bluetooth headset BT-V37 combines a high degree of carrying comfort with optimal functionality and is compatible with most mobile telephones which support Bluetooth version V2.1 + EDR. The headset, which fits in the ear, gives a plain-elegant impression with its black body and aluminium elements. The associated charger is designed accordingly and due to its suction cup can also be used in a vehicle.

Statement by the jury
This Bluetooth headset communicates its good technical properties at first glance due to the high quality of its design.

Das Bluetooth Headset BT-V37 verbindet einen hohen Tragekomfort mit optimaler Funktionalität und ist mit den meisten Mobiltelefonen, die Bluetooth-Version V2.1 + EDR unterstützen, kompatibel. Das Headset, das ins Ohr gestöpselt wird, wirkt mit seinem schwarzen Korpus und den Aluminiumelementen schlicht-elegant. Der dazugehörige Tischlader ist entsprechend gestaltet und lässt sich dank seines Saug-fußes auch im Fahrzeug verwenden.

Begründung der Jury
Dieses Bluetooth-Headset kommuniziert seine guten technischen Eigenschaften bereits auf den ersten Blick durch seine hochwertige Gestaltung.

Smart Wireless Headset pro
Bluetooth Headset

Manufacturer
Sony Mobile Communications
International AB, Lund, Sweden
In-house design
Sony Mobile UX Creative Design Team
Web
www.sonymobile.com

This elegant Bluetooth headset has an integrated MP3-Player including FM radio und superior audio properties. It can be simply clipped onto the clothing and is comfortable to wear. The surface of the plain, black-silver device comes to life as soon as an SMS or a call comes in and it gives information via display about the sender or caller, message content or missed calls.

Statement by the jury
This plainly elegant Bluetooth headset combines a high degree of user friendliness with exceptional sound properties and suitably matches every style of clothing.

Dieses elegante Bluetooth-Headset hat einen integrierten MP3-Player samt FM-Radio und verfügt über sehr gute Klangeigenschaften. Es kann einfach an die Kleidung geklemmt werden und ist komfortabel zu tragen. Die Oberfläche des schlichten schwarz-silbernen Geräts erwacht zum Leben, sobald eine Kurz-nachricht oder ein Anruf eingeht, und informiert mithilfe eines kleinen Displays über Absender, Nachrichteninhalt oder verpasste Anrufe.

Begründung der Jury
Dieses schlicht-elegante Bluetooth-Head-set vereint eine hohe Benutzerfreundlich-keit mit hervorragenden Klangeigenschaf-ten und passt zu jedem Kleidungsstil.

Vox Tube 700
Bluetooth Headset

Manufacturer
MiPow Ltd.,
Shenzhen, China
In-house design
Stanley Wai Yung Yeung
Web
www.mipow.com

Vox Tube 700 is a Bluetooth-conform and energy-saving mono headset whose elegant case of anodised aluminium is not only a headset but also a USB plug. The Vox Tube 700 can thereby be fitted into any USB port for charging without the use of an interface cable. Noise suppression of the headset is achieved via a built-in microphone and CVC technology and provides best voice quality, even excluding loud background noise. Vox Tube 700 is comfortably operated by one-key control.

Vox Tube 700 ist ein Bluetooth-konformes und energiesparendes Mono-Headset, dessen eleganter Korpus aus eloxiertem Aluminium nicht nur Headset, sondern auch USB-Stecker ist. Dadurch kann Vox Tube 700 ohne Verwendung von Adapterkabeln zum Laden in jede USB-Buchse gesteckt werden. Die Geräuschunterdrückung des Headsets wird mittels intern verbautem Mikrofon und CVC-Technologie durchgeführt und sorgt auch bei lauten Hintergrundgeräuschen für beste Sprachqualität. Vox Tube 700 lässt sich über einen einzigen Knopf komfortabel bedienen.

Statement by the jury
Vox Tube 700 surprises by its design in the form of a USB memory stick, an unusual design for a headset. This design makes it not only an eye-catcher but leads also to an intelligent, extended functionality.

Begründung der Jury
Vox Tube 700 verblüfft durch seine für ein Headset ungewohnte Gestaltung in Form eines USB-Sticks. Diese Gestaltung macht es nicht nur zum Blickfang, sondern führt auch zu einer intelligenten erweiterten Funktionalität.

Vox Tube 500
Bluetooth Headset

Manufacturer
MiPow Ltd.,
Shenzhen, China
In-house design
Stanley Wai Yung Yeung
Web
www.mipow.com

Vox Tube 500 is an energy-saving Bluetooth mono headset. The surface of the cylindrical headset is of anodised aluminium and is coated with piano lacquer in various colours. By means of integrated A2DP technology the headset can be used with the hands-free system as well as the classical headset. It supports simultaneous connection with two Bluetooth devices and informs of the present connection status via voice prompting.

Vox Tube 500 ist ein energiesparendes Bluetooth Mono-Headset. Die Oberfläche des zylindrischen Headsets ist aus eloxiertem Aluminium, die mit Klavierlack in verschiedenen Farben überzogen wurde. Durch die integrierte A2DP-Technologie lässt sich das Headset sowohl als Freisprecheinrichtung als auch als klassischer Kopfhörer nutzen. Es unterstützt die gleichzeitige Verbindung mit zwei Bluetooth-Endgeräten und informiert mit Sprachansagen über den aktuellen Verbindungsstatus.

Statement by the jury
The Vox Tube 500, with its charming cylindrical and upmarket finished aluminium case, communicates not only the high technical demands of the user but also his style feeling.

Begründung der Jury
Mit seinem anmutiger zylindrischen und hochwertig veredelter Aluminiumkörper kommuniziert Vox Tube 500 nicht nur der hohen technischen Anspruch seines Trägers, sondern auch dessen Stilgefühl.

Century SC 660/630
Headset

Manufacturer
Sennheiser Communications A/S,
Solrød Strand, Denmark
Design
Brandis Industrial Design,
Nuremberg, Germany
Web
www.senncomm.com
www.brandis-design.com

This modular designed, elegant headset system has been created for professional use in the office or call centre, where best audio quality is required, despite loud environmental noise. Because of its ergonomic design, the light construction and the large ear padding it offers a high degree of comfort when worn, even for long periods. An optimal, magnetic remote cable operation facilitates direct control of digital communication systems.

Dieses modular angelegte, elegante Headset-System ist für den professionellen Einsatz im Büro oder Call-Center konzipiert, wo beste Klangqualität trotz lauter Umgebung erforderlich ist. Durch seine ergonomische Gestaltung, die Leichtbauweise und die großen Ohrpolster bietet es auch bei Dauernutzung hohen Tragekomfort. Eine optionale, magnetische Kabelfernbedienung ermöglicht die direkte Steuerung digitaler Kommunikationssysteme.

Statement by the jury
The simple and elegant design of this headset is well-conceived and from every viewpoint is oriented towards comfortable, professional applications.

Begründung der Jury
Die schlichte und elegante Gestaltung dieses Headsets ist durchdacht und in jeder Hinsicht auf einen komfortablen professionellen Gebrauch ausgerichtet.

HBS 800
Bluetooth Headset

Manufacturer
LG Electronics Inc.,
Seoul, South Korea
In-house design
Seung-Hee Ha,
Hong-Kyu Park,
Hyun Lee
Web
www.lg.com

The Bluetooth headset HBS 800 combines a light, slim design with ease of operation. An innovative material in the neckband enables not only easy adjustment but also provides a pleasant feeling when carrying. The earphones are easily stored in the main unit by means of magnets. The weight of the main unit is supported by the shoulders; buttons which are frequently used are located in reach of the fingers.

Statement by the jury
HBS 800 impresses with its well-considered details – for instance by an innovative material use for the neckband, easing the burden for the carrier.

Das Bluetooth-Headset HBS 800 verbindet eine leichte, schlanke Gestaltung mit einfacher Bedienbarkeit. Ein innovatives Material im Halsband erleichtert nicht nur die Justierung, sondern sorgt auch für ein angenehmes Tragegefühl. Mithilfe von Magneten lassen sich die Ohrstöpsel einfach in der Haupteinheit verstauen. Das Gewicht der Haupteinheit wird von den Schultern unterstützt, häufig verwendete Tasten sind in Reichweite der Finger platziert.

Begründung der Jury
HBS 800 beeindruckt mit durchdachten Details – so etwa durch einen innovativen Materialeinsatz beim Halsband, der die Anpassung an den Träger erleichtert.

SC 30/60 USB CTRL
UC Stereo Headsets

Manufacturer
Sennheiser Communications A/S,
Solrød Strand, Denmark
Design
Brennwald Design
(Jörg Brennwald, Chang-Chin Hwang),
Kiel, Germany
Web
www.senncomm.com
www.brennwald-design.de

This headset series combines wearing comfort with very good acoustics due to its elegant and, at the same time, sturdy design which concentrates on long life and functionality. Thanks to their ergonomic quality, the headsets can be individually matched and assure fatigue-free work. The headset is provided with one or two earphones and was developed specially for professional use in UC environments.

Statement by the jury
The headset series SC 30/60 USB CTRL convinces with its classical design, good ergonomics and technically sophisticated functionality.

Diese Headset-Serie verbindet Tragekomfort und sehr gute akustische Eigenschaften durch eine elegante und zugleich robuste Gestaltung, die auf Langlebigkeit und Funktionalität ausgerichtet ist. Dank ihrer ergonomischen Qualität lassen sich die Headsets individuell anpassen und garantieren ermüdungsfreies Arbeiten. Das Headset ist mit einem oder zwei Kopfhörem erhältlich und speziell für den professionellen Einsatz in UC-Umgebungen entwickelt.

Begründung der Jury
Die Headset-Serie SC 30/60 USB CTRL überzeugt mit ihrer klassischen Gestaltung, guter Ergonomie und technisch ausgereifter Funktionalität.

WS880
Home Gateway

Manufacturer
Huawei Device Co., Ltd.,
Shenzhen, China
In-house design
Jing Wu,
Jianbo Zou,
Yang Yingzhi
Web
www.huaweidevice.com

WS880 is a wireless router with a band width of up to 1500 Mbit/s. Its design is minimalist, the surface plain and concise. A metal circle reminds one of a rising sun and is an expression of the design philosophy "Sunrise". An LED display with dot matrix gives the device a highly technical impression. When the button in the centre of the metal ring is pressed, it switches on the device and the Huawei logo lights up.

WS880 ist ein kabelloser Router mit einer Bandbreite von bis zu 1500 Mbit/s. Seine Gestaltung ist minimalistisch, die Oberfläche schlicht und prägnant. Ein Metallkreis erinnert an eine aufgehende Sonne und ist Ausdruck der Designphilosophie „Sunrise". Ein LED-Display mit Punktmatrix verleiht dem Gerät eine hochtechnologische Anmutung. Wird der Knopf im Inneren des Metallrings gedrückt, schaltet sich das Gerät ein und das Huawei-Logo leuchtet auf.

The WS880 manages in its form language to be simultaneously minimalist and concise, highly technical and emotional.

Begründung der Jury
Das WS880 schafft es, in seiner Formersprache gleichzeitig minimalistisch und prägnant, hochtechnologisch und emotional zu sein.

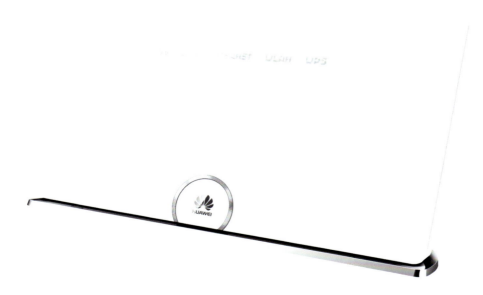

IGW 3000
Intelligent Home Gateway

Manufacturer
Icotera A/S,
Odense, Denmark
Design
Designit,
Copenhagen, Denmark
Web
www.icotera.com
www.designit.com

The home gateway IGW 3000 is characterised by a clear and unobtrusive design which allows it to merge seamlessly into the environment: all features are integrated in the housing, the symbols are discreet and the gateway does not as much as blink unless it needs attention. Installation is quick and simple; zero-touch management and remote error recognition contribute to low operating costs.

Statement by the jury
In the case of the IGW 3000, technology and a well-considered, reduced form language combine to produce a discreet, user-friendly device.

Das Home Gateway IGW 3000 ist durch eine klare und unaufdringliche Gestaltung gekennzeichnet, durch die es sich nahtlos in jede Umgebung einfügt: Sämtliche Funktionsteile sind in das Gehäuse integriert, die Symbole sind dezent und das Gateway blinkt nur, wenn es Aufmerksamkeit benötigt. Der Installationsprozess ist schnell und einfach, Zero-Touch-Management und eine Fehler-Fernerkennung tragen zu niedrigen Gesamtbetriebskosten bei.

Begründung der Jury
Beim IGW 3000 verbinden sich Technik und eine durchdachte, reduzierte Formensprache zu einem dezenten, benutzerfreundlichen Gerät.

b-box 3
Home Gateway

Manufacturer
Belgacom,
Brussels, Belgium
Design
Achilles Design,
Mechelen, Belgium
Web
www.belgacom.be
www.achilles.be

The b-box 3 home gateway can be used horizontally, vertically and by means of wall mounting. This is possible due to ventilating slits on all sides, assuring passive cooling of the internal components. The gateway has been designed in a way that it can be upgraded by interlocking additional units as new technology comes on the market. This is reflected also in its reduced form language, oriented to long life.

Statement by the jury
The b-box 3 convinces with a design which is totally geared to long life, functionality and flexibility.

Das Home Gateway b-box 3 kann horizontal, vertikal und mittels Wandbefestigung benutzt werden. Ermöglicht wird dies durch Belüftungsschlitze an allen Seiten, die eine ausreichende passive Kühlung der Innenkomponenten garantieren. Das Gateway wurde so konzipiert, dass es aufgerüstet werden kann, wenn neue Technologien auf den Markt kommen. Dies spiegelt sich auch in seiner reduzierten, auf Langlebigkeit ausgerichteten Formensprache wider.

Begründung der Jury
Das b-box 3 überzeugt mit einer Gestaltung, die ganz auf Langlebigkeit, Funktionalität und Flexibilität ausgerichtet ist.

DMG-304P
HomePlugAV PowerLine
Wi-Fi Network Adapter
Wi-Fi Netzwerkadapter

Manufacturer
D-Link Corporation,
Taipei, Taiwan
In-house design
Percent Chuang
Web
www.dlink.com

The compact, wireless network adapter DMG-304P combines special Powerline technology, Wireless-N with high band width and Gigabit Ethernet for the corresponding fastest possible network connectivity – independent of obstacles and interference. The intelligent software of the DMG-304P automatically recognises the best connection type and by means of wireless connection also the channel for the corresponding network connected device. The adaptor thus provides wall-to-wall connectivity.

Statement by the jury
This Wi-Fi network adapter provides a technically as well as elegantly designed and compact solution for network problems.

Der kompakte, kabellose Netzwerkadapter DMG-304P verbindet spezielle Powerline-Technik, Wireless-N mit hoher Bandbreite und Gigabit Ethernet für die jeweils schnellstmögliche Netzwerkverbindung – unabhängig von Hindernissen und Störungen. Die intelligente Software des DMG-304P erkennt automatisch den besten Verbindungstyp und bei einer kabellosen Verbindung auch den Kanal für das jeweilige Netzwerkgerät. So bietet der Adapter Wall-to-Wall-Connectivity.

Begründung der Jury
Dieser Wi-Fi-Netzwerkadapter bietet eine technisch wie gestalterisch elegante und kompakte Lösung bei Netzwerkproblemen.

Speedport W 724V
Router

Manufacturer
Deutsche Telekom,
Bonn, Germany
In-house design
Web
www.telekom.com

The Speedport W 724V is a universal router with ADSL/VDSL modem, powerful WLAN and an IP based telephone system for speedphones. The Speedport is characterised by its high user-friendliness: all important display and operating elements are easily accessible and located clearly visible at the front. On the back, colour coded terminals facilitate intuitive connection for network and devices. A clear, geometric and at the same time flowing form language expresses simplicity and reliability.

Der Speedport W 724V ist ein Universal-Router mit ADSL/VDSL-Modem, leistungsfähigem WLAN und einer IP-basierten Telefonanlage für Speedphones. Der Speedport zeichnet sich durch eine hohe Benutzerfreundlichkeit aus: Alle wichtigen Anzeige- und Bedienelemente befinden sich leicht zugänglich und übersichtlich auf der Vorderseite. Auf der Rückseite ermöglichen farbcodierte Anschlüsse für Netzwerk und Geräte ein intuitives Anschließen. Eine klare, geometrische und gleichzeitig fließende Formensprache drückt Einfachheit und Zuverlässigkeit aus.

Statement by the jury
The Speedport W 724V has a clear and very elegant form language and because of its well-considered design, provides surprisingly simple operation.

Begründung der Jury
Der Speedport W 724V hat eine klare und sehr elegante Formensprache und ermöglicht durch seine durchdachte Gestaltung eine überraschend einfache Bedienung.

Manufacturer
Deutsche Telekom,
Bonn, Germany
In-house design
Web
www.telekom.com

The fibreglass modem is the interface for the fibreglass network of Deutsche Telekom. The housing combines fibreglass terminal box and modem and thus reduces the space requirement on the wall. It is plain and reserved in design and fits snugly and compactly on the wall. Thanks to a wall mounting plate and a swivel-type cover, the modem is easy and flexible for the technician to install. An LED indicates correct operation.

Das Glasfaser-Modem ist die Schnitt-stelle zum Glasfaser-Netz der Deutschen Telekom. Das Gehäuse vereint Glasfaser-Anschlussdose und -Modem und reduziert so den Platzbedarf an der Wand. Es ist schlicht und zurückhaltend gestaltet und schmiegt sich kompakt an die Wand an. Dank einer Wandmontageplatte und eines schwenkbaren Deckels ist das Modem vom Techniker einfach und flexibel zu installie-ren. Eine LED-Leuchte zeigt den korrekten Betrieb an.

Statement by the jury
This fibreglass modem convinces with a well-considered functionality which eases the technician's work with many helpful details.

Begründung der Jury
Dieses Glasfaser-Modem überzeugt mit einer durchdachten Funktionalität, welche die Arbeit der Techniker mit vielen hilfrei-chen Details erleichtert.

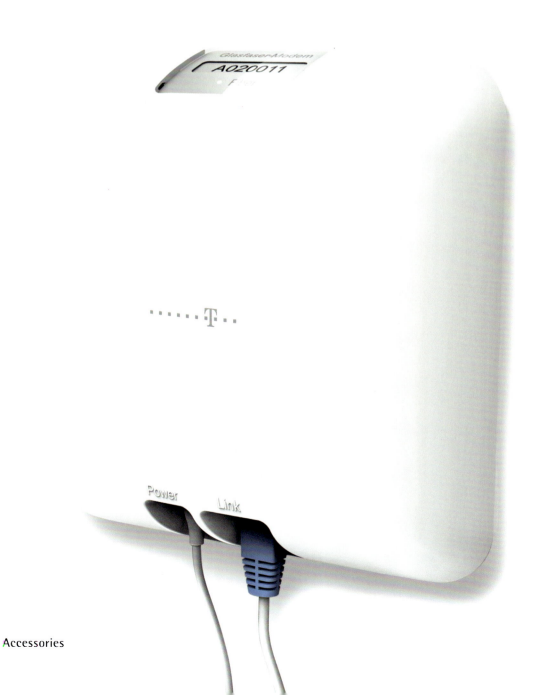

L-03E
High-Spec LTE Wi-Fi Router

Manufacturer
LG Electronics Inc.,
Seoul, South Korea
In-house design
Wataru Takahashi,
Song Seunghwan,
Tomoyuki Akutsu,
Hocheol Lee
Web
www.lg.com

L-03E is a high specification LTE Wi-Fi router with a design specially developed for the Japanese market. Thanks to a high capacity battery, the router provides many hours of operation in spite of its compact size. L-03E has a 1.8" colour LCD screen which provides all relevant information at a glance. A slide-type power switch is located on the side and replaces the power button in order to prevent accidental activation.

L-03E ist ein High-Spec LTE Wi-Fi Router mit einem speziell für den japanischen Markt entwickelten Design. Dank eines Hochleistungsakkus ermöglicht der Router trotz seiner kompakten Größe die Langzeitnutzung. L-03E hat einen farbigen 1,8"-Flüssigkristallbildschirm, der auf einen Blick alle relevanten Informationen liefert. Ein seitlich am Gehäuse platzierter Schiebeschalter ersetzt den Powerknopf, um eine versehentliche Aktivierung auszuschließen.

Statement by the jury
L-03E is a router which is optimally suitable for business people due to its high performance, its convenient form and its elegant form language.

Begründung der Jury
L-03E ist ein Router, der sich mit seiner hohen Leistung, seiner handlichen Form und seiner eleganten Formensprache optimal für Geschäftsleute eignet.

Internet Switch
Netzwerk-Schalter

Manufacturer
Huawei Device Co., Ltd.,
Shenzhen, China
In-house design
Qinna Chen,
Ding Feng
Web
www.huaweidevice.com
Honourable Mention

Internet Switch is an easily operated device for users. It converts 3G/4G signals into Wi-Fi signals and when switched on, it shares its data connection wirelessly with mobile telephones, laptops etc. It draws its energy from a battery or charger, thus it can be used anywhere in the house. Internet Switch is available in various colours and gives off a soft light due to its background illumination.

Internet Switch ist ein leicht zu bedienendes Gerät für den Endkunden, das 3G/4G-Signale in Wi-Fi-Signale umwandelt und seine Datenverbindung in eingeschaltetem Zustand drahtlos mit Mobiltelefonen, Laptops etc. teilt. Seine Energie bezieht es aus einer Batterie oder einem Ladegerät – so kann es überall im Haus genutzt werden. Internet Switch ist in verschiedenen Farben erhältlich und verbreitet dank seiner Hintergrundbeleuchtung ein sanftes Licht.

Statement by the jury
The Internet Switch is well-engineered and easy to use, at the same time it is also decorative due to its clear design and lighting.

Begründung der Jury
Der Internet Switch ist technisch ausgereift und leicht zu bedienen, mit seiner klaren Gestaltung und Beleuchtung gleichzeitig jedoch auch dekorativ.

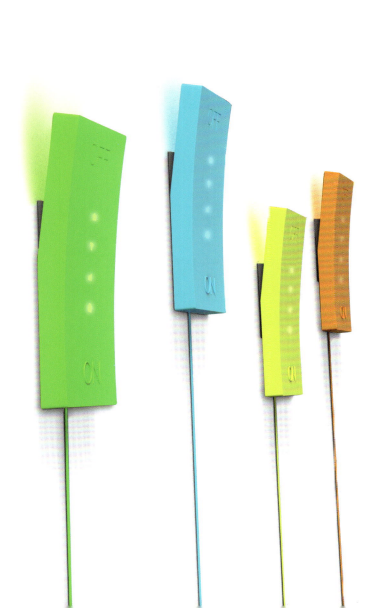

HW-02E
Mobile Wi-Fi Router
Tragbarer Wi-Fi-Router

Manufacturer
Huawei Device Co., Ltd
Shenzhen, China
In-house design
Mihoko Hotta
Web
www.huaweidevice.com

HW-02E is a 112.5 Mbps mobile high-speed communications Wi-Fi router which supports LTE cat4. With its softly rounded, compact form, it fits in every pocket. It is produced in the colours blue, white and chocolate, rather unusual colours for a router. Each colour results from a surface treatment that combines both gloss and matte. Up to ten devices can be connected to the router simultaneously.

Statement by the jury
With its compact housing and a clearly readable display, HW-02E provides a high degree of user-friendliness.

HW-02E ist ein tragbarer Wi-Fi-Router, der eine High Speed-Datenübertragung von bis zu 112,5 Mbps ermöglicht und LTE unterstützt. Mit seiner sanft geschwungenen, kompakten Form passt er in jede Tasche. Er wird in den für einen Router eher ungewöhnlichen Farbvarianten Blau, Weiß und Schokobraun produziert. Bis zu zehn Geräte können gleichzeitig mit dem Router verbunden werden.

Begründung der Jury
Mit seinem kompakten Gehäuse und einem übersichtlichen Display bietet HW-02E eine hohe Benutzerfreundlichkeit.

TL-TR861/M5350
Portable 3G WLAN Router
Tragbarer 3G-WLAN-Router

Manufacturer
TP-LINK Technologies Co., Ltd.,
Shenzhen, China
Design
Whipsaw Inc.
(Dan Harden, Elliot Ortiz),
San Jose, USA
Web
www.tp-link.com
www.whipsaw.com

TL-TR861/M5350 is a portable 3G WLAN router with a compact, very ergonomic housing. The 3G connection can be shared via Wi-Fi signal everywhere where UMTS is available. Thanks to a battery power of 2000 mAh the router operates with several Wi-Fi devices connected for up to six hours. The router is easily operated due to its OLED display and a one-touch connection.

Statement by the jury
This portable 3G WLAN router combines technical performance with organic forms, interesting exterior contrasts and ease of operation.

TL-TR861/M5350 ist ein tragbarer 3G-WLAN-Router mit einem kompakten, sehr ergonomischen Gehäuse. Die 3G-Verbindung kann via Wi-Fi-Signal überall dort für eine gemeinsame Nutzung bereitgestellt werden, wo UMTS verfügbar ist. Dank einer Akkuleistung von 2000 mAh arbeitet der Router mit mehreren gleichzeitig verbundenen Wi-Fi-Geräten bis zu sechs Stunden lang. Der Router ist mit seinem OLED-Display und einem One-Touch-Anschluss einfach zu bedienen.

Begründung der Jury
Dieser tragbare 3G-WLAN-Router vereint technische Leistung mit organischen Formen, interessanten Oberflächenkontrasten und einer guten Bedienbarkeit.

Basil (L-04D)
Mobile Wi-Fi Router
Tragbarer Wi-Fi-Router

Manufacturer
LG Electronics Inc.,
Seoul, South Korea
In-house design
Bo-Ra Choi,
Seung-Beom Park
Web
www.lg.com

This mini Wi-Fi router can be used simply with a micro-SIM card. The router weighs only 89 grams and is so compact in design, that it can be transported effortlessly and is thus ideal when on the move. The side lines and rounded forms at the front give the device a slim appearance and a good grip. The housing is obtainable in red and white; the part of the LCD display is given the colour black for each model.

Statement by the jury
Basil is easy to operate, having only three keys and an LED display and is appealing due to its slim, twin-tone design.

Dieser Mini-Wi-Fi-Router lässt sich mit einer MicroSIM-Karte einfach nutzen. Der Router wiegt nur 89 Gramm und ist so kompakt gestaltet, dass er sich mühelos transportieren lässt und damit ideal für unterwegs ist. Die Seitenlinien und abgerundeten Formen an der Vorderseite geben dem Gerät eine schlanke Erscheinung und eine gute Griffigkeit. Das Gehäuse wird in Rot oder Weiß angeboten, der Bereich der LCD-Anzeige ist jeweils in Schwarz gehalten.

Begründung der Jury
Basil ist mit nur drei Knöpfen und einem LED-Display einfach zu bedienen und begeistert mit seiner schlanken, zweifarbigen Gestaltung.

DWR-730
HSPA+ Mobile Router

Manufacturer
D-Link Corporation,
Taipei, Taiwan
In-house design
Julie Hsiau
Web
www.dlink.com

The DWR-730 converts an HSPA+-
Internet connection into a Wi-Fi hotspot, making it possible to share the
connection anywhere. For this one needs
only to insert your data-enabled SIM
card in the router and the connection
can be made available for computers
and wireless devices in the vicinity. The
DWR-730 is also equipped with a micro-
SD slot, allowing files to be shared.

Der DWR-730 verwandelt eine HSPA+-
Internetverbindung in einen Wi-Fi-
Hotspot, der eine gemeinsame Nutzung
der Verbindung an jedem Ort ermöglicht.
Hierfür muss lediglich eine SIM-Karte mit
Datentarif in den Router eingelegt werden,
und die Verbindung kann für Computer
und drahtlose Geräte in der Umgebung
freigegeben werden. Der DWR-730 ist
zudem mit einer microSD-Schnittstelle
ausgestattet, über die Dateien gemeinsam
genutzt werden können.

Statement by the jury
This mobile router gives an impressively
high level of technology, is uncomplicated in use and so compact, that it fits
in the shirt or trouser pocket.

Begründung der Jury
Dieser Mobile Router hat eine hochwertige
technische Anmutung, ist unkompliziert in
der Anwendung und so kompakt, dass er in
Hemd- oder Hosentasche passt.

Digital 9000
Digital Wireless Microphone System
Digitales Funkmikrofonsystem

Manufacturer
Sennheiser electronic GmbH & Co. KG,
Wedemark, Germany
Design
ma design GmbH & Co. KG
(Jörn Lüthe),
Kiel, Germany
Web
www.sennheiser.com
www.ma-design.de

Digital 9000 is an innovative digital wireless microphone system which can transmit totally uncompressed audio. It consists of the SKM 9000 wireless microphone, the SK 9000 bodypack transmitter and the EM 9046 receiver and is designed to meet highly professional demands. The system not only provides unprecedented audio quality but also an exceptional level of user-friendliness – both during set-up and in operation.

Digital 9000 ist ein innovatives digitales Funkmikrofonsystem, das vollkommen ohne Komprimierung übertragen kann. Es besteht aus dem Funkmikrofon SKM 9000, dem Taschensender SK 9000 und dem Empfänger EM 9046 und ist auf professionelle Anforderungen zugeschnitten. Das System bietet hervorragende Audioqualität und ist so gestaltet, dass es eine hohe Benutzerfreundlichkeit sicherstellt – sowohl bei der Systemeinrichtung als auch bei der Anwendung.

Statement by the jury
The Digital 9000 system, with its technical and formally sophisticated design, fulfils the professional requirements for a wireless microphone system.

Begründung der Jury
Das System Digital 9000 erfüllt mit seiner technisch und formal ausgereiften Gestaltung professionelle Ansprüche an ein Funkmikrofonsystem.

O2
Mobile Radio Control Head
Funkgerät-Bedienmodul

Manufacturer
Motorola Solutions,
Schaumburg, USA
In-house design
Chi Tran
Web
www.motorolasolutions.com

The O2 mobile radio control head provides a rugged vehicle mount solution for use in extreme conditions where water, dirt and vibration are prevalent. Controls are optimised for tactility and their exaggerated size allows gloved operation. Particular attention to materials and construction methods assures that it withstands the rigors of hazardous environments while providing safety and productivity for users.

Statement by the jury
For the O2 control head, materials and ergonomics have been optimised for use under extreme conditions and make the device sturdy and easy to operate.

Das O2 Funkbediengerät ist eine robuste Fahrzeuglösung, die speziell für den Einsatz unter Extrembedingungen entwickelt wurde, und Wasser, Staub und Vibrationen standhält. Das Funkbediengerät ist intuitiv und auch mit Handschuhen bedienbar. Materialien und Bauweise gewährleisten eine hohe Widerstandsfähigkeit und Zuverlässigkeit auch in anspruchsvollen Umgebungen und garantieren den Einsatzkräften Sicherheit und Produktivität.

Begründung der Jury
Beim O2 Control Head sind Materialien und Ergonomie optimal auf den Einsatz unter extremen Bedingungen abgestimmt und machen das Gerät robust und gut bedienbar.

MOTOTRBO™ SL Series
MOTOTRBO™ SL-Serie
Portable 2-Way Radio
Tragbares 2-Wege-Funkgerät

Manufacturer
Motorola Solutions,
Schaumburg, USA
In-house design
Lan Ting Garra,
Jaihar Ismail
Web
www.motorolasolutions.com

The SL Series radio is an ultra-thin, ultra-light portable two-way radio specially developed for environments where discreet, professional communication is important. It incorporates features such as hands-free, covert mode and intelligent audio where the radio's volume automatically adjusts to compensate for background noise. Weighing only 165 grams, this compact radio is unusually light and integrates voice and data applications seamlessly.

Statement by the jury
The radios of the SL Series fascinate with their extremely slim design and professional functionality.

Mit 165 Gramm bietet die SL-Serie extrem schlanke und leichte digitale Handsprechfunkgeräte, die speziell für Bereiche entwickelt wurden, in denen eine diskrete, professionelle Kommunikation wichtig ist. Die Funkgeräte verfügen über integrierte Sprach- und Datenanwendungen, Funktionen wie Freihand- und Tarnmodus sowie ein intelligentes Audio-Feature, mit dem sich die Lautstärke automatisch an das Geräuschlevel der Umgebung anpasst.

Begründung der Jury
Die Funkgeräte der SL-Serie begeistern mit ihrer extrem schlanken Gestaltung und einer professionellen Funktionalität.

Computers and information technology
Computer und Informationstechnik

Computers, notebooks, tablet PCs, PDAs, servers, keyboards, modems, USB sticks, printers, scanners, monitors, presentation technology, peripheral devices and accessories
Computer, Notebooks, Tablet-PCs, PDAs, Server, Keyboards, Modems, USB-Sticks, Drucker, Scanner, Monitore, Präsentationstechnik, Peripherie und Zubehör

5

27" iMac

Manufacturer
Apple, Inc.,
Cupertino, USA

In-house design
Apple Industrial Design Team

Web
www.apple.com

reddot design award
best of the best 2013

Brilliantly interpreted

In the late-1990s, the iMac provided an entirely new understanding of computers. It was easy to use and its all-in-one concept integrated all elements into a single housing. The new 27" iMac features an edge that is only 5mm thick, allowing for a housing of 40 per cent less volume. This was made possible by friction-stir welding, a process that is commonly used on airplane wings and rocket booster tanks. The new iMac's display was technologically refined and the cover glass is now fully laminated onto the LCD. It offers users brilliant colour definition as well as a realistic and high-contrast resolution. In addition, the display is easy on the eyes as its reflections were reduced by 75 per cent due to an anti-reflective coating. The 27" iMac features an Intel quad-core processor, improved graphics architecture by NVIDIA, and the all-new Fusion Drive, an innovative storage solution that combines the capacity of familiar drives with the performance of a flash drive. This allows even storage-intensive work to be performed quickly and efficiently. The overall design of the new 27" iMac is slender and well thought out in all of its details, elegantly carrying on the iMac line.

Brillant interpretiert

Der iMac gab einem Computer in den späten 1990er Jahren ein neues Selbstverständnis. Er war leicht zu bedienen, und sein All-in-One-Konzept integrierte alle Elemente in nur einem Gehäuse. Der Rand des neuen 27" iMacs ist lediglich 5 mm dick, weshalb das Volumen seines Gehäuses um 40 Prozent reduziert werden konnte. Ermöglicht wurde dies durch das Verfahren des Rührreibschweißens, ein Verfahren, welches üblicherweise bei Flugzeugtragflächen und Raketentreibstofftanks genutzt wird. Das Display des neuen iMacs ist technologisch ausgereift und die Glasabdeckung wurde nun direkt mit dem LCD verbunden. Die vollständige Laminierung bietet dabei eine hervorragende Farbdefinition sowie eine realistische und kontrastreiche Auflösung. Die antireflektierende Beschichtung des Displays ist zudem augenschonend, da die Blendeffekte um 75 Prozent reduziert wurden. Der 27" iMac ist mit einem Intel Quad-Core-Prozessor ausgestattet. Er bietet die fortschrittliche Grafikarchitektur von NVIDIA sowie Fusion Drive, eine innovative und wegweisende Speicheroption, welche die Kapazität herkömmlicher Festplatten mit der Leistung eines Flashspeichers verbindet. Mit dieser können speicherintensive Arbeitsschritte zügiger und auch effektiver ausgeführt werden. Die Gestaltung des 27" iMacs verleiht ihm eine schlanke Anmutung, und er ist in jedem seiner Details durchdacht. Auf elegante Weise führt er so die Kollektion des iMacs fort.

Statement by the jury

The 27" iMac effortlessly bridges the living areas of office and home and does not look like a work device. Its design successfully implements a reduction to the essential. The proportions of this iMac are well balanced and it features an elegantly flowing design appearance. Technologically convincing, it thus embodies a successful reinterpretation.

Begründung der Jury

Der 27" iMac überbrückt mühelos die Lebensbereiche von Büro und Zuhause und wirkt nicht wie ein Arbeitsgerät. Seiner Gestaltung gelingt die Reduktion auf das Wesentliche. Die Proportionen dieses iMacs sind ausgewogen und er hat eine elegant fließende Formensprache. Technologisch überzeugend, stellt er damit insgesamt eine gelungene Neuinterpretation dar.

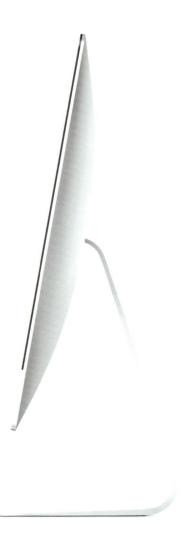

21.5" iMac

Manufacturer
Apple, Inc.,
Cupertino, USA
In-house design
Apple Industrial Design Team
Web
www.apple.com

The new iMac features a stunning design that has 40 per cent less volume than its predecessor with a thin 5 mm edge. The display has been completely reengineered for brilliant colour and contrast, with reflection reduced by 75 per cent. The cover glass is fully laminated to the LCD, and an anti-reflective coating has been applied using precision plasma deposition. Intel quad-core processors, NVIDIA graphics and innovative Fusion Drive storage make this the most advanced desktop Apple has ever made.

Der neue iMac zeigt eine beeindruckende Formgebung: Sein Volumen ist um 40 Prozent geringer ist als das seines Vorgängers, seine Kanten sind nur 5 mm dick. Zudem wurde das Display vollständig überarbeitet, damit Farben und Kontrast brillanter dargestellt und Reflexionen um 75 Prozent reduziert werden. Das Glas ist vollständig auf das LCD laminiert, eine Antireflex-Beschichtung wurde im Plasma-Abscheideverfahren aufgebracht. Dank des Intel Quad-Core-Prozessors, der NVIDIA-Grafik und des innovativen Fusion-Drive-Speichers ist dies der modernste Desktop von Apple.

Statement by the jury
The design of the iMac is classic and contemporary. It combines innovative technologies in one single unit and thus sets standards within its class.

Begründung der Jury
Der iMac ist edel und zeitgemäß gestaltet. Er vereint innovative Technologien in einem Gerät und setzt so in seiner Klasse Maßstäbe.

ASUS AiO ET2300 Series
All-in-One PC

Manufacturer
ASUSTeK Computer Inc.,
Taipei, Taiwan
In-house design
ASUSDESIGN
Web
www.asus.com
www.asusdesign.com

With its i7 processor, the AiO ET2300 offers high performance, while the built-in array speakers with subwoofer deliver strong sound quality. The PC has a double-hinge design, which allows it to be used as a traditional desktop with an upright monitor or to be placed flat on the table. Various ports, including Thunderbolt, Wireless Display and USB 3.0, provide flexible connectivity for external devices.

Der AiO ET2300 bietet mit seinem i7-Prozessor eine hohe Leistung, während die Array-Lautsprecher mit Subwoofer für eine gute Klangqualität sorgen. Der PC verfügt über ein Doppelscharnier, das es erlaubt, ihn wie ein traditionelles Desktopgerät mit aufrechtem Monitor zu verwenden oder ihn auch flach auf den Tisch zu legen. Die Schnittstellen Thunderbolt, Wireless Display und USB 3.0 bieten flexible Anschlussmöglichkeiten für externe Geräte.

Statement by the jury
This all-in-one PC is suitable for both home and office thanks to its variety of features. Depending on the circumstances, it can be used in a space-saving way

Begründung der Jury
Dank seiner Ausstattung eignet sich der All-in-one-PC sowohl für den Business- wie auch für den Heimbereich. Je nach Situation lässt er sich platzsparend nutzen.

Beta
All-in-One PC

Manufacturer
Lenovo (Beijing) Ltd.,
Beijing, China
In-house design
Yingjia Yao
Web
www.lenovo.com

Beta is an all-in-one PC with a Windows 8 operating system and a 21.5" touchscreen. It can be used as a work computer and is likewise suitable for viewing videos or gaming. The back of the unit features a foldable stand, enabling the device to be operated both in an upright and a prone position. The wedged lateral design makes the PC easy to grasp, while its rugged and durable unibody casing fosters a high-quality appearance.

Beta ist ein All-in-one-PC mit Windows 8 als Betriebssystem und einem 21,5"-Bildschirm. Er lässt sich für die Arbeit ebenso einsetzen wie für das Betrachten von Videos oder für das Spielen. An seiner Rückseite befindet sich ein ausklappbarer Standfuß, sodass das Gerät ganz nach Belieben aufgestellt oder hingelegt werden kann. Durch die abgeschrägten Kanten lässt er sich leicht greifen. Er besitzt ein Unibody-Gehäuse, das ihn robust macht und ihm eine wertige Anmutung verleiht.

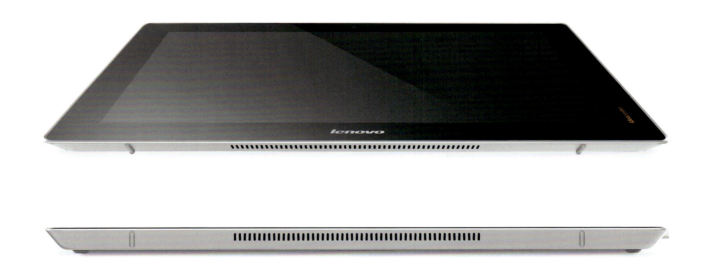

Lenovo IdeaCentre B750
All-in-One PC

Manufacturer
Lenovo (Beijing) Ltd.,
Beijing, China
In-house design
Yingjia Yao
Web
www.lenovo.com
Honourable Mention

The IdeaCentre B750 is an all-in-one PC with a 29" screen and a 2.1 surround audio system, as well as a broad range of entertainment functions. The display has a smooth glass front, focusing attention on the screen. The speakers are concealed under a subtle metal mesh so as not to detract from the overall impression. The computer rests safely on an aluminium stand and blends seamlessly with modern living spaces.

Statement by the jury
The design of the IdeaCentre helps it blend harmoniously with modern consumer electronics equipment. It is a well-conceived multimedia PC with strong sound and image display.

Das IdeaCentre B750 ist ein All-in-one-PC mit zahlreichen Entertainment-Funktionen. Er besitzt einen 29"-Bildschirm und ein 2.1-Surround-Audiosystem. Der Bildschirm zeigt eine glatte Glasfront, die die Aufmerksamkeit auf das Bild lenkt. Die Lautsprecher sind unter einem feinen Metallgewebe versteckt, um den Gesamteindruck nicht zu stören. Der Computer ruht sicher auf einem Standfuß aus Aluminium und fügt sich gut in moderne Wohnräume ein.

Begründung der Jury
Mit seiner Gestaltung passt das IdeaCentre zu aktueller Unterhaltungselektronik. Es ist ein gelungener Multimedia-PC mit einer guten Klang- und Bildwiedergabe.

ASUS Transformer AiO P1801
All-in-One PC and Tablet
All-in-one-PC und Tablet

Manufacturer
ASUSTeK Computer Inc.,
Taipei, Taiwan
In-house design
ASUSDESIGN
Web
www.asus.com
www.asusdesign.com

Transformer AiO P1801 is a combination of desktop PC and handheld tablet. The display can be detached from the base station and used as a wireless screen for the AiO PC or as an autonomous tablet. The operating systems employed are Windows 8 on the desktop PC and Android on the display so that users can choose between the proper platform, depending on the situation. The operating systems may be switched at the touch of a button.

Statement by the jury
The detachable display allows users of the desktop PC a higher degree of mobility, offering the freedom to work anywhere nearby.

Transformer AiO P1801 ist eine Kombination aus Desktop-PC und Handheld-Tablet. Der Monitor kann von der Basisstation heruntergenommen und als drahtloser Bildschirm für den AiO-PC oder als eigenständiges Tablet verwendet werden. Als Betriebssysteme kommen Windows 8 auf dem Desktop-PC und Android auf dem Monitor zum Einsatz, sodass der Benutzer je nach Situation die passende Plattform wählen kann. Per Knopfdruck wechselt er zwischen den Betriebssystemen.

Begründung der Jury
Das abnehmbare Display verschafft dem Nutzer des Desktop-PCs eine höhere Mobilität. So hat er die Freiheit, an einem beliebigen Ort weiterzuarbeiten.

Dell XPS One 27 Touch
All-in-One PC

Manufacturer
Dell Inc.,
Round Rock, USA
In-house design
Experience Design Group
Web
www.dell.com

This all-in-one PC family presents a new design language that visually reduces the unit's form to clean and minimal surfaces. The XPS One 27 Touch is the newest member of the product family. The touch-enabled 27" display with Wide Quad Hi-Def (WQHD) resolution facilitates interactive experiences with movies and applications. The stand can be tilted and its height adjusted, thus enabling a variety of positions and viewing angles.

Statement by the jury
The design of the XPS One 27 Touch is minimalist and functional. The stand allows for individual adjustment of the screen to fit various different space requirements.

Die All-in-one-PC-Familie präsentiert sich in einer neuen Formensprache, die das Produkt auf klare, minimierte Flächen reduziert. Der XPS One 27 Touch ist das neueste Mitglied der Familie. Über sein berührungsempfindliches 27"-Display mit WQHD-Auflösung können Filme und Anwendungen interaktiv erlebt werden. Der Standfuß lässt sich neigen und in der Höhe verstellen, sodass der Benutzer zahlreiche Positionen und Betrachtungswinkel einstellen kann.

Begründung der Jury
Die Gestaltung des XPS One 27 Touch ist minimalistisch und funktional. Der Standfuß ermöglicht die individuelle Ausrichtung des Bildschirms bei geringem Platzbedarf.

ThinkVision LT2323z
Monitor

Manufacturer
Lenovo,
Morrisville, USA
In-house design
Lenovo Industrial Design Group
Web
www.lenovo.com

The ThinkVision LT2323z is a VoIP- and Microsoft-Lync-certified HD display. The 23" monitor in 16:9 format has been designed for use in everyday working life, particularly for videoconferencing. The stand is height- and tilt-adjustable so that it may be aligned to an optimum position, depending on the situation at hand. The monitor also includes control buttons which allow users to answer phone calls. It is made using 80 per cent recycled plastic, thus setting an example for sustainable production.

Der ThinkVision LT2323z ist ein VoIP- und Microsoft-Lync-zertifizierter HD-Bildschirm. Der 23"-Monitor im 16:9-Format wurde für den Einsatz im beruflichen Alltag entwickelt, wo er vornehmlich bei Videokonferenzen seine Vorteile ausspielen kann. Der Standfuß ist in Höhe und Neigung verstellbar, sodass der Bildschirm je nach Situation optimal ausgerichtet werden kann. Der Monitor besitzt zudem Bedienknöpfe zum Annehmen von Telefonaten. Er besteht zu 80 Prozent aus recyceltem Kunststoff und setzt damit ein Zeichen für eine nachhaltige Produktion.

Cintiq 24HD touch
Interactive Pen Display
Interaktives Stift-Display

Manufacturer
Wacom Company Limited,
Saitama, Japan
Design
Ziba,
Portland, USA
Web
www.wacom.com
www.ziba.com

This interactive pen display allows for direct interaction with the screen, which promotes natural, creative workflows. The 24" display with full HD resolution and HDCP support presents high-quality images. Thanks to its stand design, the display can be placed in an ergonomic position, thus enabling fatigue-free work over extended periods of time. The screen is capable of identifying input by pen and multi-touch gestures.

Das interaktive Stift-Display erlaubt das direkte Arbeiten auf dem Bildschirm, so sind natürliche, kreative Arbeitsabläufe möglich. Das 24"-Display mit Full-HD-Auflösung und HDCP-Unterstützung stellt Bilder in einer hohen Qualität dar. Dank des Standfußes lässt es sich ergono-misch platzieren, sodass auch über einen längeren Zeitraum ohne Ermüdungser-scheinungen gearbeitet werden kann. Der Bildschirm erkennt sowohl die Eingabe per Stift als auch per Multi-Touch-Gesten.

Statement by the jury
The pen display convinces with its technical equipment and the variety of input options facilitating creative work.

Begründung der Jury
Das Stift-Display überzeugt durch seine technische Ausstattung und die verschiedenen Eingabemöglichkeiten, die das kreative Arbeiten erleichtern.

Acer T2
LED Touch Monitor

Manufacturer
Acer Incorporated,
New Taipei, Taiwan
In-house design
Sam CT Chen,
Mark Chang,
Eric YS Liu
Web
www.acer.com

The design of the T2 touch monitor is minimalist. A distinctive feature is its glass surface, which not only covers the display but also extends downward to form the base. This fosters a visual elongation of the display and imparts the impression of a floating image. The screen can be tilted between 30 und 80 degrees to arrive at an optimal position for each user. The monitor is available in 23" and 27" and includes HDMI, DVI and USB 3.0 connectivity.

Statement by the jury
The touch monitor rests on an elongated glass front and a metal bracket to the rear, creating an appealing and self-contained solution.

Die Gestaltung des Touchmonitors T2 ist minimalistisch. Hervorzuheben ist die Glasfläche, die nicht nur das Display bedeckt, sondern über dieses hinausreicht, um den Bildschirm vorne abzustützen. Dadurch wird zum einen das eigentliche Display optisch verlängert und zum anderen der Eindruck eines schwebenden Bildes vermittelt. Der Bildschirm lässt sich zwischen 30 und 80 Grad neigen, damit der Benutzer die beste Position einstellen kann. Er ist in 23" und 27" erhältlich und besitzt HDMI-, DVI- und USB-3.0-Schnittstellen.

Begründung der Jury
Der Bildschirm ruht auf der verlängerten Glasfront und dem Metallbügel auf der Rückseite. Diese Lösung ist ansprechend und eigenständig.

23ET63
Touch Monitor

Manufacturer
LG Electronics Inc.,
Seoul, South Korea
In-house design
Sung-Joo Cho,
Byung-Mu Huh
Web
www.lg.com

This touch monitor has a stand system with which the monitor can be tilted by up to 30 degrees. The user can individually adjust the viewing angle for ease of use. The glass front has a clear, reserved design with distinctive accents set by carefully processed decorative elements. The aluminium stand is inconspicuously welded to the monitor, matching the reserved overall impression.

Statement by the jury
The touch monitor convinces with its smooth and subtle design. With a white backside, it blends harmoniously with the respective environment.

Dieser Touchmonitor besitzt ein Standsystem, mit dem sich eine Neigung bis zu 30 Grad einstellen lässt. So kann der Benutzer den Betrachtungswinkel individuell anpassen und den Monitor mühelos bedienen. Die gläserne Front ist schlicht gestaltet, wobei sorgfältig verarbeitete dekorative Elemente Akzente setzen. Der Aluminiumständer ist unauffällig verschweißt, was gut zu dem zurückhaltenden Gesamteindruck passt.

Begründung der Jury
Der Touchmonitor überzeugt mit seiner glatten, dezenten Gestaltung. Dank seiner hellen Rückseite fügt er sich unauffällig in d e jeweilige Umgebung ein.

IdeaPad U430s
Laptop

Manufacturer
Lenovo (Beijing) Ltd.,
Beijing, China

In-house design
Yingjia Yao

Web
www.lenovo.com

reddot design award
best of the best 2013

Inspiration of the new

Laptops mark the individual lifestyles of their users as they are companions in everyday life. The design inspiration of the IdeaPad U430s is derived from the language of a dynamic journal, a characteristic that the laptop manages to translate into a product with a fascinatingly iconic appearance. The design language of this laptop is marked by carefully harmonised materials and a U-shaped cover surrounded by a distinctively coloured body. Complemented by an innovatively concave shape, this provides an overall outstanding user experience. The laptop is easy to open using just one finger, and the 45-degree angle on the back-end allows it to be held comfortably and securely. The IdeaPad U430s stands up to the rigours of everyday life with its full-metal high-strength body, which is made of aluminium alloy and thus makes the laptop weigh only 1.5 kg. In addition, the laptop features a thickness of only 14.8 mm and thus impresses with an overly slim appearance. This 14" laptop is equipped with a multi-touch screen in FHD resolution. Its large glass multi-touch touchpad and backlit keyboard are intuitive and convenient to use. The IdeaPad U430s also offers users a pleasing tactile experience due to a special surface finish that makes it feel soft and natural to the touch. In a highly inspired manner, the design of the IdeaPad U430s places the user centre-stage – it has emerged as a laptop with fascinatingly new qualities.

Die Inspiration des Neuen

Das Laptop prägt den persönlichen Lebensstil, denn es ist ein stetiger Begleiter im Alltag. Die Inspiration für die Gestaltung des IdeaPad U430s lag in der dynamischen Anmutung eines Journals, und es gelingt ihr, deren Charakter in ein beeindruckend ikonografisch anmutendes Produkt zu überführen. Die Formensprache dieses Laptops wird geprägt von sorgfältig abgestimmten Materialien und einer U-förmigen Abdeckung, die auffällig farbig umrandet ist. Diese bietet durch ihre innovative konkave Form Nutzererlebnisse der besonderen Art. Das Laptop lässt sich leicht mit nur einem Finger öffnen, und ein 45-Grad-Winkel an der Rückseite erlaubt ein komfortables Festhalten. Das IdeaPad U430s trotzt dem Alltag durch ein gänzlich aus Metall gefertigtes Gehäuse. Da eine Aluminiumlegierung eingesetzt wird, ist es mit 1,5 kg Gewicht dennoch leicht. Es hat zudem eine Dicke von nur 14,8 mm und beeindruckt insgesamt durch seine schlanke Anmutung. Ausgestattet ist dieses 14"-Laptop mit einem Multi-Touch-Screen in einer FHD-Auflösung. Über ein gläsernes Touchpad und eine hintergrundbeleuchtete Tastatur lässt es sich unkompliziert und intuitiv bedienen. Das IdeaPad U430s bietet dem Nutzer eine ansprechende Haptik, denn durch ein besonderes Finish fühlt es sich an der Oberfläche natürlich und weich an. Auf eine inspirierte Art und Weise stellte die Gestaltung des IdeaPad U430s den Nutzer in den Mittelpunkt – es entstand ein Laptop mit faszinierend neuen Qualitäten.

Statement by the jury

The intelligent design of the IdeaPad U430s redefines laptops with a new and refreshing approach. It looks slim and emotionalises its users. The colour concept of this laptop is well matched to the graphics. It also fascinates with its sophisticated functionality. It is easy to use with only one hand and offers a high degree of user-friendly convenience.

Begründung der Jury

Die intelligente Gestaltung des IdeaPad U430s definiert das Laptop auf neue und erfrischende Weise. Es wirkt leicht und emotionalisiert seinen Besitzer. Das Farbkonzept dieses Laptops wurde gut mit seiner Grafik abgestimmt. Es begeistert auch durch seine ausgefeilte Funktionalität. Man kann es leicht mit nur einer Hand bedienen und es bietet ein hohes Maß an nutzerfreundlichem Komfort.

Laptops

Aspire R7
Convertible Laptop

Manufacturer
Acer Incorporated,
New Taipei, Taiwan
In-house design
Gary Chang, Joe SB Chen,
CY Liu, Gavin Sung
Web
www.acer.com

Aspire R7 is a slim convertible laptop with a special connection between casing and display. The latter is based on the patented Ezel hinge, allowing users to position it either as a notebook, pad or a display. To this end, the hinge provides two different resistance levels, which can be adjusted according to the intended use. In pad mode, resistance is less so that the display can easily be brought into four different positions with just one hand. In notebook mode, resistance is higher, and the display remains in a fixed position. The notebook also includes four high-performance speakers.

Aspire R7 ist ein schlanker Convertible-Laptop mit einer besonderen Verbindung von Gehäuse und Bildschirm. Dieser ruht auf dem patentierten Ezel-Gelenk, das es ermöglicht, ihn als Notebook, Tablet oder Bildschirm zu positionieren. Das Gelenk bietet dafür zwei unterschiedliche Widerstände, die sich dem Verwendungszweck anpassen. Im Tablet-Modus ist der Widerstand geringer, sodass der Bildschirm mit einer Hand leicht in vier verschiedene Stellungen gebracht werden kann. Im Notebook-Modus ist der Widerstand hoch, und der Bildschirm bleibt in einer festen Position. Das Notebook besitzt zudem vier leistungsfähige Lautsprecher.

Dell XPS 12
Convertible Laptop

Manufacturer
Dell Inc.,
Round Rock, USA
In-house design
Experience Design Group
Web
www.dell.com

The Dell XPS 12 is a dual-function device which can seamlessly change from laptop to tablet and back again. It thus offers both the high performance of a laptop and the convenience and user-friendliness of a tablet. The display is protected by rugged, rimless Gorilla Glass with integrated buttons. The carbon fibre back saves weight without compromising durability.

Der Dell XPS 12 ist ein Gerät mit Doppel-funktion, das sich mit wenigen Handgrif-fen von einem Laptop in ein Tablet und wieder zurück verwandeln lässt. Es bietet damit sowohl die Leistungsfähigkeit eines Laptops als auch die Annehmlichkeiten und Benutzerfreundlichkeit eines Tablets. Das Display wird durch robustes, randlo-ses Gorilla-Glas mit integrierten Tasten geschützt. Die Rückseite aus Carbonfaser spart Gewicht, ohne Zugeständnisse an die Haltbarkeit zu machen.

Statement by the jury
The XPS 12 features a unique display solution: the screen can be rotated within its frame by 180 degrees so that the back of the screen comes to rest on the keyboard.

Begründung der Jury
Die Display-Lösung des XPS 12 ist ungewöhnlich: Der Bildschirm lässt sich innerhalb seines Rahmens um 180 Grad drehen, sodass er mit der Rückseite auf der Tastatur liegt.

HP Envy x2
Convertible Laptop

Manufacturer
Hewlett-Packard,
Houston, USA
In-house design
HP PPS Industrial Design Team
Web
www.hp.com

This very slim, high-performance convertible laptop unites the advantages of two devices in one. With the detachable screen inserted, it is an 11.6" notebook with a Windows 8 operating system, 2 GB internal memory, HDMI and USB connectivity and an 8 MP camera. When the screen is detached from the magnetic latch, it turns into a tablet with touch functionality. When the notebook is opened, the hinge in which the screen is inserted moves slightly downward. This lifts the back of the keyboard to a slightly inclined position so as to offer a comfortable position for typing.

Statement by the jury
Envy x2 convinces with its versatility. The anodised aluminium case lends the device a premium appearance.

Der leistungsstarke, sehr flache Convertible Laptop vereint die Leistung von zwei Geräten. Mit eingestecktem Bildschirm ist er ein 11,6"-Notebook mit Windows 8, 2 GB Arbeitsspeicher, HDMI- und USB-Anschlüssen sowie einer 8-MP-Kamera. Wird der Bildschirm aus der Magnethalterung gelöst, wird er zum Tablet mit Touchfunktion. Wenn das Notebook geöffnet wird, fährt das Scharnier nach unten aus. Dadurch wird die Tastatur hinten etwas angehoben und steht dann leicht schräg, was für eine angenehme Position beim Tippen sorgt.

Begründung der Jury
Envy x2 überzeugt durch seine Wandlungsfähigkeit. Zudem vermittelt das Gehäuse aus gebürstetem Aluminium eine wertige Anmutung.

Dell Latitude 6430u
Ultrabook

Manufacturer
Dell Inc.,
Round Rock, USA
In-house design
Experience Design Group
Web
www.dell.com

The Latitude 6430u was designed for
mobile professionals. With a weight of
only 1.69 kg and a thickness of 20.9 mm,
it is very lightweight and flexible.
The durable enclosure is capable of
withstanding the rigours of everyday
business life. The spill-resistant keyboard
and protective LCD seal help protect
the laptop from damage and adverse
weather conditions. Reinforced metal-
plated hinges ensure durability and also
the easy opening and closing of the
cover.

Statement by the jury
This ultrabook is designed to withstand
the most adverse conditions, down to
the last detail. Particularly noteworthy
are the spill-resistant keyboard and the
reinforced hinges.

Das Latitude 6430u wurde für mobile Be-
rufstätige konzipiert. Es ist entsprechend
leicht und wiegt nur 1,69 kg bei einer
Dicke von 20,9 mm. Das stabile Gehäuse
meistert den Business-Alltag problemlos.
Die spritzwassergeschützte Tastatur und
die Schutzversiegelung des LCDs bewahren das Laptop vor Beschädigungen und
Witterungseinflüssen. Verstärkte Metall-
blech-Scharniere sorgen für Langlebigkeit
und einfaches Öffnen und Schließen des
Deckels.

Begründung der Jury
Das Ultrabook ist bis ins Detail auf hohe
Belastungen ausgelegt. Zu nennen sind
hier insbesondere die spritzwasserge-
schützte Tastatur und die verstärkten
Scharniere.

13" MacBook Pro
with Retina Display

Manufacturer
Apple, Inc.,
Cupertino, USA
In-house design
Apple Industrial Design Team
Web
www.apple.com

The 13" MacBook Pro features a stunning Retina display and all-flash storage in a new, compact design. At a mere 0.75 inches and 3.57 pounds, the remarkably portable 13" MacBook Pro is 20 per cent thinner and almost a pound lighter than the current 13" MacBook Pro.

Das MacBook Pro 13" zeigt eine neue, kompakte Formgebung. Es besitzt ein brillantes Retina-Display und ist mit Flash-Speicher ausgestattet. Es ist mit 1,9 cm sehr dünn und wiegt lediglich 1,62 kg. Da es 20 Prozent dünner und fast 500 Gramm leichter als das aktuelle 13" MacBook Pro ist, bietet es eine noch größere Mobilität.

Statement by the jury
Display, memory solution, and design – the MacBook Pro sets standards for compact notebooks with regard to performance and equipment.

Begründung der Jury
Display, Speicherlösung, Gestaltung – das MacBook Pro setzt Maßstäbe bei kompakten Notebooks hinsichtlich Leistung und Ausstattung.

15" MacBook Pro with Retina Display

Manufacturer
Apple, Inc.,
Cupertino, USA
In-house design
Apple Industrial Design Team
Web
www.apple.com

Featuring a precision engineered aluminium unibody design and an all flash storage architecture, the new 15" MacBook Pro is the lightest MacBook Pro to date and nearly as thin as a MacBook Air, measuring a mere 0.71 inches and weighing only 4.46 pounds.

Das neue MacBook Pro 15" besitzt ein präzisionsgefertigtes Unibody-Gehäuse aus Aluminium und ist mit Flashspeicher ausgestattet. Das bisher leichteste MacBook Pro ist annähernd so schlank wie das MacBook Air. Es ist nur 1,8 cm dünn und wiegt lediglich 2,02 kg.

Statement by the jury
Despite its low weight, the notebook renders a convincing performance. The high-quality design, including the unibody housing, underscores this in an impressive way.

Begründung der Jury
Trotz seines geringen Gewichts bietet das Notebook eine überzeugende Leistung. Die wertige Gestaltung mit dem Unibody-Gehäuse unterstreicht dies eindrucksvoll.

HP Envy Sleekbook
Notebook

Manufacturer
Hewlett-Packard,
Houston, USA
In-house design
HP PPS Industrial Design Team
Web
www.hp.com

The Sleekbook is a very lightweight and slim 15" notebook. It weighs 1.75 kg and is only 19.88 mm thick, enabling it to be easily taken along and used anywhere. The cover is made of anodised aluminium, while the dark red sides set a bright accent. The notebook has one HDMI and three USB ports, as well as a card reader for SD cards. Its slip-resistant base guarantees a firm stand and safe grip when carrying the notebook with both hands.

Das Sleekbook ist ein sehr leichtes und dünnes 15"-Notebook. Es wiegt 1,75 kg und ist nur 19,88 mm dick. So lässt es sich leicht mitnehmen und überall einsetzen. Der Gehäusedeckel besteht aus gebürstetem Aluminium, während die Seiten in Dunkelrot einen farblichen Akzent setzen. Das Notebook besitzt eine HDMI- und drei USB-Schnittstellen sowie einen Kartenleser für SD-Karten. Seine rutschfeste Unterseite gewährleistet einen festen Stand und sorgt dafür, dass das Notebook beim Tragen sicher in der Hand liegt.

Z360
Ultrabook

Manufacturer
LG Electronics Inc.,
Seoul, South Korea
In-house design
Yi-Hyun Moon, Tae-Jin Lee,
Byung-Mu Huh
Web
www.lg.com

The Ultrabook Z360 is merely 13.8 mm thick and weighs only 1.16 kg. It has a 13.3" display and connections for HDMI, Ethernet and micro SD cards. Its design is characterised by the seamless cover and the slightly slanted bottom side. The cover can be opened with just one finger, while hotkeys for specific Windows 8 functions are located on the left-hand side of the keyboard.

Das Ultrabook Z360 ist lediglich 13,8 mm dick und wiegt nur 1,16 kg. Es besitzt einen 13,3"-Bildschirm sowie Anschlüsse für HDMI, Ethernet und Micro-SD-Karten. Seine Gestaltung ist von dem nahtlosen Gehäusedeckel und der leicht abgeschrägten Unterseite geprägt. Der Deckel lässt sich mit einem Finger öffnen, Hotkeys für die Nutzung spezifischer Windows-8-Funktionen befinden sich am linken Rand der Tastatur.

Statement by the jury
This ultrabook captivates with its slim and seamless contemporary design, which imparts a high degree of mobility.

Begründung der Jury
Das Ultrabook beeindruckt durch seine schlanke und nahtlose Gestaltung. Sie ist zeitgemäß und vermittelt eine hohe Mobilität.

ASUS U38 Series
Notebook

Manufacturer
ASUSTeK Computer Inc.,
Taipei, Taiwan
In-house design
ASUSDESIGN
Web
www.asus.com
www.asusdesign.com

The notebooks of the U38 series consist entirely of a metal body with hairline finish, highlighting its plain and elegant style. To further underscore this simplicity, the design has deliberately dispensed with a locking mechanism. The U38 configuration features special core technology, individual VGA, a Full HD touchscreen, a 500 GB hard drive, many different connection options and a backlit keyboard.

Die Notebooks der Serie U38 bestehen rundum aus Metall mit einer strichge-schliffenen Oberfläche, die ihren einfachen und eleganten Stil hervorhebt. Um diese Einfachheit zu unterstreichen, wurde bewusst auf einen Verriegelungsmecha-nismus verzichtet. Die U38-Konfiguration verfügt über eine spezielle Core-Techno-logie, individuelles VGA, einen Full-HD-Touchscreen, eine 500-GB-Festplatte, zahlreiche Anschlüsse sowie eine beleuch-tete Tastatur.

Statement by the jury
The high-performance notebook is lightweight and slim. With a battery life of seven hours, it is also very mobile.

Begründung der Jury
Das leistungsfähige Notebook ist leicht und flach und darüber hinaus dank seiner Akkuleistung von sieben Stunden sehr mobil.

ASUS Commercial
Notebook Series
Notebookserie

Manufacturer
ASUSTeK Computer Inc.,
Taipei, Taiwan
In-house design
ASUSDESIGN
Web
www.asus.com
www.asusdesign.com

The Commercial notebook series was conceptualised for use in small and medium businesses. It offers high performance, a reliable system and durable construction. With a design that is minimalist and functional, the notebooks blend harmoniously with any working environment. The devices in this series not only render high performance but also convey user-friendliness.

Statement by the jury
This notebook series convinces with its objective and functional design. The devices are durable and therefore well suited for business use.

Die Notebookserie Commercial wurde für den Einsatz in kleinen und mittleren Unternehmen entwickelt. Die Geräte bieten eine hohe Leistung, ein zuverlässiges System und sind stabil konstruiert. Ihre Gestaltung ist minimalistisch und funktional, sodass sie sich in eine Arbeitsumgebung harmonisch einfügen. Die Notebooks der Serie sind nicht nur leistungsfähig, sondern auch benutzerfreundlich.

Begründung der Jury
Die Notebookserie überzeugt durch ihre sachliche Gestaltung. Die Geräte sind robust und eignen sich daher gut für den geschäftlichen Einsatz.

Toshiba Satellite U840W/U800W/ Dynabook R542
Ultrabook

Manufacturer
Toshiba Corporation,
Digital Products & Services Company,
Tokyo, Japan
In-house design
Yusuke Kawai
Web
www.toshiba.co.jp

This ultrabook is equipped with an LC display in 21:9 format. Multiple windows can be arranged in a proper size next to each other, facilitating the work experience. The central design focus in developing this device was to achieve maximum mobility. Accordingly, the ultrabook is merely 21 mm thick, weighs 1.6 kg and has a rechargeable battery capable of powering the unit for up to eight hours. With integrated high-quality speakers, it also delivers good sound.

Statement by the jury
The two-part display cover already indicates that the wide-format screen of this ultrabook is its distinguishing feature.

Das Ultrabook besitzt ein LC-Display im 21:9-Format. So können mehrere Fenster in einer ordentlichen Größe nebeneinander angeordnet werden, was das Arbeiten erleichtert. Das Hauptaugenmerk bei der Entwicklung dieses Geräts lag auf größtmöglicher Mobilität. Entsprechend ist das Ultrabook nur 21 mm dick, wiegt lediglich 1,6 kg und hat eine Akkulaufzeit von etwa acht Stunden. Dank hochwertiger Lautsprecher bietet es zudem einen guten Klang.

Begründung der Jury
Schon der zweigeteilte Displaydeckel weist darauf hin, dass der Bildschirm dieses Ultrabooks aufgrund seines breiten Formats das hervorstechendste Merkmal ist.

ASUS Taichi Series
Ultrabook/Tablet

Manufacturer
ASUSTeK Computer Inc.,
Taipei, Taiwan
In-house design
ASUSDESIGN
Web
www.asus.com
www.asusdesign.com
Honourable Mention

The Taichi series offers a combination of ultrabook and tablet. When opened, the device can be operated like a notebook. When closed, it turns into a tablet, as it features a second display on the back of the notebook. Both screens are also capable of simultaneously displaying the same content, which is advantageous in the case of presentations. The device is lightweight and has a very slim aluminium case.

Die Geräte der Serie Taichi sind eine Kombination aus Ultrabook und Tablet. Ist das Gerät geöffnet, lässt es sich wie ein Notebook bedienen. In geschlossenem Zustand wird es zum Tablet, da sich an der Rückseite des Notebooks ein zweites Display befindet. Beide Bildschirme können auch gleichzeitig dieselben Inhalte zeigen, was beispielsweise bei Präsentationen von Vorteil ist. Das Gerät ist leicht und besitzt ein sehr flaches Aluminiumgehäuse.

Statement by the jury
With the second display, this hybrid device demonstrates high flexibility. The aluminium case moreover gives it a high-quality appearance.

Begründung der Jury
Durch den zweiten Bildschirm lässt sich das Hybridgerät flexibel einsetzen. Das Aluminiumgehäuse verleiht ihm eine wertige Anmutung.

Computers and information technology

HP Elitebook Revolve
Notebook

Manufacturer
Hewlett-Packard,
Houston, USA
In-house design
HP PPS Industrial Design Team
Web
www.hp.com

The Elitebook Revolve is a business notebook with touchscreen and Windows 8. Its cover, which includes the screen, can be rotated and inclined downward so that it rests on the keyboard, turning the notebook into a tablet PC. The 11.6" display is made of scratch-resistant Gorilla Glass and a durable magnesium frame. The notebook weighs 1.4 kg and includes two USB 3.0 connections and a DisplayPort. In addition, it features Wi-Fi, Bluetooth and NFC wireless technologies.

Das Elitebook Revolve ist ein Business-Notebook mit Touchscreen und Windows 8. Der Gehäusedeckel und damit der Bildschirm des Geräts lässt sich drehen und nach unten klappen, sodass er auf der Tastatur zum Liegen kommt. So wird aus dem Notebook ein Tablet-PC. Das 11,6"-Display besteht aus kratzfestem Gorilla-Glas und einem robusten Magnesiumgehäuse. Das Gerät wiegt 1,4 kg und besitzt zwei USB-3.0-Schnittstellen sowie einen DisplayPort. Außerdem verfügt es über die kabellosen Technologien WLAN, Bluetooth und NFC.

Statement by the jury
This high-performance notebook is particularly suited for business applications. The rotatable display not only turns it into a tablet but also allows it to be flexibly aligned.

Begründung der Jury
Das leistungsfähige Notebook eignet sich gut für den Business-Bereich. Der drehbare Bildschirm verwandelt es nicht nur in ein Tablet, sondern kann auch flexibel ausgerichtet werden.

iPad mini

Manufacturer
Apple, Inc.,
Cupertino, USA
In-house design
Apple Industrial Design Team
Web
www.apple.com

The iPad mini displays a completely new design that is 23 per cent thinner and 53 per cent lighter than the third generation iPad. With a stunning 7.9" Multi-Touch display, two cameras, ultrafast wireless and ten hours of battery, the iPad mini is every inch an iPad, yet in an innovative design that can be held in one hand. It comes in a silver aluminium housing with white glass and a slate aluminium housing with black glass.

Das iPad mini wurde komplett neu gestaltet. Es ist um 23 Prozent dünner und um 53 Prozent leichter als die dritte iPad-Generation. Es besitzt ein beeindruckendes 7,9"-Multi-Touch-Display, zwei Kameras, schnelles WLAN und einen Akku, der eine Betriebsdauer von zehn Stunden ermöglicht. Das iPad mini ist ein vollwertiges iPad; es kann jedoch aufgrund seiner innovativen Formgebung mit einer Hand gehalten werden. Es ist mit einem silberfarbenen Aluminiumgehäuse mit weißem Glas oder einem schiefergrauen Aluminiumgehäuse mit schwarzem Glas erhältlich.

Padfone Infinity
Smartphone/Tablet

Manufacturer
ASUSTeK Computer Inc.,
Taipei, Taiwan
In-house design
ASUSDESIGN
Web
www.asus.com
www.asusdesign.com

The Padfone Infinity consists of an Android smartphone and a dock. When the smartphone is inserted into the well of the dock, it turns into a fully fledged tablet with a 10.1" display. The smartphone is recharged in the process since the dock includes an integrated battery. The unibody enclosure is made mainly of aluminium. With its rounded edges, it provides comfortable haptics and presents a high-quality appearance.

Das Padfone Infinity besteht aus einem Android-Smartphone und einem Dock. Wird das Smartphone in den Schacht des Docks gesteckt, wird es zu einem voll-wertigen Tablet mit einem 10,1"-Display. Gleichzeitig wird das Smartphone aufge-laden, da das Dock einen integrierten Akku besitzt. Das Unibody-Gehäuse besteht überwiegend aus Aluminium. Dank seiner abgerundeten Kanten bietet es eine ange-nehme, wertige Haptik.

Statement by the jury
The Padfone Infinity combines the advantages of two devices: it offers the mobility of a smartphone and the large display of a tablet.

Begründung der Jury
Das Padfone Infinity vereint die Vorteile von zwei Geräten: Es bietet die Mobilität eines Smartphones und das große Display eines Tablets.

Folder Pad
Tablet

Manufacturer
Lenovo (Beijing) Ltd.,
Beijing, China
In-house design
Yingjia Yao
Web
www.lenovo.com

With its integrated wireless technology, the Folder Pad tablet PC offers new freedom for the user. It features a special stand which is folded out backward like an easel. This enables the tablet to be positioned upright like a monitor instead of always held in the hand. When wirelessly connected to the keyboard, the device can be used like a desktop PC.

Dank der integrierten kabellosen Technologie bietet der Tablet-PC Folder Pad neue Freiheiten für den Benutzer. Er besitzt zudem einen besonders gestalteten Standfuß, der wie eine Bildstütze nach hinten weggeklappt wird. Dadurch muss das Tablet nicht immer in der Hand gehalten werden, sondern lässt sich wie ein Monitor aufstellen. Wird das Gerät kabellos mit einer Tastatur gekoppelt, kann es wie ein Desktop-PC verwendet werden.

Statement by the jury
This tablet PC impresses with an intelligent design solution for the integrated support. Notable is also the appealing interplay of different colours.

Begründung der Jury
Der Tablet-PC überzeugt durch die intelligente Gestaltung der Stütze. Sie ist in das Produkt integriert und bietet darüber hinaus ein ansprechendes Spiel mit unterschiedlichen Farben.

Kobo Arc
Tablet

Manufacturer
Kobo Inc.,
Toronto, Canada
In-house design
James Wu
Web
www.kobo.com

Kobo Arc was designed with the reader and multimedia consumer in mind. The 7" tablet features a fast dual-core 1.5 GHz processor and 1 GB of low-power RAM. The durable HD screen is highly resistant, making it ideal to withstand the challenges of everyday life. With more than 1 million pixels on the HD display, all words and images are crisp and clear. The new interface Tapestries allows the user to organise content thematically and pin media into customised groups. The device is available in black or white in 16, 32, and 64 GB models.

Statement by the jury
Kobo Arc is a high-performance tablet, offering a large number of functions. The rubberised back contributes to a comfortable and safe grip.

Kobo Arc wurde auf die Bedürfnisse von Lesern und Multimedianutzern abgestimmt. Das 7"-Tablet besitzt einen schnellen Dual-Core-1,5-GHz-Prozessor und 1 GB RAM mit geringem Leistungsverbrauch. Der langlebige HD-Bildschirm ist sehr widerstandsfähig, sodass er alltägliche Herausforderungen problemlos meistert. Dank mehr als einer Million Pixel werden Schrift und Bilder klar und brillant dargestellt. Die neue Benutzeroberfläche „Galerien" erlaubt dem Benutzer, Inhalte nach Themen zusammenzustellen und Medien in benutzerdefinierten Gruppen anzuordnen. Das Gerät ist in Schwarz oder Weiß mit 16, 32 oder 64 GB erhältlich.

Begründung der Jury
Kobo Arc ist leistungsfähig und bietet umfangreiche Funktionen. Dank seiner gummierten Rückseite lässt er sich angenehm und sicher in der Hand halten.

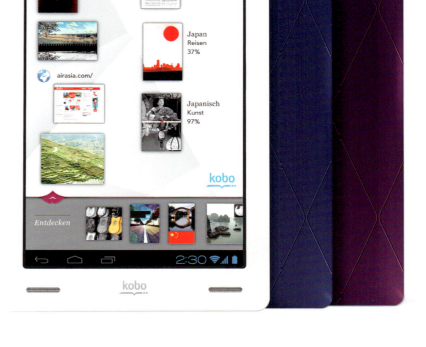

Xperia Tablet Z
Tablet

Manufacturer
Sony Mobile Communications
International AB,
Lund, Sweden
In-house design
Sony Mobile UX Creative Design Team
Web
www.sonymobile.com

The Xperia Tablet Z, thanks to the use of fibreglass reinforced plastic, is extremely light and flat. At the same time it is sturdy and, in spite of its slim silhouette, provides an 8 megapixel camera as well as NFC and infrared interfaces. The reduced form language directs the attention to the significant characteristics of the tablet: the seamlessly fitted full HD screen, high-grade, contrasting materials and precise engineering.

Statement by the jury
The design of the Xperia Tablet Z, with its contrasting surfaces, the clear linearity and the pleasant haptics is well conceived to the last detail.

Das Xperia Tablet Z ist dank des Einsatzes von glasfaserverstärktem Kunststoff extrem leicht und flach. Gleichzeitig ist es robust und bietet trotz seiner schlanken Silhouette eine 8-Megapixel-Kamera sowie NFC- und Infrarot-Schnittstellen. Die reduzierte Formensprache lenkt die Aufmerksamkeit auf die wesentlichen Merkmale des Tablets: den nahtlos eingepassten Full HD-Bildschirm, hochwertige, kontrastierende Materialien und eine präzise Fertigung.

Begründung der Jury
Die Gestaltung des Xperia Tablet Z mit seinen kontrastreichen Oberflächer, der klaren Linienführung und der angenehmen Haptik ist bis ins Detail durchdacht.

Ultrathin Keyboard Cover for iPad
Ultradünne Tastatur und Abdeckung für das iPad

Manufacturer
Logitech Europe S.A.,
Lausanne, Switzerland
Design
Design Partners
(Mathew Bates, John Moriarty),
Bray, Ireland
Web
www.logitech.com
www.designpartners.com

The Ultrathin Keyboard Cover is a very thin Bluetooth keyboard made of aluminium designed specifically for use with an iPad. It attaches to the iPad with a magnetic hinge to form a cover or into a magnetic, custom slot for a naturally ergonomic typing position. Bluetooth connectivity automatically deactivates when the unit is not in use as a keyboard in order to conserve battery power.

Statement by the jury
This keyboard harmonises beautifully with the design of the iPad. It also impresses with its dual functionality as keyboard and protective cover.

Das Ultrathin Keyboard Cover ist eine sehr dünne Bluetooth-Tastatur aus Aluminium, die speziell für die Verwendung mit einem iPad konzipiert wurde. Sie kann wie ein Deckel mit einem magnetischen Scharnier am iPad befestigt werden. Zudem besitzt die Tastatur eine magnetische Nut, in die das iPad eingesteckt werden kann. So ermöglicht sie ergonomisches Tippen. Wird die Tastatur nicht verwendet, schaltet sich Bluetooth automatisch ab, um Akkuleistung zu sparen.

Begründung der Jury
Diese Tastatur harmoniert hervorragend mit der Gestaltung des iPads. Zudem überzeugt das Produkt durch seine Doppelfunktion als Tastatur und Schutz.

S-Fun
iPad Series Leather Case
Lederhülle für die iPad-Serie

Manufacturer
Sound Young Originality,
Taipei, Taiwan
In-house design
Alec Wu
Web
www.sound-young.com

The S-Fun leather case is designed to make the iPad more comfortable and fun to use. It encloses the iPad, yet all operating elements are left freely accessible. In this way, the user can hold and handle the device comfortably. The variable backrest allows the iPad to be placed on a smooth surface at different tilted angles, for instance for conveniently viewing pictures or videos.

Statement by the jury
The leather case provides good functional protection for the iPad series and also convinces with its pleasant haptics.

Mit der Lederhülle S-Fun soll das Benutzen des iPads noch komfortabler werden und gleichzeitig Spaß machen. Die Lederhülle umschließt das iPad, lässt aber die Bedienelemente frei. So kann es der Anwender sicher in der Hand halten und bequem bedienen. Die variable Rückenstütze sorgt dafür, dass sich das iPad in unterschiedlicher Neigung auf einen glatten Untergrund stellen lässt, sodass beispielsweise Bilder oder Videos bequem betrachtet werden können.

Begründung der Jury
Die Lederhülle bietet einen guten, funktionalen Schutz für die iPad-Serie und überzeugt darüber hinaus durch ihre angenehme Haptik.

I-Line

Industrial Panel PC
Industrie-Panel PC

Manufacturer
CRE Rösler Electronic GmbH,
Hohenlockstedt, Germany
In-house design
Frank Michaelsen
Web
www.cre-electronic.com

I-Line is a multitouch panel PC with a screen size of 21.5" in portrait or landscape format which is particularly suitable for chemistry and hygiene areas. The durable stainless-steel case has a depth of 40 mm and includes a hardened glass surface meeting all requirements of the food, pharmaceutical and chemical industries, and it also attains tightness classes up to IP65. Moreover, the capacitive touchscreen can be activated when wearing gloves.

I-Line ist ein Multi-Touch-Panel-PC mit einem 21,5"-Bildschirm im Hoch- oder Querformat, der besonders für Chemie- und Hygienebereiche geeignet ist. Das robuste Edelstahlgehäuse ist 40 mm tief und hat eine gehärtete Oberfläche aus Glas. Sie erfüllt die Anforderungen der Lebensmittel-, Pharma- und Chemie-industrie und erreicht Dichtheitsklassen bis IP65. Der kapazitive Touchscreen lässt sich auch mit Handschuhen bedienen.

Statement by the jury
This panel PC exhibits a high-grade design, both internally and externally. The stainless-steel case underscores its capacity for high performance and suitability for sensitive areas within industrial settings.

Begründung der Jury
Der Panel-PC ist innen wie außen hoch-wertig gestaltet. Das Edelstahlgehäuse unterstreicht seine Leistungsfähigkeit und die Eignung für sensible Bereiche in der Industrie.

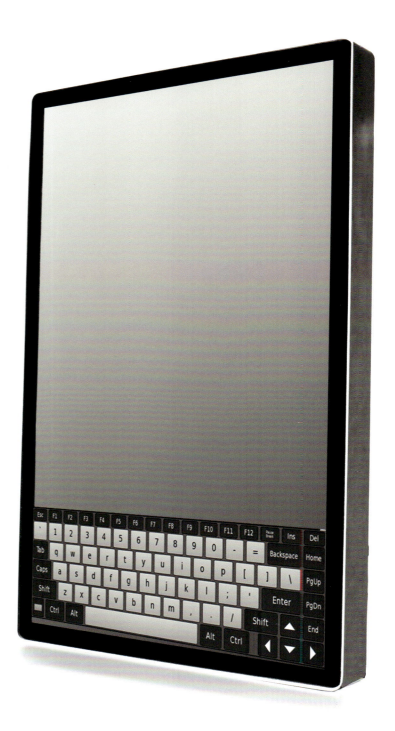

HP Envy 120 e-All-in-One
Inkjet All-in-One Printer
All-in-One-Tintenstrahldrucker

Manufacturer
Hewlett-Packard,
San Diego, USA

In-house design
Mylene Tjin

Web
www.hp.com

See-through

The daily handling of data and documents is a complex process, a reason why the printer is also used a lot in the home office. The HP Envy 120 e-All-in-One merges newly thought-through functionality with an aesthetic that is fascinating for this type of devices. This inkjet all-in-one printer features an elegantly compact and slim form factor. The subtle design combination of glass and metal is complemented by a careful finish, lending it a high-value and noble ambient aesthetic that makes the printer blend in well in a home environment. This inkjet all-in-one printer is very quiet in operation and provides sharp high-resolution print results. In addition, its innovative design is also exemplary as it reflects high levels of ecological awareness. The printer is made without PVC and offers the option of consuming up to 50 per cent less paper by using the built-in automatic two-sided printing function. The HP Envy 120 e-All-in-One is capable of wireless setup. It features Internet connectivity and offers easy printing from smartphones and tablet computers. A surprising feature of this device is that it allows users to see directly into the scanner, which adds a particular sense of depth. Due to its innovative design, the HP Envy 120 e-All-in-One also turns the scanning of documents into a new experience in terms of form and function.

Gut gesichtet

Der tägliche Umgang mit Daten und Schriftstücken ist komplex, weshalb der Drucker auch im Homeoffice viel genutzt wird. Der HP Envy 120 e-All-in-One verbindet eine neu durchdachte Funktionalität mit einer für ein solches Gerät beeindruckenden Ästhetik. Dieser All-in-One-Tintenstrahldrucker ist auf elegante Weise dünn und kompakt gestaltet. Eine feinsinnige Kombination der Materialien Glas und Metall sowie ein sorgfältiges Finish lassen ihn wertig und edel anmuten. Auf diese Weise bereichert er das Ambiente und kann gut im Wohnzimmer stehen. Mit diesem All-in-One-Tintenstrahldrucker geht das Drucken sehr leise vonstatten und er bietet hochaufgelöste Druckergebnisse. Seine innovative Gestaltung verbindet sich zudem mit einem vorbildlichen ökologischen Denken. Er ist ohne den Einsatz von PVC gefertigt, und durch die Option des beidseitigen Druckens können bis zu 50 Prozent des Papiers eingespart werden. Der Einsatz des HP Envy 120 e-All-in-One geschieht kabellos. Er verfügt über einen Internetzugang und kann leicht auch mit dem Smartphone oder einem Tablet-Computer angesteuert werden. Überraschend ist es bei diesem Gerät, dass man direkt in den Scanner hineinsehen kann und dabei eine besondere Tiefe erlebt. Durch diese innovative Art der Gestaltung wird auch das Scannen mit dem HP Envy 120 e-All-in-One zu einem neuen Erlebnis von Form und Funktion.

Statement by the jury

The HP Envy 120 e-All-in-One integrates all functions in a fascinating manner. A familiar product thus acquires a new form and functionality. The innovative design with a glass lid that allows users to see into the device lends this inkjet all-in-one printer a high sense of lightness. Its well-arranged touch screen provides an intuitive user experience.

Begründung der Jury

Bei dem HP Envy 120 e-All-in-One sind alle Funktionen auf beeindruckende Weise integriert. Ein bekanntes Produkt erhält hier eine neue Form und Funktionalität. Die innovative Gestaltung mit einem gläsernen Deckel, durch den man hineinsehen kann, verleiht diesem All-in-One-Tintenstrahldrucker Leichtigkeit. Sein übersichtlich gestaltetes Touchscreen ermöglicht dem Nutzer eine intuitive Bedienung.

iScan Air
Portable Wireless Scanner
Tragbarer kabelloser Scanner

Manufacturer
Mustek Systems, Inc.,
Hsinchu, Taiwan
In-house design
Huizong Chen
Web
www.mustek.com.tw

The iScan Air is a Wi-Fi-enabled scanner capable of communicating with iOS, Android, Mac OS and Windows 8 devices via its own dedicated app. The scanned image can be viewed with the respective mobile device and subsequently stored or processed further. The built-in rechargeable battery allows the scanner to be used on the go. Since it is connected via Wi-Fi, there is no need for interfering cables. A graphical interface makes the scanner simple to operate.

Der iScan Air ist ein WLAN-Scanner, der über eine eigene App mit iOS-, Android-, Mac OS- und Windows-8-Geräten kommunizieren kann. Das gescannte Bild lässt sich dann im jeweiligen Mobilgerät betrachten und speichern oder weiterverarbeiten. Der Scanner kann dank seines integrierten Akkus unterwegs eingesetzt werden. Da er über WLAN verbunden wird, kommt er ohne störende Kabel aus. Über eine grafische Benutzeroberfläche ist er leicht zu bedienen.

Statement by the jury
This scanner is lightweight, handy and mobile. Its reserved design in terms of form and colour turns it into an inconspicuous companion.

Begründung der Jury
Der Scanner ist leicht, handlich und mobil. Seine zurückhaltende Farb- und Formgebung machen ihn zu einem unauffälligen Begleiter.

LSM-300
Mouse Scanner
Maus-Scanner

Manufacturer
LG Electronics Inc.,
Seoul, South Korea
In-house design
Bum-Sang Lee,
Sang-Ik Lee
Web
www.lg.com

LSM-300 is a mouse with integrated scanner functionality. It also scans areas difficult to access, such as books or walls, something that is not possible with flatbed scanners. The cable is detachable so that the mouse can also be used in wireless mode, and its plain design ensures an optimal grip. The device has been optimised in such a way that – despite the built-in scan module and batteries – it still displays a size typical of a conventional computer mouse.

LSM-300 ist eine Maus mit integrierter Scanner-Funktion. Sie scannt auch an schwierigen Stellen wie in Büchern oder an Wänden, was mit Flachbettscannern nicht möglich ist. Durch das abnehmbare Kabel kann sie auch als Funkmaus genutzt werden. Ihre schlichte Form sorgt zudem für eine optimale Griffigkeit beim Halten. Sie wurde so optimiert, dass sie trotz des eingebauten Scan-Moduls und der Batterien eine Maus-typische Größe hat.

Statement by the jury
This mouse offers new scanning options. It likewise displays an ergonomic design and is easily handled.

Begründung der Jury
Der LSM-300 bietet beim Scannen neue Möglichkeiten. Zudem ist er ergonomisch gestaltet und leicht zu bedienen.

HP Z4000 Ultrabook Mouse
Computer Mouse
Computer-Maus

Manufacturer
Hewlett-Packard,
Houston, USA
In-house design
HP PPS Industrial Design Team
Web
www.hp.com

The HP Z4000 Ultrabook Mouse is
so slim and lightweight that it is ideally
suited for mobile use. It also offers
optimised ergonomics for the user. The
seamless design of the mouse underscores its high utility value, visible
in both its precise control features and
its compactness for flexible use. The
computer mouse is only 25 mm thick
and exhibits a distinctive profile.

Die HP Z4000 Ultrabook Mouse ist so
schlank und leicht, dass sie sich gut für
den mobilen Einsatz eignet. Darüber hinaus
bietet sie eine sehr gute Ergonomie für den
Nutzer. Die nahtlose Form der Maus unterstreicht ihren hohen Gebrauchswert, der
sich in der präzisen Steuerung wie auch
in ihrer Kompaktheit zeigt. Die Computer-
maus ist nur 25 mm dick und besitzt ein
markantes Profil.

Statement by the jury
This computer mouse convinces with
its smooth, reserved design. Despite its
reduced form, it offers a high degree of
user-friendliness.

Begründung der Jury
Die Computermaus überzeugt mit ihrer
zurückhaltenden glatten Gestaltung. Trotz
ihrer reduzierten Form ist sie sehr benutzerfreundlich.

Computers and information technology

GIGABYTE Aivia Neon
Air Presenter Mouse
Freihand-Präsentationsmaus

Manufacturer
Gigabyte Technology Co., Ltd.,
New Taipei City, Taiwan
In-house design
Eddie Lin
Web
www.gigabyte.com

The air presenter mouse combines the advantages of mouse, freehand mouse and presenter in one single, compact device. It is equipped with a gyroscope so that the cursor movement can be detected in the air. In addition, the mouse can be used as a remote control for PC-based media centres, extending the usage from conference room to living room. With a mini charging device integrated into the USB nano receiver, the Aivia Neon can be quickly and easily recharged on the go, making it always ready for use during a presentation.

Die Freihand-Präsentationsmaus kombiniert die Vorzüge von Maus, Freihandmaus und Presenter in einem einzigen, kompakten Gerät. Sie ist mit einem Gyroskop ausgestattet, das Mauszeigerbewegungen im Freihandbetrieb ohne Unterlage ermöglicht. Darüber hinaus lässt sich die Maus auch als Fernbedienung für PC-basierte Mediencenter verwenden und damit nicht nur im Konferenzraum, sondern auch im Wohnzimmer einsetzen. Über das in den USB-Nanoempfänger integrierte Miniladegerät wird die Aivia Neon schnell und einfach zwischendurch aufgeladen. So ist sie auch während einer Präsentation stets einsatzbereit.

Statement by the jury
The Aivia Neon air presenter mouse not only offers a wide variety of functions but also convinces with its well-conceived ergonomic design, thus facilitating easy and comfortable use.

Begründung der Jury
Die Freihand-Präsentationsmaus Aivia Neon bietet nicht nur zahlreiche Funktionen, sondern auch eine durchdachte ergonomische Gestaltung. Dadurch lässt sie sich bequem und einfach handhaben.

SpaceMouse Pro
3D Mouse
3D-Maus

Manufacturer
3D Connexion GmbH,
Munich, Germany
In-house design
Antonio Pascucci,
Benedict Brandt
Design
Design Partners
(Eugene Canavan, David Fleming),
Bray, Ireland
Web
www.3dconnexion.com
www.designpartners.com

The innovative SpaceMouse Pro is designed to facilitate intuitive 3D navigation. Digital content can be panned, rotated or zoomed to allow viewing in any position. The mouse includes a soft palm rest for the wrist, which enables comfortable working over longer periods of time. The user's hand is positioned optimally on the controller cap to improve usability and comfort. 15 programmable keys are customisable to include frequently used commands.

Statement by the jury
SpaceMouse Pro exhibits a sophisticated ergonomic design. With its well-conceived functions, it facilitates and speeds up workflow.

Die innovative SpaceMouse Pro ermöglicht eine intuitive 3D-Navigation. Digitale Inhalte lassen sich kippen, drehen oder zoomen und so in jeder Position betrachten. Die Maus besitzt eine weiche Stütze für das Handgelenk, die das komfortable Arbeiten über längere Zeiträume ermöglicht. Zudem liegt die Hand optimal auf der Controller-Kappe, was die Bedienung und den Komfort erheblich verbessert. 15 programmierbare Tasten können mit oft verwendeten Befehlen belegt werden.

Begründung der Jury
Die SpaceMouse Pro ist in hohem Maße ergonomisch gestaltet. Dank ihrer durchdachten Funktionen erleichtert und beschleunigt sie Arbeitsabläufe.

G600 MMO
Gaming Mouse
Gaming-Maus

Manufacturer
Logitech,
Newark, USA
Design
Design Partners
(Cormac Ó Conaire, James Lynch),
Bray, Ireland
Web
www.logitech.com
www.designpartners.com

This gaming mouse is designed for massive-multiplayer online (MMO) PC gaming, with a focus on optimum control. The 12-button thumb array is easily accessible and formed in such a way that it can be precisely handled without looking. In addition, the mouse includes the three classic main buttons and a scroll wheel that can be tilted sideward. All buttons are already preconfigured for MMO.

Statement by the jury
The gaming mouse offers all functions required by MMO gamers. Its design is pleasantly reserved, with numerous control buttons unobtrusively arranged on the side.

Die Gaming-Maus wurde für Massive-Multiplayer-Online (MMO)-Spieler entwickelt. Ihre Gestaltung ist auf eine bestmögliche Kontrolle während des Spiels ausgelegt. So sind die zwölf Daumentasten leicht zu erreichen und so geformt, dass sie sich auch blind präzise bedienen lassen. Zusätzlich besitzt die Maus die drei klassischen Haupttasten und ein Scrollrad, das sich auch zur Seite neigen lässt. Alle Tasten sind bereits für MMO vorkonfiguriert.

Begründung der Jury
Die Gaming-Maus bietet alle Funktionen, die MMO-Spieler brauchen. Ihre Gestaltung ist dabei angenehm zurückhaltend, da die zahlreichen Kontrolltasten unauffällig an der Seite angeordnet sind.

Rotatable mouse
Computer Mouse
Computer-Maus

Manufacturer
Lenovo (Beijing) Ltd.,
Beijing, China
In-house design
Yingjia Yao
Web
www.lenovo.com

This computer mouse features a two-part housing. Thanks to a folding mechanism, the smaller section can be angled downward, enabling the device to be operated like a mouse. At the same time, the mechanism serves as an on/off switch for the mouse. When not in use, it can be folded back to its flat state and easily transported or used as a laser pointer.

Statement by the jury
The Rotatable mouse is a practical device with dual functionality. Thanks to its unique shape, it is comfortable and easy to operate.

Das Gehäuse der Computermaus ist zweigeteilt. Durch einen Drehmechanismus lässt sich der kleinere Teil abwinkeln, sodass das Gerät wie eine Maus bedient werden kann. Gleichzeitig wird die Maus über den Drehmechanismus an- oder abgeschaltet. Wird sie nicht benötigt, kann sie wieder zur Kartenform gedreht werden. Dann ist sie so flach und handlich, dass sie einfach mitgenommen oder als Laserpointer verwendet werden kann.

Begründung der Jury
Die Rotatable mouse ist ein praktisches Gerät mit zwei Funktionen. Dank ihrer besonderen Formgebung liegt sie angenehm und leicht in der Hand.

RollerMouse Re:d
Computer Mouse
Computer-Maus

Manufacturer
Contour Design Inc.,
Windham, USA
In-house design
Steve Wang
Web
www.rollermouse.com

The operating concept of the Roller-Mouse Re:d differs from a conventional computer mouse: the mouse pointer is controlled with a roller bar, thus reducing hand motion to a minimum. The roller bar has a soft, non-slip surface that offers optimum control. It utilises seven sensors for tracking the user's hand movements and for allowing precise control of the mouse pointer. The device is fitted with snap-on wrist rests that support the wrists and hands of the user. The aluminium body of the mouse is placed centrally in front of the user to guarantee an optimised working position.

Die RollerMouse Re:d wird anders als eine herkömmliche Computermaus bedient: Der Mauszeiger wird mit einem Rollstab gesteuert, wodurch die Bewegung der Hand auf ein Minimum beschränkt wird. Der Rollstab hat eine weiche und griffige Oberfläche und bietet damit die bestmögliche Kontrolle. Dazu tragen auch die sieben Sensoren bei, die die Handbewegungen verfolgen und die präzise Steuerung des Mauszeigers ermöglichen. Handgelenkstützen, die Hände und Handgelenke ergonomisch unterstützen, können einfach an das Gerät eingeklickt werden. Das Aluminiumgehäuse der Maus liegt mittig vor dem Benutzer und sorgt so für eine optimale Arbeitshaltung.

HC1
Headset Computer

Manufacturer
Motorola Solutions,
Schaumburg, USA
In-house design
Ian Jenkins, Nicole Tricoukes
Design
Thinkable Studio (Jorg Schlieffers),
Abington, GB
Web
www.motorolasolutions.com
www.thinkable.co.uk

The HC1 is a wireless, rugged hands-free mobile headset computer designed for data, video and voice communication in harsh environments or remote locations. It has a small display equivalent to that of a 15" virtual screen and the ability to stream video via a camera for real-time collaboration. It enables access to existing wireless and network infrastructure via Bluetooth or Wi-Fi. It is controlled via voice commands or tilting of the head.

Der HC1 ist ein kabelloser, robuster Headset-Computer, der für die Daten-, Video- und Sprachkommunikation in rauen Umgebungen entwickelt wurde. Er besitzt einen kleinen Bildschirm, dessen Sichtfeld mit dem eines 15"-Bildschirms vergleichbar ist. Eine Webcam an der anderen Seite des Headsets überträgt per WLAN oder Bluetooth Bilder der Umgebung, in der sich der Anwender befindet. Gesteuert wird der Computer über Sprache oder Kopfbewegungen.

Statement by the jury
With its well-conceived design, the HC1 offers maximum mobility, creating new possibilities when working with a computer.

Begründung der Jury
Der HC1 bietet dank seiner durchdachten Gestaltung größtmögliche Mobilität. Er eröffnet so neue Möglichkeiten beim Arbeiten mit einem Computer.

Presentair™

Presenter, Laser Pointer and Stylus
Presenter, Laserpointer und Eingabestift

Manufacturer
Kensington Computer Products Group,
Redwood Shores, USA
In-house design
Kensington Industrial Design Team
(Dominic Peralta, Todd Robinson, Kee Wilcox)
Web
www.kensington.com

Presentair is a compact presenter in ergonomic stylus form. It can be used as a Bluetooth presenter, laser pointer or stylus. It features an intuitive four-button control disk providing haptic and audible feedback when pressed. Thanks to its ergonomic design, Presentair fits well in the hand and is easily stowed away. The rechargeable battery supplies up to nine hours of power during normal use, and up to ten days in standby mode. It is charged by plugging the included cable into the integrated micro-USB port.

Presentair ist ein kompaktes Präsentationsgerät in ergonomischer Stiftform. Er kann sowohl als Bluetooth-Presenter, Laserpointer oder Eingabestift verwendet werden. Die Bedienung erfolgt intuitiv über eine 4-Tasten-Steuerungsscheibe, die beim Drücken eine haptische und hörbare Rückmeldung gibt. Dank seiner ergonomischen Gestaltung liegt der Presentair gut in der Hand und lässt sich zudem problemlos verstauen. Seine Akkulaufzeit beträgt bei normaler Nutzung bis zu neun Stunden und zehn Tage im Stand-by-Modus. Aufgeladen wird er per Kabel über einen Mikro-USB-Anschluss.

Statement by the jury
Presentair is a true all-rounder that convinces with high functionality and easy handling. Its design is pleasantly self-contained.

Begründung der Jury
Presentair ist ein Multitalent, das durch seine Funktionsvielfalt und einfache Bedienung überzeugt. Seine Gestaltung ist dabei angenehm zurückhaltend.

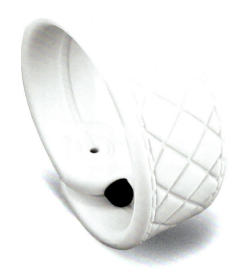

Flaxus
Stylus
Eingabestift

Manufacturer
Inno Lifestyles Ltd.,
Hong Kong
In-house design
Jack Chau
Web
www.aeglo.com
Honourable Mention

Flaxus is a flexible stylus for touch-screens. The dual-tip feature allows the user to quickly bring the stylus into a writing position. Made of surgical-grade silicone, the stylus is so slim and flexible that it can curl and be worn on the wrist when not in use. When needed for writing or typing, it is simply bent straight. The stylus is available in a variety of different colours.

Statement by the jury
Since styluses have a way of getting lost, Flaxus addresses this problem with a clever solution: it can be converted into a fashionable accessory.

Flaxus ist ein flexibler Eingabestift für Touchscreens. Er besitzt an jedem Ende eine Spitze, sodass der Nutzer den Stift schnell in eine Schreibposition bringen kann. Aus medizinischem Silikon gefertigt, ist er so flach und flexibel, dass er sich um das Handgelenk legen lässt. Wenn er zum Schreiben oder Tippen verwendet werden soll, wird er einfach geradegebogen. Er ist in verschiedenen Farben erhältlich.

Begründung der Jury
Eingabestifte gehen gerne verloren. Flaxus begegnet diesem Umstand mit einer cleveren Lösung, da er sich zu einem modischen Accessoire umgestalten lässt.

Jot Touch
Stylus
Eingabestift

Manufacturer
Adonit Corp. Ltd.,
Taipei, Taiwan
In-house design
Zach Zeliff,
Kris Perpich
Web
www.adonit.net
Honourable Mention

Jot Touch is a stylus for the iPad. With its optimal weight, it is comfortable and enjoyable to use. Aluminium and stain-less-steel elements communicate quality and durability. A pressure-sensitive tip is the special feature of this stylus, which uses Bluetooth to transfer the contact pressure to the iPad. The ballpoint with disk enables precise input.

Statement by the jury
The stylus has a high-quality, durable appearance. Its integrated spring mechanism offers the user an authentic writing experience.

Jot Touch ist ein Eingabestift für das iPad. Dank seines optimalen Gewichts liegt er gut in der Hand und ist angenehm zu verwenden. Elemente aus Aluminium und Edelstahl zeugen von Qualität und Beständigkeit. Das Besondere an diesem Stift ist seine drucksensitive Spitze, die über Bluetooth den aktuellen Anpressdruck auf das iPad überträgt. Die Kugelspitze mit Scheibe ermöglicht eine sehr präzise Eingabe.

Begründung der Jury
Der Eingabestift wirkt wertig und robust. Sein Federmechanismus vermittelt ein authentisches Schreibgefühl.

Bamboo Stylus feel carbon
Stylus
Eingabestift

Manufacturer
Wacom Company Limited,
Saitama, Japan
In-house design
Takaaki Nakata
Web
www.wacom.com
Honourable Mention

The Bamboo Stylus feel carbon is a pressure-sensitive, intuitively operated active touchscreen stylus for Windows 8 and Android tablet users. It facilitates the precise input of notes, thoughts and creative ideas on the smartphone or tablet. The high-quality stylus is made of carbon fibre and is thus lightweight and durable. With its well-conceived ergonomic design, it rests comfortably in the hand.

Statement by the jury
This stylus evinces a self-contained design in terms of colour and materials, which lend it a high-quality appearance.

Der Bamboo Stylus feel carbon ist ein drucksensitiver und intuitiv zu bedienender aktiver Touchscreen-Stift. Er richtet sich an Nutzer aktueller Windows-8- und Android-Tablets. Notizen, Gedanken und kreative Ideen können damit auf dem Smartphone oder Tablet exakt eingegeben werden. Der hochwertige Stift aus Carbonfaser ist leicht und strapazierfähig. Er bietet eine gute Ergonomie und liegt optimal in der Hand.

Begründung der Jury
Der Eingabestift zeigt eine eigenständige Farb- und Materialanmutung, die ihm ein wertiges Aussehen verleiht.

Cyber Clean® Stylus-Pro
Stylus
Eingabestift

Manufacturer
JOKER AG/SA,
Switzerland
In-house design
Royce Yu
Web
www.joker-group.net

Stylus-Pro is not only a stylus pen; it also removes dirt, dust and stains from the sensitive screens of mobile devices, such as tablet PCs and smartphones. For this, it integrates a cartridge of alcohol-free cleaning liquid, a fine spray nozzle and a hydrophilic, multilayer cloth. It can be attached to any tablet cover with the functional clip.

Statement by the jury
The Stylus-Pro is an intelligently designed stylus with integrated cleaning function. It is available in several different colours and thus fits well with all current mobile devices.

Stylus-Pro ist nicht nur ein Eingabestift, sondern beseitigt zudem Schmutz, Staub und Flecken von berührungsempfindlichen Bildschirmen mobiler Geräte wie Tablet-PCs und Smartphones. Zu diesem Zweck besitzt er eine integrierte Patrone mit einer alkoholfreien Reinigungslösung, einen fein zerstäubenden Sprühkopf sowie ein hydro-philes, mehrlagiges Reinigungsvlies. Mit dem praktischen Clip lässt er sich an jedem Tablet-Cover befestigen.

Begründung der Jury
Der Stylus-Pro ist ein durchdacht gestalteter Eingabestift mit integrierter Reinigungsfunktion. Das Produkt ist in zahlreichen Farben erhältlich und passt dadurch zu vielen aktuellen Mobilgeräten.

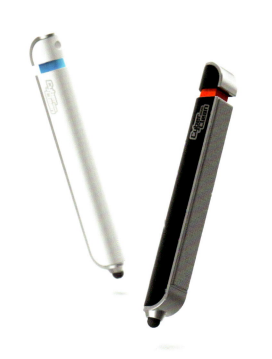

NU: Stylus
Stylus
Eingabestift

Manufacturer
OO Factory,
Seoul, South Korea
In-house design
Jongwon Park
Web
www.mobile-tail.com

This stylus weighs merely four grams and is thus lighter than a piece of A4 paper. Made of aluminium, it is robust and durable. For transport, the rounded silicone tip can simply be bent and safely inserted into the metal body. Even when pushed down hard on a screen, the display surface is never scratched or damaged. The stylus allows for precise input on tablet and mobile phone displays.

Statement by the jury
The stylus is very lightweight and rests comfortably in the hand. Another advantage is that its tip gives way under excessive pressure, thus avoiding damage to the display surface.

Der Touchscreen-Stift wiegt lediglich vier Gramm und ist somit leichter als ein DIN-A4-Blatt. Da er aus Aluminium besteht, ist er trotzdem widerstandsfähig und stabil. Die abgerundete Silikonspitze kann für den Transport einfach gebogen werden und verschwindet sicher in der Metallhülle. Display-Oberflächen werden selbst bei starkem Druck nicht beschädigt, dafür ermöglicht der Stift präzise Eingaben auf Tablets und Mobiltelefonen.

Begründung der Jury
Der Eingabestift ist sehr leicht und liegt dadurch angenehm in der Hand. Dass die Spitze auf Druck nachgibt und so Beschädigungen der Oberfläche verhindert, ist ein weiterer Pluspunkt.

Keyboard
Tastatur

Manufacturer
Keyview,
Yarkona, Israel
Design
prime.total product design
(David Keller), Tel Aviv, Israel
Web
www.thesmartype.com
www.prime-do.com
Honourable Mention

The smartype keyboard includes a display window along the top, which shows what is currently being typed. This does away with the need to move one's eyes back and forth between the screen and the keyboard, for users can see their hands and the typed results at the same time. With constant eye and head movement no longer necessary, typing errors are reduced considerably. Alternatively, the keyboard may display various widgets like a calendar, weather data or e-mails.

Am oberen Rand der smartype Tastatur ist ein Display integriert, das zeigt, was der Anwender gerade tippt. So braucht er seinen Blick nicht abwechselnd auf einen Bildschirm und die Tastatur zu richten, sondern kann seine Hände und das Tippergebnis gleichzeitig sehen. Da die ständige Bewegung von Kopf und Augen entfällt, werden Tippfehler reduziert. Alternativ können im Display verschiedene Widgets wie Kalender, Wetter oder E-Mail dargestellt werden.

Statement by the jury
This keyboard exhibits an innovative approach: the integrated display enhances comfort and efficiency.

Begründung der Jury
Die Tastatur zeigt einen innovativen Ansatz: Durch das integrierte Display werden der Komfort für den Benutzer und die Effizienz erhöht.

VZ-C6 / VZ-C3D
Document Camera
Dokumentenkamera

Manufacturer
WolfVision GmbH,
Klaus, Austria
In-house design
Johannes Fraundorfer
Design
VIEWSDESIGN GmbH
(Lutz Gebhardt, Stefan Diethelm),
Zug, Switzerland
Web
www.wolfvision.com
www.viewsdesign.ch

Belonging to a range of document cameras, the VZ-C6 visualiser was designed specifically for use in medium-sized rooms and small videoconferencing installations. The VZ-C3D stereoscopic visualiser, in turn, is a document camera capable of three-dimensional imaging. The Full HD cameras, the new image processor YSOP1 and the energy-efficient LED light projector all guarantee high-quality images. Fields of application include education, research or medicine.

Statement by the jury
These document cameras contain high-performance technology in a discreetly designed enclosure. When mounted to the ceiling, they are hardly perceivable.

Der Visualizer VZ-C6 wurde speziell für den Einsatz in mittelgroßen Räumen und für kleinere Videokonferenzanlagen entwickelt; der stereoskopische Visualizer VZ-C3D ist eine Dokumentenkamera für dreidimensionale Aufnahmen. Die Full-HD-Kameras, der neue Bildbearbeitungsprozessor YSOP1 und der energieeffiziente LED-Lichtprojektor garantieren Aufnahmen in hoher Qualität. Anwendungsgebiete finden sich in Lehre, Forschung oder im medizinischen Bereich.

Begründung der Jury
Die Dokumentenkameras beherbergen hochwertige Technik in einem dezent gestalteten Gehäuse. An der Decke angebracht, werden sie kaum wahrgenommen.

VZ-8plus4 / light4
Document Camera
Dokumentenkamera

Manufacturer
WolfVision GmbH,
Klaus, Austria
Design
VIEWSDESIGN GmbH
(Lutz Gebhardt, Philipp Maurer),
Zug, Switzerland
Web
www.wolfvision.com
www.viewsdesign.ch

This swivelling document camera is mounted on a folding support and delivers high-quality images from the whiteboard projection base. Pictures can be taken from materials like books, photos or three-dimensional objects and transmitted to beamers, monitors, digital whiteboards or videoconferencing systems. It also allows for colour and image corrections, as well as for the saving and sending of pictures. The camera also includes a preview monitor.

Statement by the jury
The document camera offers a high degree of functionality. The swivel arm and foldable stand enable individual alignment for optimal imaging.

Die schwenkbare Dokumentenkamera am Klappstativ liefert Bilder in hoher Qualität von der auch als Whiteboard nutzbaren Arbeitsplatte. Bilder werden von Vorlagen wie Büchern, Fotos oder dreidimensionalen Gegenständen aufgenommen und an Beamer, Monitore, digitale Whiteboards oder Videokonferenz-Systeme übertragen. Zusätzlich ermöglicht die Kamera Farb- und Bildkorrekturen sowie das Speichern und Versenden von Bildern. Ein Vorschaumonitor ist ebenfalls integriert.

Begründung der Jury
Die Dokumentenkamera bietet eine hohe Funktionalität. Der Schwenkarm und das Klappstativ ermöglichen die individuelle Ausrichtung am Objekt.

Donut

Cable Organiser

Kabelhalter

Manufacturer
Just Mobile Ltd.,
Taichung, Taiwan
Design
Tools Design
(Claus Jensen, Henrik Holbæk),
Copenhagen, Denmark
Web
www.just-mobile.com
www.toolsdesign.com

Donut is a handy cable box that simplifies cable management on the desktop and also the transport of cables and headsets. It offers space for multiple cables and adapters, which can be safely stored together in one place. The solid aluminium lid, which screws into a black plastic case, features 12 holes around the rim for optimum cable management.

Donut ist eine handliche Kabelbox, die das Kabelmanagement auf dem Schreibtisch und den Transport von Kabeln und Headsets erleichtert. Sie bietet Platz für mehrere Kabel und Adapter, die sicher an einem Ort aufbewahrt werden können. Das Oberteil aus solidem Aluminium wird auf das Gegenstück aus schwarzem Plastik geschraubt, an dem sich zwölf Löcher befinden, durch die die Kabel geführt werden können.

Statement by the jury
With its aluminium lid, the cable box provides a pleasant tactile experience. It is an effective option for intelligently storing and routing cables.

Begründung der Jury
Die Kabelbox bietet dank ihres Aluminiumdeckels eine angenehme Haptik. Sie ist eine effektive Möglichkeit, Kabel inte ligent aufzubewahren und zu führen.

iCB13, iCB17, iCB19, iCB21, iCB27, iCB55, iCB110, iCB112, iCB117, iCB360
Audio/Charging Cable Series
Audio-/Ladekabel-Serie

Manufacturer
iLuv Creative Technology,
Port Washington, USA
In-house design
iLuv Design Team
Web
www.iluv.com

iLuv cables are suitable for a wide variety of application areas in different environments. Each cable is constructed with double-shielded insulation, making it durable while minimising signal loss. The connectors are designed using two different materials with varying grades of hardness for additional durability and compactness. The cable necks are integrated into the cable for a clean, aesthetic look.

Statement by the jury
The cables have been designed for different uses, yet they are still recognisable as belonging to one single product family due to uniform design language.

iLuv-Kabel eignen sich für zahlreiche Verwendungszwecke in unterschiedlichen Umgebungen. Jedes Kabel bietet eine doppelte Isolierung, die es langlebig macht und den Signalverlust minimiert. Die Verbindungsstücke wurden aus zwei Materialien mit unterschiedlichen Härtegraden gefertigt. Das sorgt für zusätzliche Haltbarkeit und macht die Kabel sehr kompakt. Die Übergänge wurden in das Kabel integriert und bieten dadurch eine klare, saubere Ästhetik.

Begründung der Jury
Die Kabel wurden zwar für unterschiedliche Anwendungen konzipiert, sind aber aufgrund ihrer einheitlichen Formensprache als Vertreter einer Produktfamilie zu erkennen.

Topp USB Hub SPUH01

Manufacturer
MiPow Ltd.,
Shenzhen, China
In-house design
Stanley Wai Yung Yeung
Web
www.mipow.com
Honourable Mention

The Topp USB Hub is a compact USB 2.0 port with four outputs. It is compatible with all operating systems and features a flat USB cable that can be wrapped around the hub for transport so that all ports are neatly covered. The top is fashioned from anodised aluminium, lending the hub an elegant look.

Statement by the jury
The Topp USB Hub solves the problem of stowing cables in an intelligent way, resulting in a compact and tidy appearance.

Der Topp USB-Hub ist ein kompakter USB-2.0-Hub mit vier Buchsen, der mit allen Betriebssystemen kompatibel ist. Das flache USB-Kabel kann für den Transport um den Hub gewickelt werden, sodass alle Buchsen verdeckt sind. Seine Oberseite besteht aus eloxiertem Aluminium, was dem Hub ein elegantes Aussehen verleiht.

Begründung der Jury
Der Topp USB-Hub löst das Verstauen des Kabels auf eine intelligente Weise. So erscheint er stets kompakt und aufgeräumt.

AluCube Mini
Cable Organiser
Kabelhalter

Manufacturer
Just Mobile Ltd.,
Taichung, Taiwan
Design
WE+ design studio (Lu-Wei Chen),
Taipei, Taiwan
Web
www.just-mobile.com
www.weplus-design.com

The AluCube Mini is a simple and elegant option for avoiding cable chaos. It is crafted from a single piece of aluminium and has a rubber-lined interior for protecting the cables and keeping them securely in position. Two adhesive patches, which can be removed residue-free, may be used to fixate the cable organiser on the desktop if needed. It can be comfortably used at home, in the office or on the go.

Statement by the jury
Despite its plain design, the AluCube Mini offers a high degree of functionality, including well-conceived details like the rubber-lined interior.

Der AluCube Mini ist eine elegante und schlichte Möglichkeit, Kabelsalat zu vermeiden. Er ist aus einem Stück Aluminium gefertigt und besitzt eine gummierte Innenseite, die die Kabel schützt und zuverlässig in ihrer Position hält. Zwei Klebestreifen, die sich rückstandslos wieder entfernen lassen, helfen, den Kabelhalter bei Bedarf auf dem Schreibtisch zu fixieren. So lässt er sich zu Hause, unterwegs oder im Büro bequem verwenden.

Begründung der Jury
Der AluCube Mini bietet trotz seiner schlichten Gestaltung eine hohe Funktionalität, insbesondere durch Details wie beispielsweise die gummierte Innenseite.

Personal Cloud Storage
Persönlicher Cloud-Speicher

Manufacturer
Acer Incorporated,
New Taipei, Taiwan
In-house design
Tzuhsiang Chang
Web
www.acer.com

Acer Orbe is a storage device that can be installed in interiors such as living and office spaces. It wirelessly integrates into a local network via Wi-Fi or Ethernet. The device can then be used as a personal cloud and save or retrieve data from anywhere. Data access is established with a compatible device based on iOS, Android or Windows.

Acer Orbe ist ein Speichermedium, das sich in Innenräumen wie Wohnungen oder Büros installieren lässt. Eingebunden wird es in ein lokales Netzwerk über WLAN oder Ethernet. Der Anwender kann dieses Gerät dann wie eine persönliche Cloud nutzen und von jedem beliebigen Ort aus Daten darauf speichern oder abrufen. Dazu ist ein kompatibles Gerät nötig, das auf iOS, Android oder Windows basiert.

Statement by the jury
The storage solution has a friendly and unobtrusive form and colour design. It can smoothly blend with different living and office environments.

Begründung der Jury
Die Speicherlösung besitzt eine sympathische und unauffällige Form- und Farbgebung. So kann sie problemlos in unterschiedlichen Wohn- oder Büroräumen platziert werden.

ASUS VariDrive
Optical Disk Drive
Optisches Laufwerk

Manufacturer
ASUSTeK Computer Inc.,
Taipei, Taiwan
In-house design
ASUSDESIGN
Web
www.asus.com
www.asusdesign.com

VariDrive is a versatile optical disk drive. It connects to a device via the fast USB 3.0 port and expands its functions to include VGA, HDMI and Ethernet connectivity. It also reads and burns DVDs with up to 8x speed. A portion of the disk drive can be angled to form an L-shape, enabling placement in an upright, space-saving position. Alternatively, it can be situated horizontally.

Statement by the jury
VariDrive represents an appealing way to complement a notebook with a disk drive. Its housing conveys high value and professionalism.

VariDrive ist ein vielseitiges optisches Laufwerk. Es wird über die schnelle USB-3.0-Schnittstelle an ein Gerät angeschlossen und erweitert dieses um VGA-, HDMI- und Ethernet-Funktionalität. Zusätzlich liest und brennt es DVDs mit achtfacher Geschwindigkeit. Ein Teil des Laufwerks lässt sich zu einer L-Form abwinkeln, sodass es platzsparend hochkant aufgestellt werden kann. Alternativ kann es auch liegend platziert werden.

Begründung der Jury
VariDrive ist eine ansprechende Möglichkeit, ein Notebook mit einem Laufwerk zu ergänzen. Sein Gehäuse vermittelt Wertigkeit und Professionalität.

STOR.E Slim
External Hard Drive
Externe Festplatte

Manufacturer
Toshiba Corporation,
Storage Products Division,
Tokyo, Japan
In-house design
Masaaki Kurata
Web
www.toshibastorage.com

The STOR.E Slim is a lightweight and slender external hard drive with a memory capacity of 500 GB. With small dimensions of 107 x 75 x 9 mm, it is only slightly larger than an iPhone 4 and therefore represents a handy memory solution for on the go. Its casing consists of brushed aluminium in black or silver. Featuring plug and play, the hard drive is simply connected via USB a port to a PC or Mac. It supports the fast USB 3.0 connection and is downward compatible with USB 2.0. Password protection ensures the security of the stored data.

Statement by the jury
The aluminium casing gives the hard drive an elegant, valuable appearance. Since it is compact and light, the drive can easily be taken anywhere.

Die STOR.E Slim ist eine leichte und dünne externe Festplatte mit einer Speicherkapazität von 500 GB. Mit Abmessungen von 107 x 75 x 9 mm ist sie kaum größer als ein iPhone 4 und damit eine handliche Speicherlösung für unterwegs. Ihr Gehäuse besteht aus gebürstetem Aluminium in Schwarz oder Silber. Die Festplatte ist mit Plug-and-play ausgestattet und wird über USB einfach an einen PC oder Mac angeschlossen. Sie unterstützt das schnelle USB 3.0 und ist abwärtskompatibel mit USB 2.0. Ein Passwortschutz gewährleistet die Sicherheit der gespeicherten Daten.

Begründung der Jury
Das Aluminiumgehäuse verleiht der Festplatte eine elegante, wertige Anmutung. Da sie so kompakt und leicht ist, lässt sie sich bequem mitnehmen.

brinell Stick single-action
USB Flash Drive
USB-Stick

Manufacturer
brinell gmbh,
Karlsruhe, Germany
In-house design
Michael Föhrenbach
Web
www.brinell.net

The brinell Stick single-action is a USB flash drive that features a concealed plug connector which snaps out at the push of a button. It is available in a variety of materials, such as wood, leather, carbon, stainless steel and with Swarovski crystals, each enclosed in a stainless-steel frame. The stick has a memory capacity of 64 GB and supports the fast USB 3.0 standard, allowing for data transfer speeds of up to 90 MB/s.

Statement by the jury
Not only the concealed plug connector but also the different high-quality surfaces give this stick a high degree of design independence.

Der brinell Stick single-action ist ein USB-Stick mit einem Stecker, der im Gehäuse verborgen ist und auf Knopfdruck heraus-schnellt. Das Gehäuse ist in den Ausfüh-rungen Holz, Leder, Carbon, Edelstahl und mit Swarovski-Kristallen erhältlich, jeweils eingefasst in einen Edelstahlrahmen. Der Stick fasst ein Datenvolumen von 64 GB und unterstützt den schnellen USB-3.0-Standard. So lassen sich bis zu 90 MB/s übertragen.

Begründung der Jury
Nicht nur der im Gehäuse verborgene Stecker, sondern auch die verschiedenen wertigen Oberflächen verleihen dem Stick eine hohe Eigenständigkeit.

Firma F80
USB Flash Drive
USB-Stick

Manufacturer
Silicon Power Computer & Communications Inc., Taipei, Taiwan
In-house design
Ming-Sung Lin,
Chia-Chen Wu
Web
www.silicon-power.com

This USB flash drive has a metal case with a circular ring at the upper end, so that it may be attached to a key ring and is ready for use at any time. The one-piece storage medium is lightweight and slim and can thus be taken virtually anywhere. The metal case is fashioned from a zinc alloy, lending the flash drive both durability and an elegant appearance. Its sand-blasted surface prevents scratches and fingerprint marks. With the integrated chip-on-board technology, it is water-, dust- and vibration-proof.

Statement by the jury
The USB flash drive has a clean and functional design that places emphasis on practicality: it features a built-in ring for versatile use and also does without a cap.

Der USB-Stick besitzt ein Metallgehäuse, dessen oberes Ende in einen Ring über-geht. So kann er beispielsweise an einem Schlüsselring befestigt werden und ist damit jederzeit einsatzbereit. Der einteilige Speicher ist leicht und flach und lässt sich überallhin mitnehmen. Das Metallgehäuse besteht aus einer Zinklegierung, die den Stick robust macht und ihm eine elegante Anmutung verleiht. Die Oberfläche ist sandgestrahlt und so unempfindlich gegen Kratzer und Fingerabdrücke. Dank der Chip-on-Board-Technologie ist er wider-standsfähig gegen Staub, Erschütterungen und Wasser.

Begründung der Jury
Der USB-Stick ist sauber und funktional gestaltet. Er besitzt keine Kappe, die verlo-ren gehen kann, lässt sich aber stattdessen an seinem praktischen Ring befestigen.

Armor A15
External Hard Drive
Externe Festplatte

Manufacturer
Silicon Power Computer & Communications
Inc., Taipei, Taiwan
In-house design
Mei-Ling Chiu, Wen-Te Shen
Web
www.silicon-power.com
Honourable Mention

The Armor A15 external hard drive is designed to securely house large amounts of data. It features a rubber-coated case, which is practical while simultaneously protecting the hard drive from impact. A reinforced, wave-shaped frame and an internal suspension system likewise play a protective role. The unit can withstand accidental drops from a height of up to 1.2 metres without any damage. It is equipped with a USB 3.0 port, allowing for the fast transmission of large amounts of data. Data backup can be accomplished at just the touch of a button.

Die externe Festplatte Armor A15 eignet sich für das sichere Speichern von großen Datenmengen. Ihr Gehäuse besteht aus Gummi, der es nicht nur griffig macht, sondern die Festplatte auch vor Erschütterungen schützt. Für Schutz sorgen außerdem der verstärkte wellenförmige Rahmen und ein Federungsdämpfer im Inneren des Gehäuses. So übersteht die Festplatte schadlos auch Stürze aus bis zu 1,2 Metern Höhe. Ausgestattet ist sie mit einer USB-3.0-Schnittstelle, die die schnelle Übertragung auch von größeren Datenmengen ermöglicht. Die Datensicherung lässt sich mit einer Taste durchführen.

Huawei OceanStor HVS
IT Storage System
IT-Speichersystem

Manufacturer
Huawei Technologies Co., Ltd.,
Shenzhen, China
In-house design
Biaoke Zhong,
Lingyan Chen
Design
Design 3
(Jan-Michael von Lewinski, Philipp Christ),
Hamburg, Germany
Web
www.huawei.com
www.design3.de

The OceanStor HVS storage system is capable of intelligently managing large amounts of data, as is required in dynamically growing data centres. The system is flexible and scalable so that it can easily be expanded if needed. This flexibility also shows up in the design, the units can be seamlessly merged into one. The storage solution provides indispensable basic technical prerequisites like security and reliability. These criteria are expressed by its restrained appearance, featuring precise manufacturing and the use of high-grade materials. The design focuses on a few fine-tuned, functional details like the lamellate profiles, which ensure efficient air flow and determine the visual structure.

Mit dem Speichersystem OceanStor HVS können große Datenmengen, wie sie in dynamisch wachsenden Rechenzentren anfallen, intelligent gespeichert werden. Das System ist flexibel und skalierbar, sodass es sich bei Bedarf leicht erweitern lässt. Diese Flexibilität zeigt sich ebenso in der Gestaltung, denn die Einheiten werden nahtlos miteinander verbunden. Dabei bietet die Speicherlösung die unabdingbaren technischen Grundvoraussetzungen wie Sicherheit und Zuverlässigkeit. Diese Kriterien prägen das reduziert seriöse Erscheinungsbild mit seiner präzisen Formgebung und den hochwertigen Materialien. Die Gestaltung beschränkt sich auf wenige, fein abgestimmte, funktionale Details wie die profilierten Lamellen, die einen effizienten Luftstrom gewährleisten und die visuelle Struktur bestimmen.

Statement by the jury
The storage system conveys its most important characteristics in a discreet and reserved way. It is easily expanded yet maintains a unified, harmonious overall image.

Begründung der Jury
Das Speichersystem vermittelt seine wichtigsten Eigenschaften dezent und zurückhaltend. Es lässt sich einfach erweitern, wobei das einheitliche, harmonische Gesamtbild erhalten bleibt.

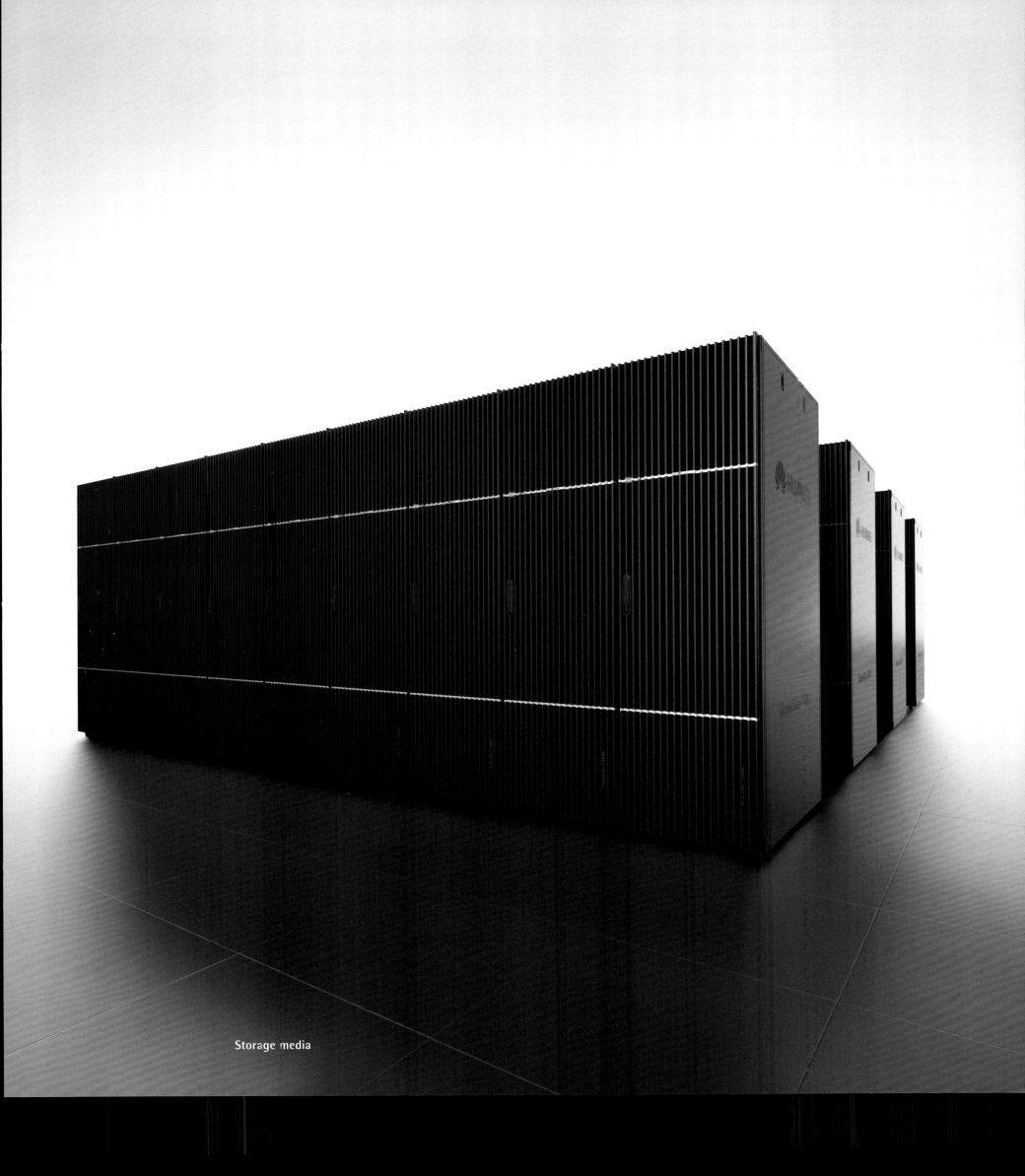

Storage media

ThinkServer Family
Server

Manufacturer
Lenovo,
Morrisville, USA
In-house design
Lenovo Industrial Design Group
Web
www.lenovo.com

The ThinkServer family is a comprehensive system with a well-conceived design. The servers are available as racks or towers and can be modularly configured, based on the space available. The flush-faced design makes it easy to locate problems and to securely maintain the units in the crowded data centres of medium- and large-sized companies. Hard drives are hot-swappable, and the vented release handles guarantee optimum airflow. User interfaces are highlighted in red for easy recognition.

Die ThinkServer-Familie ist ein umfangreiches System mit einer durchdachten Formgebung. Die Server sind als Rack oder Tower erhältlich und lassen sich modular je nach Platzangebot konfigurieren. Die bündige Gestaltung erleichtert das Auffinden von Problemen und macht die Wartung in überfüllten Datencentern in mittleren und großen Unternehmen sicherer. Festplatten lassen sich im laufenden Betrieb austauschen, die belüfteten Entriegelungsgriffe gewährleisten eine optimale Luftzufuhr. Die rot hervorgehobenen Benutzerschnittstellen sind leicht zu erkennen.

ASUS ROG
Maximus V Formula (M5F)
Motherboard

Manufacturer
ASUSTeK Computer Inc.,
Taipei, Taiwan
In-house design
ASUSDESIGN
Web
www.asus.com
www.asusdesign.com

The ROG Maximus V Formula (M5F) motherboard is equipped with both air and water cooling. The cooling parts have been enhanced with additional heat-dissipating layers, and the radial groove design is capable of efficiently diverting heat under full load as well. Heat dissipation technology components are special water channels and heat pipes made of 100 per cent copper. Gaming PCs are thus provided with optimal cooling and also extensive overclocking options.

Das Motherboard ROG Maximus V Formula (M5F) wird sowohl mit Luft als auch mit Wasser gekühlt. Die Kühlteile wurden mit zusätzlichen wärmeableitenden Schichten ausgestattet, und das radiale Rillendesign kann Wärme auch unter Volllast effizient ableiten. Bestandteile der Technologie zur Wärmeableitung sind spezielle Wasser-kanäle und Heatpipes aus 100 Prozent Kupfer. Dadurch erhalten Gaming-PCs eine optimale Kühlung und umfangreiche Möglichkeiten zum Übertakten.

Statement by the jury
The motherboard shows a well-conceived design, down to the last detail, to ensure optimal cooling.

Begründung der Jury
Das Motherboard bietet eine bis ins Detail durchdachte Gestaltung, um bestmögliche Kühlung zu gewährleisten.

V8 GTS
CPU Cooler
Prozessorkühler

Manufacturer
Cooler Master Co., Ltd.,
New Taipei, Taiwan
In-house design
Kino Chen
Web
www.coolermaster.com

The design of the high-performance CPU coolers in the V-series is inspired by race car engines. The V8 GTS cooler is equipped with a vapour chamber sealed under a partial vacuum. This causes the coolant to boil at room temperature, and heat can dissipate up to eight times faster than with heat pipes. This approach considerably reduces the heat generation common to modern multi-core processors.

Die Hochleistungs-CPU-Kühler der V-Serie sind von Rennmotoren inspiriert. Der V8 GTS basiert auf einer Wärmeleitkammer, die unter einem partiellen Vakuum versiegelt wurde. Dadurch verkocht die Kühlflüssigkeit bereits bei Raumtemperatur, und Wärme kann bis zu achtmal schneller als mit Heatpipes abtransportiert werden. So wird die Hitzeentwicklung bei modernen Multi-Core-Prozessoren deutlich reduziert.

Statement by the jury
The powerful design of this CPU cooler underscores its high performance and gives it a dynamic appearance.

Begründung der Jury
Die kraftvolle Gestaltung des Prozessorkühlers unterstreicht seine hohe Leistungsfähigkeit und verleiht ihm darüber hinaus eine dynamische Anmutung.

Chaser A71
Computer Chassis
Computergehäuse

Manufacturer
Thermaltake Technology Co., Ltd.,
Taipei, Taiwan
In-house design
TtDesign
Web
www.thermaltakecorp.com

The Chaser A71 was developed for computer gamers in need of high-performance hardware. Its design is characterised by numerous fluorescent stripes on the front panel. Three 5.25" and six 3.5" drive bays are accessible without tools. The chassis offers ample space and intelligent cable management. In addition, the power supply is situated at the bottom so as to free up enough space for particularly large mother-boards and graphic cards.

Statement by the jury
The chassis convinces with its user-friendly approach. Drives can be easily integrated, while power supply, cooling system and overall layout are optimised for gamers.

Das Chaser A71 wurde für Computerspieler entwickelt, die leistungsfähige Hardware benötigen. Seine Gestaltung ist durch fluoreszierende Streifen auf der gesamten Vorderseite gekennzeichnet. Drei 5,25"- und sechs 3,5"-Laufwerksschächte sind ohne Werkzeug zugänglich. Das Gehäuse bietet viel Platz und ein intelligentes Kabelmanagement. Zudem ist das Netzteil an der Unterseite platziert, sodass auch größere Motherboards und Grafikkarten Platz finden.

Begründung der Jury
Das Gehäuse überzeugt durch seine Nutzerfreundlichkeit. Laufwerke lassen sich einfach integrieren, Stromversor-gung, Kühlung und Layout sind für Gamer optimiert.

Level 10 Limited
Computer Chassis
Computergehäuse

Manufacturer
Thermaltake Technology Co., Ltd.,
Taipei, Taiwan
Design
BMW Group DesignworksUSA
(Sonja Schiefer),
Munich, Germany
Web
www.thermaltakecorp.com
www.designworksusa.com

The Level 10 Limited computer chassis has a modern and innovative appear-ance. Compactness and performance are reinterpreted by means of its open design. A modular structure highlights the built-in hardware components, such as the drives, which are stored in individual casing. This is complemented by the sandblasted aluminium surface accented in red, thus lending the chassis a high-grade appeal.

Statement by the jury
The design of this computer chassis is very self-contained. Due to its chamber-type construction, the interplay of components is visible for the user.

Das Computergehäuse Level 10 Limited zeigt sich modern und innovativ. Kom-paktheit und Leistung werden durch die offene Bauweise neu interpretiert. Seine modulare Struktur betont die verbauten Hardwarekomponenten, wie beispielsweise die Laufwerke, die in einzelnen Gehäusen untergebracht werden. Dazu passt die sandgestrahlte Aluminiumoberfläche, die um rote Akzente ergänzt wird und dem Gehäuse eine wertige Anmutung verleiht.

Begründung der Jury
Die Gestaltung des Computergehäuses ist sehr eigenständig. Das Zusammenspiel der Komponenten wird durch die Kammerbau-weise für den Nutzer sichtbar.

Jury 2013
International orientation and objectivity
Internationalität und Objektivität

The jurors of the Red Dot Award: Product Design
All members of the Red Dot Award: Product Design jury are appointed on the basis of independence and impartiality. They are independent designers, academics in design faculties, representatives of international design institutions, and design journalists.

The jury is international in its composition, which changes every year. These conditions assure a maximum of objectivity. The members of this year's jury are presented in alphabetical order on the following pages.

Die Juroren des Red Dot Award: Product Design
In die Jury des Red Dot Award: Product Design wird als Mitglied nur berufen, wer völlig unabhängig und unparteiisch ist. Dies sind selbstständig arbeitende Designer, Hochschullehrer der Designfakultäten, Repräsentanten internationaler Designinstitutionen und Designfachjournalisten.

Die Jury ist international besetzt und wechselt in jedem Jahr ihre Zusammensetzung. Unter diesen Voraussetzungen ist ein Höchstmaß an Objektivität gewährleistet. Auf den folgenden Seiten werden die Jurymitglieder des diesjährigen Wettbewerbs in alphabetischer Reihenfolge vorgestellt.

reddot design award
product design 2013

01

Prof. Masayo Ave
Japan

Jury member since 2004
Appointed six times
Jurymitglied seit 2004
Berufen sechs Mal

Professor Masayo Ave, born in 1962 in Tokyo, graduated in architecture and design from Hosei University in Tokyo. She worked with I. Ebihara Architect and Associates before moving to Milan in 1990 to study and graduate with a Master of Arts in industrial design from the Domus Academy. In 1992, she founded her own design studio under the name "Ave design corporation" with headquarters in Tokyo and Milan, received a scholarship in 1996 from the Akademie Schloss Solitude in Germany, and, in 2000, launched her own collection called "MasayoAve creation". She has received numerous international awards, including the ICFF 2000 Editors Award, as well as the A&W Mentor Award 2006, and has been working as a designer for companies such as Authentics, DuPont Corian, Yamaha, and Panasonic.

Since 2001, Masayo Ave has been conducting research into the design value of tactile sensitivity and, while a professor at the Berlin University of the Arts from 2004 to 2007, founded a design institute for haptic interface design, which she integrated into her own office in 2010 when it moved from Milan to Berlin.

Professorin Masayo Ave, 1962 in Tokio geboren, graduierte an der Hosei-Universität in Tokio in Architektur und Design. Sie arbeitete bei I. Ebihara Architect and Associates, bevor sie 1990 nach Mailand zog und einen M. A. in Industriedesign an der Domus Academy ablegte. 1992 eröffnete sie ihr eigenes Designstudio „Ave design corporation" mit Sitz in Tokio und Mailand, 1996 erhielt sie ein Stipendium der Akademie Schloss Solitude in Deutschland und im Jahr 2000 führte sie ihre eigene Kollektion „MasayoAve creation" ein. Sie erhielt zahlreiche internationale Auszeichnungen, u. a. den ICFF 2000 Editors Award sowie den A&W Mentor Award 2006, und ist als Designerin für Unternehmen wie Authentics, DuPont Corian, Yamaha und Panasonic tätig.

Seit 2001 erforscht Masayo Ave auf taktiler Sensitivität beruhende Gestaltungen und gründete während ihrer Professur an der Universität der Künste Berlin von 2004 bis 2007 ein Designinstitut für Haptic Interface Design, das sie 2010 in ihr von Mailand nach Berlin übersiedeltes Büro integrierte.

01 **GENESI**
Table light designed with
a cover made from a
washable open-cell polyester
and a body in chromed steel,
launched in her own collection "MasayoAve creation",
1998
Tischleuchte, die mit einem
Lampenschirm aus wasch-
barem, offenporigem Poly-
ester und einem Körper aus
verchromtem Stahl entworfen
wurde; erschienen in ihrer
eigenen Kollektion „Masayo-
Ave creation", 1998

02 **"Quadrato",
"Circolo", "Triangolo"
by Bruno Munari**
Book design and translation,
published by Heibonsha
Limited, Japan, 2010
Buchgestaltung und Über-
setzung, herausgegeben von
Heibonsha Limited, Japan,
2010

02

»I consider the sensorial value of a product as a signifier of design quality.«

»Ich erachte den sinnlichen Wert eines Produktes als Zeichen von Designqualität.«

What current trends do you see in the category "Offices"?
I have noticed that the "living office" is a new trend direction in this category. Office spaces today have to be designed not only for efficiency, but also for their living quality. I have already seen this reflected in many new products.

What are the important criteria for you as a juror in the assessment of a product?
A fine balance between advanced technology and sophisticated craftsmanship.

What challenges do you see for the future in design?
The continuous challenge that I foresee for the future of design consists in cultivating the sensorial qualities of industrial products.

Do you have a philosophy toward life?
Keeping alive my innate sense of wonder.

Welche Trends konnten Sie im Bereich „Büro" in den letzten Jahren ausmachen?
Ich sehe das „Wohnbüro" als den neuen Trend in diesem Bereich. Die Büroräume von heute müssen nicht nur der Effizienz zu liebe gestaltet werden, sondern auch Wohnqualität ausweisen, und dies schlägt sich bereits in vielen neuen Produkten nieder.

Worauf achten Sie als Jurorin, wenn Sie ein Produkt bewerten?
Eine ausgezeichnete Balance zwischen fortschrittlicher Technologie und ausgefeilter Handwerkskultur.

Welche Herausforderungen sehen Sie für die Zukunft im Design?
Die anhaltende Herausforderung, die ich für die Zukunft im Design sehe, besteht darin, die sinnliche Qualität von Industrieprodukten zu kultivieren.

Haben Sie ein Lebensmotto?
Mir meinen angeborenen Sinn des Staunens lebendig zu erhalten.

01

Marcell von Berlin
Poland/Germany

Jury member since 2013
Appointed for the first time
Jurymitglied seit 2013
Berufen zum ersten Mal

Marcell von Berlin is founder and creative director of the fashion label "Marcell von Berlin". In early 2012, the designer presented his first collection "Born for Fame" consisting of a series of high-class basics. In November 2012, he added the couture line "Atelier Marcell von Berlin". Glamorous cuts and luxurious materials – the creations are targeted at cosmopolitan customers all around Europe, including celebrities such as Judith Rakers, Anja Kling and Nazan Eckes. Before Marcell von Berlin founded his own label, he studied fashion design in New York and worked as a consultant and stylist for various fashion companies. With his own brand, he advocates the belief that collections can be successful regardless of the season, as long as they follow their own style and are further developed according to customer needs.

Marcell von Berlin ist Gründer und Creative Director des Fashion-Labels „Marcell von Berlin". Anfang 2012 präsentierte der Designer seine erste Kollektion „Born for Fame" mit einer Reihe hochwertiger Basics, im November 2012 kam die Couture-Linie „Atelier Marcell von Berlin" hinzu. Glamouröse Schnitte und luxuriöse Materialien – die Kreationen richten sich an kosmopolitische Kundinnen in ganz Europa, darunter Persönlichkeiten wie Judith Rakers, Anja Kling und Nazan Eckes. Bevor Marcell von Berlin sein eigenes Label gründete, studierte er Modedesign in New York und stand diversen Modeunternehmen beratend und als Stylist zur Seite. Mit seiner eigenen Marke vertritt er die Überzeugung, dass Kollektionen saisonunabhängig erfolgreich sein können, wenn sie einem eigenen Stil folgen und entsprechend den Kundenbedürfnissen weiterentwickelt werden.

02

»My philosophy of life is:
Do what you love and love
what you do.«

»Mein Lebensmotto lautet:
Tue, was du liebst, und liebe,
was du tust.«

What is, in your opinion, the significance of design
quality in the product categories you evaluated?
The category "Fashion, lifestyle and accessories"
demands products that are sold on the worldwide
fashion market. It is essential that these products are
targeted at a broad customer base but are unique
in their designs. They should offer innovative aspects
that are reflected in new, high-quality materials or
cuts that are thematically linked by look and usability.

What aspects of the jury process for the
Red Dot Award: Product Design 2013
have stayed in your mind in particular?
It was an incredible experience to meet, work with
and exchange views with the other jurors. Not
only were decisions difficult due to the high number
of outstanding product submissions, but it was
also challenging because every juror has a different
personality, taste and ideas on design quality.

Wie schätzen Sie den Stellenwert der Designqualität
in den von Ihnen beurteilten Produktkategorien ein?
Die Kategorie „Mode, Lifestyle und Accessoires" verlangt
nach Produkten, die auf dem weltweiten Modemarkt
vertrieben werden. Wesentlich ist, dass sie sich an eine
breite Konsumentenbasis richten, in ihrer Gestaltung
aber einzigartig sind. Sie sollten innovative Aspekte
aufweisen, die sich in neuen, wertigen Materialien oder
Schnitten widerspiegeln, in denen sich Aussehen und
Gebrauchswert thematisch miteinander verbinden.

Was ist Ihnen von der Jurierung des
Red Dot Award: Product Design 2013
besonders im Gedächtnis geblieben?
Es war eine unglaubliche Erfahrung, die anderen Juroren
zu treffen, zusammen zu arbeiten und uns auszutauschen. Aufgrund der großen Anzahl herausragender
Produkteinreichungen waren nicht nur die Entscheidun-
gen schwierig; es war auch eine Herausforderung,
da jeder Juror eine andere Persönlichkeit, Vorliebe und
Vorstellungen von Designqualität hat.

01

Gordon Bruce
USA

Jury member since 2008
Appointed six times
Jurymitglied seit 2008
Berufen sechs Mal

Gordon Bruce is the managing director of Gordon Bruce Design LLC and has been a design consultant for 40 years working with many international corporations in Europe, Asia and the USA. He has worked with numerous multinational corporations on a wide range of different products – from aeroplanes to computers and medical equipment. From 1991 to 1994, Gordon Bruce was a consulting vice-president for the Art Center College of Design's Kyoto programme and, from 1995 to 1999, chairman of Product Design for the Innovative Design Lab of Samsung (IDS) in Seoul, Korea. In 2003, he played a crucial role in helping to establish Porsche Design's office in the USA. Gordon Bruce is a visiting professor at several universities in the USA and in China. For many years, he served as head design consultant for Lenovo's Innovative Design Center (IDC) in Beijing and for Changhong in China, and he is presently working with Bühler in Switzerland and Huawei Technologies, Co., LTD in China.

Gordon Bruce ist Direktor der Gordon Bruce Design LLC und seit mittlerweile 40 Jahren als Designberater für zahlreiche internationale Unternehmen in Europa, Asien und den USA tätig. Er arbeitete mit zahlreichen multinationalen Unternehmen an unterschiedlichsten Produkten in allen Größenlagen – von Flugzeugen über Computer bis hin zu medizinischen Geräten. Von 1991 bis 1994 war Gordon Bruce beratender Vorstand des Kioto-Programms am Art Center College of Design sowie von 1995 bis 1999 Vorsitzender für Produktdesign beim „Innovative Design Lab of Samsung" (IDS) in Seoul, Korea. Im Jahr 2003 war er wesentlich daran beteiligt, das Büro von Porsche Design in den USA aufzubauen. Gordon Bruce ist Gastprofessor an zahlreichen Universitäten in den USA und in China. Über viele Jahre war er leitender Designberater für Lenovos „Innovative Design Center" (IDC) in Beijing und für Changhong in China, und derzeit arbeitet er für Bühler in der Schweiz und Huawei Technologies, Co., LTD in China.

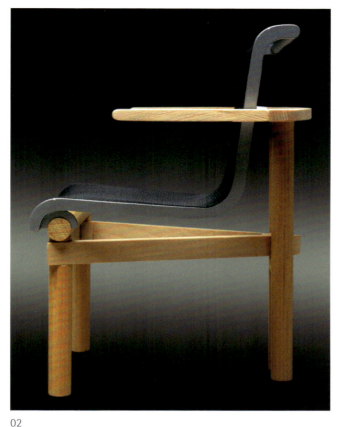

02

»Design quality is a result of experience over time with various attributes of the product and this must be considered when judging a product.«

»Designqualität zeigt sich im Erfahrungswert der verschiedenen Eigenschaften des Produkts über die Zeit hinweg, und dies muss mit in seine Beurteilung einfließen.«

What is, in your opinion, the significance of design quality in the product categories you evaluated?
From my point of view, design is all that matters in all the product categories. I think of design as a verb, not a noun. As a verb, design has multi-dimensional qualities over time. Empiric design performance takes place at every stage of interaction.

What current trends do you see in the category "Computers and information technology"?
In the laptop sector, the trend seems to be towards simplicity, refined materials and finishes, while a few unique hinge details as well as some clever transformations from laptop to tablet configurations have evolved. In desktop computers, the "all in one" classification is becoming simpler, more sophisticated and more prevalent. Thanks to novel curved-display technology, new directions have opened up in design.

Wie schätzen Sie den Stellenwert der Designqualität in den von Ihnen beurteilten Produktkategorien ein?
Meiner Ansicht nach spielt das Design des Produkts die einzige Rolle in allen Kategorien. Ich betrachte Design als ein Verb, nicht als ein Nomen. Als Verb verweist „Gestalten" auf multidimensionale Qualitäten mit zeitlichem Ablauf. In jedem Moment der Interaktion vollzieht sich eine empirische Gestaltungsleistung.

Welche Trends konnten Sie im Bereich „Computer und Informationstechnik" in den letzten Jahren ausmachen?
In der Kategorie „Laptops" hat es den Anschein, dass Schlichtheit und edle Materialien sowie Finishes immer noch im Trend liegen, während sich ein paar einzigartige Scharnierdetals sowie clevere Umwandlungen vom Laptop zur Tablet-PC-Konfiguration herausbilden. Bei Desktop-Computern wird die „All-in-one"-Klassifizierung zusehends einfacher, ausgeklügelter und verbreiteter. Mit dank neuer Technologien gewölbten Displays haben sich neue Wege in der Gestaltung eröffnet.

01

Tony K. M. Chang
Taiwan

Jury member since 2007
Appointed seven times
Jurymitglied seit 2007
Berufen sieben Mal

Tony K. M. Chang, born in 1946, studied architecture at Chung Yuan Christian University in Chung Li, Taiwan. Since 2004 he has been chief executive officer of the Taiwan Design Center as well as editor-in-chief of DESIGN magazine. Chang has made tremendous contributions to industrial design, both in his home country and across the entire Asia-Pacific region. As an expert in design management and design promotion, he has served as a consultant for governments and the corporate sector for decades. From 2005 to 2007 and from 2009 to 2011 he was an executive board member of the Icsid and master-minded the 2011 IDA Congress in Taipei. In 2008, he was elected founding chairman of the Taiwan Design Alliance, a consortium of government-supported and private design entities aimed at promoting Taiwanese design. Chang has lectured in Europe, the United States and Asia, and he frequently serves as a juror in prestigious international design competitions.

Tony K. M. Chang, 1946 geboren, studierte Architektur an der Chung Yuan Christian University in Chung Li, Taiwan. Seit 2004 ist er Chief Executive Officer des Taiwan Design Centers und zudem Chefredakteur des Magazins DESIGN. Chang hat im Bereich Industrie-design Erhebliches geleistet, sowohl in seinem Heimat-land als auch im gesamten Asien-Pazifik-Raum. Als Experte in Designmanagement und Designförderung ist er seit Jahrzehnten als Berater in Regierungs- und Unternehmenskreisen tätig. Von 2005 bis 2007 sowie 2009 bis 2011 war er Vorstandsmitglied des Icsid und federführend in der Planung des IDA Congress 2011 in Taipeh. Im Jahr 2008 wurde er zum Founding Chairman der Taiwan Design Alliance gewählt, einem Konsortium staatlich geförderter und privater Designorgane mit dem Ziel der Förderung taiwanesischen Designs. Chang hält Vorträge in Europa, den USA und Asien und fungiert häufig als Juror hochrangiger internationaler Designwettbewerbe.

02

03 / 04

»The challenges for the future of design lie in creating maximised values through minimised designs.«

»Die Herausforderung für die Zukunft des Designs besteht im Schaffen maximaler Wertigkeit durch minimale Gestaltung.«

What is, in your opinion, the significance of design quality in the product categories you evaluated?
Using the right materials and attaching importance to detailed design.

What trends have you noticed in the field of "Tableware and decoration" in recent years?
Increasing numbers of designers integrate cultural elements into their creations in order to enrich their design concepts. Silicone has been widely applied in various fields.

Do you see a correlation between the design quality of a company's products and the economic success of this company?
Undoubtedly, for a design company, design quality is a basic prerequisite in this industry. However, when it comes to the assessment of success, it is not easy to come up with a universal rule since the key factors vary depending on the situation.

Wie schätzen Sie den Stellenwert der Designqualität in den von Ihnen beurteilten Produktkategorien ein?
Sie liegt im Einsatz der richtigen Materialien und der Konzentration auf eine detaillierte Gestaltung.

Welche Trends konnten Sie im Bereich „Tableware und Dekoration" in den letzten Jahren ausmachen?
Immer mehr Gestalter integrieren kulturelle Elemente in ihre Entwürfe, um ihre Gestaltungskonzepte zu bereichern. Silikon kam in verschiedenen Bereichen ausgiebig zum Einsatz.

Sehen Sie einen Zusammenhang zwischen der Designqualität, die sich in den Produkten eines Unternehmens äußert, und dem wirtschaftlichen Erfolg dieses Unternehmens?
Zweifelsfrei ist Designqualität für ein Designunternehmen eine Grundvoraussetzung in dieser Branche. Wenn es aber um die Beurteilung von Erfolg geht, gibt es keine schnell heranziehbare universelle Regel, da die Hauptfaktoren gemäß den jeweiligen Umständen variieren.

Tony K. M. Chang

01

02

Vivian Wai-kwan Cheng
Hong Kong

Jury member since 2006
Appointed eight times
Jurymitglied seit 2006
Berufen acht Mal

Vivian Wai-kwan Cheng, born in 1962 in Hong Kong, studied industrial design at the Swire School of Design, Hong Kong Polytechnic University. In 1987, she graduated with a Bachelor of Arts degree and, in the same year, was awarded a special prize by the Federation of Hong Kong Industries in the "Young Designers of the Year" competition. She began her career as a watch designer and later continued as an industrial designer of fashion accessories. She went on to join Lambda Industrial Limited where she won the "Governor's Award for Industry: Consumer Product Design" in 1989. Vivian Wai-kwan Cheng has had a major impact on the Hong Kong design scene; she has been a member of numerous important design boards and is a senior lecturer in product design at the Hong Kong Design Institute (HKDI), where she is also in charge of international networking and collaboration.

Vivian Wai-kwan Cheng, 1962 in Hongkong geboren, studierte Industriedesign an der Swire School of Design, Hong Kong Polytechnic University. Sie schloss ihr Studium 1987 mit dem Bachelor of Arts ab und erhielt im gleichen Jahr von der Federation of Hong Kong Industries einen Sonderpreis im Wettbewerb „Young Designers of the Year". Ihre Karriere begann sie als Gestalterin für Uhren. Sie entwarf als Industriedesignerin Modeaccessoires und ging später zu Lambda Industrial Limited. Für ihre Arbeit dort gewann sie 1989 den „Governor's Award for Industry: Consumer Product Design". Vivian Wai-kwan Cheng hat in der Designszene Hongkongs viel in Bewegung gesetzt; so gehörte und gehört sie zahlreichen wichtigen Designgremien an und unterrichtet als Dozentin für Produktdesign am Hong Kong Design Institute (HKDI), wo sie zudem die Verantwortliche für internationale Zusammenarbeit und Networking ist.

03

»Everything I do today is a little better than what I did yesterday – this has been my philosophy of life since the age of 35.«

»Alles, was ich heute mache, ist etwas besser als das, was ich gestern gemacht habe – das ist meine Lebensphilosophie, seitdem ich 35 bin.«

What current trends have you noticed in the category "Fashion, lifestyle and accessories" in recent years?
This category is very much dominated by the way we live and the clothes we wear. Mobile phones are no longer just communication tools, but are also equipped with state-of-the-art technology. As such, they have also become a means for people to express their personal style and taste. There are therefore many products that have been designed to meet this rapidly growing demand.

What current trends do you see in the category "Watches and jewellery"?
We were surprised to find products that are made of recycled or environmentally friendly materials and that are produced with social responsibility in mind.

Welche Trends konnten Sie im Bereich „Mode, Lifestyle und Accessoires" in den letzten Jahren ausmachen?
Diese Kategorie wird stark von unserer Lebensart und der Kleidung, die wir tragen, dominiert. Mobiltelefone sind nicht mehr lediglich Kommunikationsinstrumente, sondern verfügen über modernste Technologie und sind zugleich Mittel, mit denen Menschen ihren persönlichen Stil und Geschmack ausdrücken. Daher gibt es eine große Anzahl an Produkten, die gestaltet worden sind, um diese schnell wachsende Nachfrage zu befriedigen.

Welche aktuellen Trends sehen Sie im Bereich „Uhren und Schmuck"?
Wir waren überrascht Produkte vorzufinden, die aus recycelten oder umweltfreundlichen Materialien hergestellt sind und deren Herstellung soziale Verantwortlichkeit berücksichtigt.

Vivian Wai-kwan Cheng

01

02

Dato' Prof.
Jimmy Choo OBE
Great Britain/
Malaysia

Jury member since 2013
Appointed for the first time
Jurymitglied seit 2013
Berufen zum ersten Mal

Dato' Professor Jimmy Choo Yeang Keat OBE is descended from a family of Malaysian shoemakers and studied at Cordwainers College, which is today part of the London College of Fashion. After graduating in 1983, he founded his own couture label and opened a show store in London's East End in the late 1980s. Choo's regular customers included Diana, the Princess of Wales. In 1996, building upon his growing international reputation, Choo launched his ready-to-wear line with Tom Yeardye. He sold his share of the ready-to-wear business in 2001 to Equinox Luxury Holdings Ltd., while continuing to run his own couture line. Choo now spends his time as ambassador for footwear education at the London College of Fashion and as spokesperson for the British Council in their promotion of British education to foreign students. In 2003, Jimmy Choo was honoured for his contribution to fashion by Queen Elizabeth II, who appointed him "Officer of the Order of the British Empire".

Dato' Professor Jimmy Choo Yeang Keat OBE entstammt einer malaysischen Schuhmacher-Familie und studierte am Cordwainers College, heute Teil des London College of Fashion. Nach seinem Abschluss 1983 gründete er Ende der 1980er Jahre sein eigenes Couture-Label und eröffnete ein Schuhgeschäft im Londoner East End. Zu seiner Stammkundschaft gehörte auch Lady Diana, Prinzessin von Wales. Aufbauend auf seiner wachsenden internationalen Reputation führte er 1996 gemeinsam mit Tom Yeardye seine Konfektionslinie ein. 2001 verkaufte Jimmy Choo seine Anteile an dem Unternehmen für Konfektionskleidung an die Equinox Luxury Holdings Ltd. und kümmerte sich weiter um seine eigene Couture-Linie. Heute engagiert sich Dato' Professor Jimmy Choo als Botschafter für Footwear Education am London College of Fashion sowie als Sprecher des British Council für die Förderung der Ausbildung ausländischer Studenten in Großbritannien. Für seine Verdienste für die Mode wurde er 2003 von Königin Elisabeth II. mit dem Titel „Officer of the Order of the British Empire" geehrt.

03

»The design quality at the Red Dot Design Award 2013 was outstanding.«

»Die Gestaltungsqualität im Red Dot Design Award 2013 war herausragend.«

What trends have you noticed in the field of "Fashion, lifestyle and accessories" in recent years?
Since the recession started in 2008, there have been a few interesting trend developments in consumer behaviour. Some consumers have chosen to focus on spending more on one special piece. Others have chosen to make do and mend, which has meant an increase in crafts, as well as an interest in products that are made with traditional skills.

Do you see a correlation between the design quality of a company's products and the economic success of this company?
If the design is of a good quality and you have the right marketing for the product it will sell well. The marketing that the Red Dot Design Award offers winners is invaluable.

Do you have a philosophy toward life?
Always move forward. Work hard. Believe in yourself.

Welche Trends konnten Sie im Bereich „Mode, Lifestyle und Accessoires" in den letzten Jahren ausmachen?
Mit der Rezession seit 2008 haben sich ein paar interessante Trends im Käuferverhalten herausgebildet. Manche Verbraucher geben für ein besonderes Stück bewusst mehr aus. Andere wollen mit wenig auskommen und reparieren lieber, was zu einem Anstieg im Handwerk und einem Interesse an traditionell gefertigten Produkten geführt hat.

Sehen Sie einen Zusammenhang zwischen der Designqualität, die sich in den Produkten eines Unternehmens äußert, und dem wirtschaftlichen Erfolg dieses Unternehmens?
Wenn die Gestaltung von guter Qualität ist und man das richtige Marketing für das Produkt hat, wird es sich gut verkaufen. Das Marketing, das der Red Dot Design Award den Gewinnern bietet, ist unschätzbar.

Haben Sie ein Lebensmotto?
Immer nach vorne gehen. Hart arbeiten. An sich selber glauben.

Dato' Prof. Jimmy Choo OBE

01

Mårten Claesson
Sweden

Jury member since 2004
Appointed eight times
Jurymitglied seit 2004
Berufen acht Mal

Mårten Claesson was born in 1970 in Lidingö, Sweden. After studying at the Vasa Technical College in Stockholm in the department of Construction Engineering and at the Parsons School of Design in New York in the departments of Architecture and Product Design, he graduated in 1994 with an MFA degree from Konstfack, the University College of Arts, Crafts and Design in Stockholm. He is co-founder of the Swedish design partnership "Claesson Koivisto Rune", which is multidisciplinary in the classic Scandinavian way and pursues the practice of both architecture and design. Mårten Claesson is also a writer and lecturer in the field of architecture and design.

Mårten Claesson wurde 1970 in Lidingö, Schweden, geboren. Nach einem Studium am Vasa Technical College in Stockholm im Bereich „Construction Engineering" und an der Parsons School of Design in New York in den Bereichen „Architecture" und „Product Design" schloss er 1994 seine Ausbildung an der Konstfack, dem University College of Arts, Crafts and Design, in Stockholm ab. Er ist Mitbegründer der Design-Partnerschaft „Claesson Koivisto Rune", die sich nach klassischer skandinavischer Art multidisziplinär sowohl mit Architektur als auch mit Design beschäftigt. Mårten Claesson ist darüber hinaus als Autor und Dozent im Bereich „Architektur und Design" tätig.

02

**»A company can survive
for a while without design,
but not in the long run.«**

**»Ein Unternehmen kann ohne
Design kurzfristig überleben,
aber nicht auf Dauer.«**

**What is, in your opinion, the significance of design
quality in the product categories you evaluated?**
The difference between "home" and "contract"
furniture is blurred today. This means that you may
well choose to furnish your living room or office
with the same type of products. From the contract
side comes the tradition of quality in manufacturing
and professional design development and from the
home side the demand for aesthetic, poetic and
elegant design solutions.

**What current trends do you see
in the category "Offices"?**
The trend in offices is a kind of resurrection of the
cubicle office of the 1960s. Although it is described
as room-within-room furniture, it still creates
privacy.

**What are the important criteria for you
as a juror in the assessment of a product?**
Good design is like natural beauty. Not-so-good
design is like make-up or plastic surgery: it always
reveals itself. I prefer the real thing.

**Wie schätzen Sie den Stellenwert der Designqualität
in den von Ihnen beurteilten Produktkategorien ein?**
Die Unterscheidung zwischen ausgesprochenen Wohn-
und Büromöbeln verwischt zusehends. Das bedeutet,
dass man sich eben entscheiden kann, das Wohnzimmer
oder das Büro mit den gleichen Produkttypen auszu-
statten. Aus dem Bürosegment stammt die Tradition
der Qualität in professioneller Designentwicklung und
Herstellung. Und aus dem Wohnsegment die Forderung
nach eleganten, poetischen und ästhetischen Gestal-
tungslösungen.

**Welche Trends konnten Sie im Bereich
„Büro" in den letzten Jahren ausmachen?**
Der Trend in Büros ist eine Art Wiederauferstehung
des Zellenbüros aus den 1960ern. Obwohl als Raum-
im-Raum-Mobiliar beschrieben, entsteht so doch eine
Abgrenzung.

**Worauf achten Sie als Juror,
wenn Sie ein Produkt bewerten?**
Gute Gestaltung ist wie natürliche Schönheit. Nicht so
gute Gestaltung ist wie Makeup oder plastische Chirurgie,
sie verrät sich stets. Ich mag das Natürliche.

Mårten Claesson

435

01

Guto Indio da Costa
Brazil

Jury member since 2011
Appointed three times
Jurymitglied seit 2011
Berufen drei Mal

Guto Indio da Costa, born in 1969 in Rio de Janeiro, studied product design and graduated from the Art Center College of Design in Switzerland in 1993. He is design director of Indio da Costa A.U.D.T., a consultancy based in Rio de Janeiro, which develops architectural, urban planning, design and transportation projects. It works with a multidisciplinary strategic-creative group of designers, architects and urban planners, supported by a variety of other specialists.

Guto Indio da Costa is a member of the Design Council of the State of Rio de Janeiro, former Vice President of the Brazilian Design Association (Abedesign) and founder of CBDI (Brazilian Industrial Design Council). He is an active speaker for and contributor to the "ArcDesign" magazine in São Paulo and has been a jury member of many design competitions in Brazil and abroad.

Guto Indio da Costa, geboren 1969 in Rio de Janeiro, studierte Produktdesign und machte 1993 seinen Abschluss am Art Center College of Design in der Schweiz. Er ist Gestaltungsdirektor von Indio da Costa A.U.D.T., einem in Rio de Janeiro ansässigen Beratungsunternehmen, das Projekte in Architektur, Stadtplanung, Design- und Transportwesen entwickelt und mit einem multidisziplinären, strategisch-kreativen Team aus Designern, Architekten und Stadtplanern sowie mit der Unterstützung weiterer Spezialisten operiert.

Guto Indio da Costa ist Mitglied des Design Councils des Bundesstaates Rio de Janeiro, ehemaliger Vize-Präsident der brasilianischen Designvereinigung (Abedesign) und Gründer des CBDI (Industrial Design Council Brasilien). Er ist aktiver Sprecher und Mitarbeiter der Zeitschrift „ArcDesign" in São Paulo und ist als Jurymitglied in vielen Designwettbewerben in Brasilien und im Ausland tätig.

02

»Design quality has become an extremely important factor for economic success – not only for companies but also for the countries they represent.«

»Designqualität ist für den wirtschaftlichen Erfolg mittlerweile von enormer Bedeutung – nicht nur für Unternehmen, sondern auch für die Länder, die sie repräsentieren.«

What is, in your opinion, the significance of design quality in the product categories you evaluated?
As long as high quality is universal and technology is used by different manufacturers in the same way, design is the only way to differentiate products and experiences. By that I mean design in the broader sense; not only form and aesthetics, but also the way in which a product functions, the way it is manufactured, distributed and used by people. This is why design has become highly significant in every field and category.

What challenges do you see for the future in design?
I believe the world population will have to change how it lives, manufactures, uses and consumes products. To promote the correct balance between economy, energy and the environment will be the greatest challenge design has to face in the near future.

Wie schätzen Sie den Stellenwert der Designqualität in den von Ihnen beurteilten Produktkategorien ein?
Solange hohe Qualität ein allgemeines Gut ist und Technologie von verschiedenen Herstellern gleichermaßen benutzt wird, ist Design der einzige Weg, um Produkte und Erfahrungen voneinander abzuheben. Design im weiteren Sinne: nicht nur die Form und die Ästhetik, sondern auch die Art, wie ein Produkt funktioniert, die Art, wie es hergestellt, vertrieben und von den Menschen benutzt wird. Deshalb hat Design in allen Kategorien und Bereichen einen sehr hohen Stellenwert erreicht.

Welche Herausforderungen sehen Sie für die Zukunft im Design?
Ich glaube, die Weltbevölkerung wird die Art, wie sie lebt, produziert und Dinge benutzt und konsumiert, verändern müssen. In der nahen Zukunft besteht die größte Herausforderung für das Design darin, die richtige Balance zwischen Wirtschaft, Energie und Umwelt voranzubringen.

01

Vincent Créance
France

Jury member since 2008
Appointed three times
Jurymitglied seit 2008
Berufen drei Mal

Vincent Créance, born in 1961, graduated from the Ecole Supérieure de Design Industriel. He began his career in 1985 at the Plan Créatif Agency, where he became design director in 1990. He joined Alcatel in 1996 as design director for all telephone activities and became vice-president Brand for Alcatel Mobile Phone, in his role as design/UI and communication director, in 1999. In 2006, Vincent Créance was appointed president of MBD Design, one of the principal design agencies in France, providing design solutions in transportation and product design. MBD Design is also a globally leading agency for railway transportation design. He is a member of APCI (Agency for the Promotion of Industrial Creation) and the ENSCI (National College of Industrial Design) board of directors, as well as a member of the scientific advisory board of Strate College.

Vincent Créance, 1961 geboren, machte seinen Abschluss an der Ecole Supérieure de Design Industriel. Seine berufliche Laufbahn begann er 1985 bei Plan Créatif Agency, wo er 1990 zum Design Director aufstieg. 1996 ging er als Design Director für sämtliche Telefonaktivitäten zu Alcatel und wurde in seiner Funktion als Design/UI und Communication Director 1999 zum Vice-President Brand für Alcatel Mobile Phone ernannt. Vincent Créance wurde 2006 President von MBD Design, einer der wichtigsten Designagenturen in Frankreich, und entwickelte Designlösungen für Verkehrs- und Produktdesign. MBD Design ist zudem eine weltweit führende Agentur für schienengebundenes Verkehrsdesign. Er ist Mitglied von APCI (Agency for the Promotion of Industrial Creation), Vorstand von ENSCI (National College of Industrial Design) und Mitglied des wissenschaftlichen Beirats des Strate College.

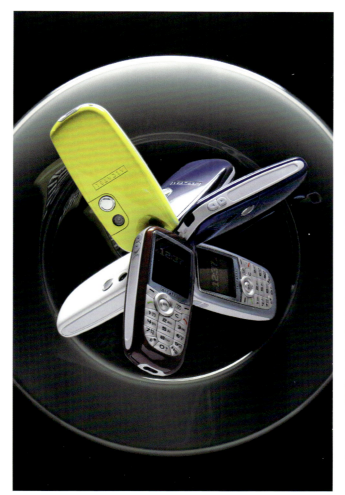

02

»Above all, design must arouse emotions and desire and then link them intimately to the brand.«

»Gestaltung muss vor allem Emotion und Verlangen hervor-rufen, und diese dann aufs Engste mit der Marke verknüpfen.«

What are the important criteria for you as a juror in the assessment of a product?
As always: simplicity, immediate understandability, finishing and detailing. However, I would also add beauty. It is a pity that many designers don't dare to use this word more frequently.

Do you see a correlation between the design quality of a company's products and the economic success of this company?
Yes, obviously. Good design is much more powerful than technical or marketing efficiency, because design talks primarily to the heart. It plays with desire instead of rationality, and desire always wins out over reason... If there is passion, the price is no longer so important!

Do you have a philosophy toward life?
Don't play safe! Seek out intense moments! Accept that taking risks is necessary, and be brave. In other words, conduct your own life like a design project!

Worauf achten Sie als Juror, wenn Sie ein Produkt bewerten?
Wie immer: Einfachheit, unmittelbare Verständlichkeit, Finishing und Details. Allerdings würde ich außerdem noch Schönheit hinzufügen. Es ist schade, dass sich viele Designer nicht trauen, dieses Wort öfter zu gebrauchen.

Sehen Sie einen Zusammenhang zwischen der Designqualität, die sich in den Produkten eines Unternehmens äußert, und dem wirtschaftlichen Erfolg dieses Unternehmens?
Ja, ganz offensichtlich. Gute Gestaltung ist sehr viel wirkungsmächtiger als Technik- oder Vermarktungseffizienz, denn Gestaltung richtet sich zunächst ans Herz. Es spielt mit dem Verlangen anstelle der Rationalität, und Verlangen gewinnt immer gegen Vernunft ... Wenn das Verlangen da ist, ist der Preis nicht mehr so wichtig!

Haben Sie ein Lebensmotto?
Gehe nicht auf Nummer sicher! Suche intensive Momente! Akzeptiere, dass es nötig ist, Risiken einzugehen, und sei tapfer. Anders gesagt, führe dein eigenes Leben wie ein Gestaltungsprojekt!

Vincent Créance

439

01

Martin Darbyshire
Great Britain

Jury member since 2008
Appointed six times
Jurymitglied seit 2008
Berufen sechs Mal

Martin Darbyshire is founder and managing director of the design studio "tangerine" in London and Seoul and has nearly 30 years of experience in the design sector. Before founding "tangerine" in 1989, he joined Moggridge Associates and worked in San Francisco at ID TWO (now IDEO). Most notably, Darbyshire led the multidisciplinary team that created both generations of the "Club World" business-class aircraft seating for British Airways – the world's first fully flat bed in the business class which, since its launch in 2000, has remained the profit engine of the airline. Martin Darbyshire also worked as visiting professor at the University of the Arts, Central Saint Martins; he is an industry spokesperson on design and innovation and was board member of the Icsid from 2007 to 2012.

Martin Darbyshire ist Gründer sowie Geschäftsführer des Designbüros „tangerine" mit Standorten in London und Seoul und kann mittlerweile auf eine fast dreißigjährige Erfahrung in der Designbranche zurückblicken. Bevor er „tangerine" 1989 gründete, arbeitete er bei Moggridge Associates sowie bei ID TWO (heute IDEO) in San Francisco. Darbyshire leitete das multidisziplinäre Team, das beide Generationen der „Club World", der so bezeichneten Business-Class-Flugzeugsitze für British Airways, entwickelte. Das weltweit erste komplett flache Bett in einer Business Class hat der Airline seit seiner Markteinführung im Jahr 2000 enorme Umsatzzahlen beschert. Martin Darbyshire arbeitete darüber hinaus als Gastprofessor an der University of the Arts, Central Saint Martins; er ist Wortführer für Design und Innovation und war von 2007 bis 2012 Vorstandsmitglied des Icsid.

01　Hyundai Heavy Industries
　　(HHI)
　　"tangerine" created a future
　　vision for HHI construction
　　vehicles – the exterior view
　　of a futuristic excavator
　　„tangerine" schuf eine Zu-
　　kunftsvision für die HHI-
　　Baumaschinen – die Außen-
　　ansicht eines futuristischer
　　Baggergeräts

02　British Airways
　　business class seats
　　Sitze für die Business
　　Class von British Airways
　　For more than a decade,
　　this design by "tangerine"
　　has remained the profit
　　engine of British Airways
　　Seit mehr als einem Jahrzehnt
　　ist diese Gestaltung von
　　„tangerine" bereits der
　　Gewinnmotor von British
　　Airways

02

»Money can't buy you happiness but good design can!«

»Geld kann nicht glücklich machen, gutes Design dagegen schon!«

What trends have you noticed in the field of "Automotive and transportation" in recent years?
In automotive design, there is a continuing trend to concentrate design energy and investment on vehicle exteriors rather than what is inside and particularly on the user-interface. In transportation, notably in contract vehicles, there is some interesting use of materials with clever functionality for the user, producing "more for less". That is what clever design is all about and is one of the reasons I love judging Red Dot.

How do you assess the significance of design quality for the success of a company?
The basis of good design has to be that the product must deliver groundbreaking innovation and be commercially viable. As a jury, we always weigh up the commercial potential of a design as well as considering how it enhances and underpins the brand's reputation in the market it has been designed for.

Welche Trends konnten Sie im Bereich „Automotive und Transport" in den letzten Jahren ausmachen?
Im Fahrzeugdesign besteht ein anhaltender Trend, Investition und gestalterische Energie auf die Außengestaltung zu konzentrieren anstatt auf das, was im Inneren passiert, und vor allem auf die Bedienoberfläche. Im Bereich „Transport", insbesondere bei Vertragsfahrzeugen, zeigt sich ein interessanter Einsatz von Materialien mit ausgeklügelter Funktionalität für den Nutzer, wodurch „mehr für weniger" entsteht. Das ist, worum es bei intelligentem Design geht, und einer der Gründe, warum ich gerne Juror bei Red Dot bin.

Wie schätzen Sie den Stellenwert der Designqualität in den von Ihnen beurteilten Produktkategorien ein?
Die Grundlage guten Designs muss sein, dass das Produkt bahnbrechende Innovation aufweist und finanziell realisierbar ist. Als Jury wägen wir immer auch das kommerzielle Potenzial einer Gestaltung ab sowie die Frage, wie es die Markenreputation in dem Markt, für den es entworfen wurde, stärken und untermauern kann.

Martin Darbyshire

01

Prof. Stefan Diez
Germany

Jury member since 2011
Appointed three times
Jurymitglied seit 2011
Berufen drei Mal

Professor Stefan Diez, born 1971 in Freising, Germany, studied industrial design at the Stuttgart State Academy of Art and Design. In 2003 he opened his own studio in Munich and became professor at the Karlsruhe University of Applied Sciences in 2010. Together with two partners, he took over the art direction of Authentics from 2008 to 2009.

Stefan Diez specialises in product and exhibition design, working for Bree, e15, Established & Sons, Merten, Moroso, Rosenthal, Thonet, Wilkhahn and others; his designs have garnered many awards and are on display in various exhibitions. He is considered to be one of the most innovative and promising designers.

Professor Stefan Diez, 1971 in Freising geboren, studierte Industriedesign an der Staatlichen Akademie der Bildenden Künste Stuttgart. 2003 eröffnete er sein eigenes Studio in München und seit 2010 ist er Professor an der Staatlichen Hochschule für Gestaltung Karlsruhe. Zusammen mit zwei Partnern übernahm er von 2008 bis 2009 die Art Direction der Firma Authentics.

Stefan Diez ist im Produkt- und Ausstellungsdesign u. a. für Bree, e15, Established & Sons, Merten, Moroso, Rosenthal, Thonet und Wilkhahn tätig; seine Arbeiten wurden vielfach ausgezeichnet und in zahlreichen Ausstellungen präsentiert. Er gilt als einer der innovativsten und vielversprechendsten Designer.

01 **THIS THAT OTHER**
Seating series for e15 with
side chair, lounge chair and
barstool, made of moulded
oak-veneered plywood,
lacquered
Sitzserie für e15 mit Beistell-
stuhl, Sessel und Barhocker
aus geformten, eichenfurnierten
Schichtholzplatten, lackiert

02 **TROPEZ**
Collection of outdoor
furniture for Saskia Diez,
made of thermo-lacquered
aluminium profiles and
water-repellent fabric
Kollektion von Outdoor-
Möbeln für Saskia Diez,
gefertigt aus thermolackier-
ten Aluminiumprofilen und
wasserabweisendem Stoff

02

»Over the years the Red Dot jury has almost become a kind of circle of friends, even though there are new jurors every year and others take a break.«

»Die Red Dot-Jury ist über die Jahre fast eine Art Freundeskreis geworden, obwohl jedes Jahr neue Juroren hinzukommen und andere Pause machen.«

What is, in your opinion, the significance of design quality in the product categories you evaluated?
Besides fashion, design is currently paying most attention to furniture and lamps. Thus the jury's expectations were high. What inspired me was also the large number of innovative projects, above all in the area of architectural lighting.

What are the important criteria for you as a juror in the assessment of a product?
This always concerns the question of necessity, the logical consistency and the distinctiveness of a design. And it of course also concerns first impressions. It goes without saying that products are complex and multi-faceted, but they have to function at all levels and they have to fascinate. On top of that, I believe that the relationship between price and effort must be balanced.

Wie schätzen Sie den Stellenwert der Designqualität in den von Ihnen beurteilten Produktkategorien ein?
Neben der Mode bekommen Möbel und Leuchten gerade die meiste Aufmerksamkeit im Design. Von daher waren die Erwartungen der Jury entsprechend hoch. Was mich begeistert hat, war aber auch der hohe Anteil an innovativen Projekten, vor allem im Bereich des Architectural Lighting.

Worauf achten Sie als Juror, wenn Sie ein Produkt bewerten?
Es geht immer um die Frage nach der Notwendigkeit, um die Schlüssigkeit und um die Prägnanz eines Ent-wurfs. Und natürlich um den ersten Eindruck. Natürlich sind Produkte komplex und haben viele Schichten, aber sie müssen auf allen Ebenen funktionieren und einen fesseln. Ich finde, dass auch Preis und Aufwand in einem guten Verhältnis stehen müssen.

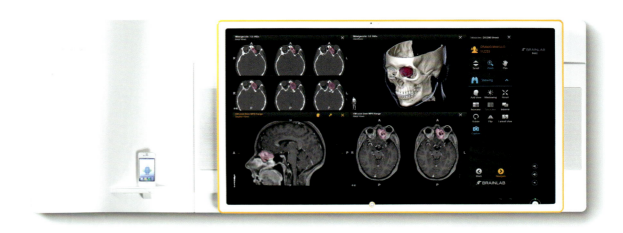

01

Stefan Eckstein
Germany

Jury member since 2013
Appointed for the first time
Jurymitglied seit 2013
Berufen zum ersten Mal

Stefan Eckstein, born 1961 in Stuttgart, studied industrial design at the Muthesius Academy of Fine Arts and Design in Kiel and attended courses at the Anthropological Institute of the Christian Albrecht University of Kiel. After holding a position as an assistant in Hamburg, he founded his own design studio in 1989 in Munich.

Since 2012, Stefan Eckstein has been the president of the Association of German Industrial Designers (VDID), giving highest priority to authenticity, responsibility and ethics. As project leader, he and his team developed the VDID Codex, which lays out the ethical values of the profession of industrial designers. His studio has received many design awards in national and international award competitions. Today, Stefan Eckstein is recognised as a renowned designer in industrial design, in particular in the fields of medical technology, as well as production and consumer goods.

Stefan Eckstein,1961 in Stuttgart geboren, studierte an der Muthesius-Hochschule in Kiel Industriedesign und besuchte Seminare am Anthropologischen Institut der Christian-Albrechts-Universität zu Kiel. Nach einer Assistenz in Hamburg gründete er 1989 sein eigenes Designstudio in München.

Seit 2012 ist Stefan Eckstein Präsident des Verbandes Deutscher Industrie Designer (VDID) und gibt Authentizität, Verantwortung und Ethik oberste Priorität. Als Projektleiter entwickelte er mit einem Team den VDID Codex, der die ethischen Werte des Berufsstandes der Industriedesigner beschreibt. Sein Büro erhielt zahlreiche Auszeichnungen in nationalen wie internationalen Wettbewerben. Stefan Eckstein zählt heute zu den renommierten Designern für Industriedesign in der Medizintechnik sowie Investitions- und Konsumgütergestaltung.

02

»In future, industrial designers will have to prove themselves with new role models and fascinating product strategies.«

»Industriedesigner müssen in Zukunft mit neuen Leitbildern und faszinierenden Produkt- strategien überzeugen.«

What trends have you noticed in the field of "Bathrooms, spas and air conditioning" in recent years?
High-quality real materials will continue to be a trend. Alongside the traditional reduced language of form, which has visibly come to the fore over the years, the bionic language of form will in future also increasingly establish itself.

How do you assess the significance of design quality for the success of a company?
The responsibility that is placed on industrial design and individual designers today is enormous. The success of all people involved in the development, production and marketing of a product, as well as the product support over its entire lifecycle, is often heightened disproportionly by the achievements of the industrial designers. Industrial design is therefore essential for companies and is clearly a critical success factor.

Welche Trends konnten Sie im Bereich „Bad, Wellness und Klima" in den letzten Jahren ausmachen?
Ein Trend werden weiter hochwertige Echt-Materialien sein. Neben der schon tradierten reduzierten Formsprache, die sich hier über Jahre sichtbar in den Vordergrund gespielt hat, wird sich in Zukunft auch die bionische Formsprache mehr und mehr etablieren.

Wie schätzen Sie den Stellenwert des Designs für den Erfolg eines Unternehmens ein?
Die Verantwortung, die Industriedesign bzw. Designer für Unternehmen heute tragen, ist enorm. Der Erfolg aller, die an der Entwicklung, Herstellung und Vermarktung eines Produkts und seiner Pflege über den gesamten Lebenszyklus mitwirken, wird häufig überproportional durch die Leistungen der Industriedesigner gesteigert. Hieraus ergibt sich, dass das Industriedesign elementar für die Zukunft von Unternehmen steht und ihren Erfolg sichtbar kennzeichnet.

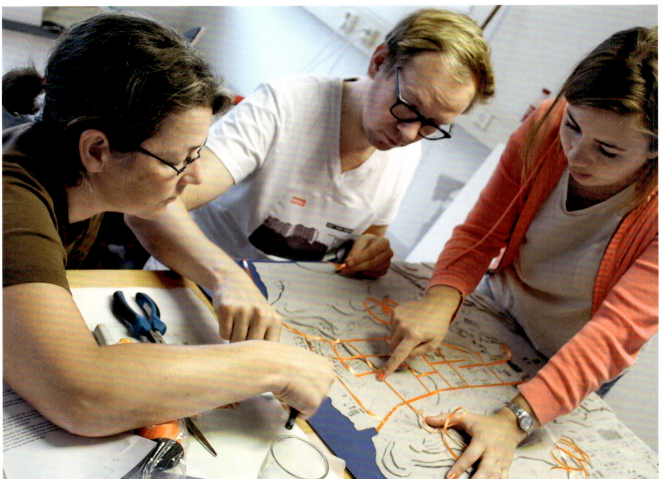

01

Robin Edman
Sweden

Jury member since 2007
Appointed six times
Jurymitglied seit 2007
Berufen sechs Mal

Robin Edman, born in 1956 and raised in Sweden, studied industrial design at the Rhode Island School of Design in Providence, USA. After graduating in 1981, he started as an industrial designer and later advanced to assistant director of Industrial Design at AB Electrolux in Stockholm. In 1989, he moved to Columbus, USA, as vice president of Industrial Design for the Frigidaire Company, where he also initiated and ran the Electrolux Global Concept Design Team for future forecasting of user needs.

In 1997, Edman moved back to Stockholm as vice president of Electrolux Global Design and was appointed chief executive of the Swedish Industrial Design Foundation (SVID) in 2001. From 2003 to 2007, Robin Edman was a board member of the International Council of Societies of Industrial Design, and from 2005 to 2007, he served as its treasurer. Since 2012 he has been a board member of the BEDA, the Bureau of European Design Associations.

Robin Edman, geboren 1956 und aufgewachsen in Schweden, studierte Industriedesign an der School of Design in Providence, Rhode Island, USA. Nach seinem Abschluss 1981 arbeitete er zunächst als Industriedesigner, später als Assistant Director für Industriedesign bei AB Electrolux in Stockholm. 1989 zog er nach Columbus, USA, um bei der Frigidaire Company die Position des Vizepräsidenten für Industriedesign zu übernehmen sowie das Electrolux Global Concept Design Team zur Vorhersage der Verbraucherbedürfnisse ins Leben zu rufen und zu leiten.

1997 kehrte Edman als stellvertretender Geschäftsführer bei Electrolux Global Design nach Stockholm zurück und wurde 2001 zum Geschäftsführer der Stiftung Schwedisches Industriedesign (SVID) ernannt. Von 2003 bis 2007 war Robin Edman Mitglied im Vorstand des International Council of Societies of Industrial Design, wobei er zwischen 2005 und 2007 das Amt des Schatzmeisters innehatte. Seit 2012 ist er Vorstandsmitglied des BEDA, Bureau of European Design Associations.

01 The Swedish Industrial Design Foundation's destination programme focuses on how the design process can develop attractive regions, locations, environments or events
Das Zielprogramm der Stiftung Schwedisches Industriedesign fokussiert sich darauf, wie der Design-prozess attraktive Regionen, Orte, Umgebungen oder Veranstaltungen entwickeln kann

02 Swedish Design Research Journal
The publication of the Swedish Industrial Design Foundation publishes research-based articles that explore how design can contribute to the sustainable development of industry, public sector and society
Die Publikation der Stiftung Schwedisches Industriedesign veröffentlicht forschungsba-sierte Artikel, die darstellen, was Design zu einer nachhal-tigen Entwicklung in Industrie, dem öffentlichen Sektor und der Gesellschaft beitragen kann

02

»I would like to see design become one of the obvious drivers of sustainable development and have it included in all innovation and change processes.«

»Ich würde Design gerne als eine der klaren Antriebskräfte für nachhaltige Entwicklung in allen Innovations- und Wandlungs-prozessen einbezogen sehen.«

What trends have you noticed in the field of "Gardens" in recent years?
Across the board there is a noticeable improvement in quality levels, a better use of materials and a "make my life easy" approach by manufacturers.

What current trends do you see in the category "Outdoor, leisure and sports"?
Multi-functionality and combinations instead of a new product for every use. Take for example clothing, strollers and toys. The improved use of new materials and clever solutions opens up new opportunities and result in better products for end-users.

What aspects of the jury process for the Red Dot Award: Product Design 2013 have stayed in your mind in particular?
Well-organised, tight schedules and 100 per cent support by the Red Dot team. A friendly and lively atmosphere and, of course, the networking and presence of so many great minds from around the world.

Welche Trends konnten Sie im Bereich „Garten" in den letzten Jahren ausmachen?
Auf ganzer Linie lässt sich ein höherer Qualitätsstan-dard, ein verbesserter Einsatz von Materialien und ein „Mach mein Leben leicht"-Ansatz durch die Hersteller beobachten.

Welche Trends konnten Sie im Bereich „Outdoor, Freizeit und Sport" in den letzten Jahren ausmachen?
Multifunktionalität und Kombinationen anstelle eines neuen Produkts für jeden Gebrauch. Beispielhaft ist dies bei Kleidung, Kinderwagen und Spielzeugen zu beobachten. Der verbesserte Einsatz neuer Materialien und intelligenter Lösungen schafft neue Möglichkeiten und bessere Produkte für die Nutzer.

Was ist Ihnen von der Jurierung des Red Dot Award: Product Design 2013 besonders im Gedächtnis geblieben?
Gut durchorganisierte, straffe Zeitpläne und 100-pro-zentige Unterstützung durch das Red Dot-Team. Eine freundliche und animierte Atmosphäre und selbstver-ständlich das Netzwerk und die Präsenz so vieler kluger Köpfe aus aller Welt.

Robin Edman

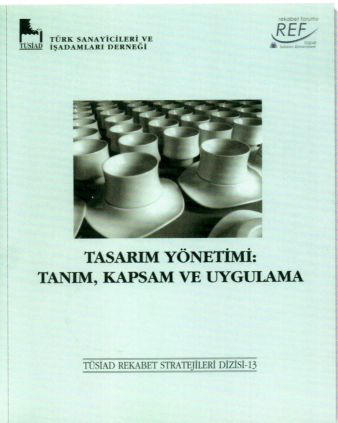

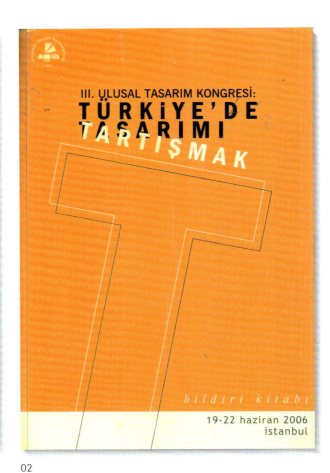

01

02

Prof. Dr. Alpay Er
Turkey

Jury member since 2012
Appointed twice
Jurymitglied seit 2012
Berufen zwei Mal

Professor Dr. Alpay Er studied industrial design at the Middle East Technical University (METU) in Ankara, Turkey, and completed his Ph.D. at Manchester Metropolitan University in the UK in 1994. He joined the Department of Industrial Product Design at Istanbul Teknik Universitesi (ITU) in 1997, where he served as chairperson from 2006 to 2013. Currently, he is founding chair of the Industrial Design Department at Ozyegin University in Istanbul. Alpay Er is also active in the Industrial Designers' Society of Turkey (ETMK), serving as an executive committee member as well as chairman of the ETMK Istanbul Branch. In cooperation with ITU and the Istanbul Chamber of Industry (ISO) he initiated the Industrial Design for SMEs project, the first programme for the education and promotion of design aimed at small and medium-sized enterprises (SMEs) in Turkey. In addition, he works as a consultant for various institutions and companies, is a member of the Executive Hosting Committee for the 2013 International Design Alliance (IDA) Congress hosted in Istanbul, and is a member of the Design Research Society (DRS). Alpay Er is also a member of the Icsid Executive Board (2011–2013).

Professor Dr. Alpay Er studierte Industriedesign an der Middle East Technical University (METU) in Ankara, Türkei, und promovierte 1994 in England an der Manchester Metropolitan University. 1997 trat er der Fakultät für Industriedesign an der Istanbul Teknik Universitesi (ITU) bei, wo er von 2006 bis 2013 als Vorsitzender tätig war. Derzeit ist er mitbegründender Vorsitzender der Fakultät für Industriedesign an der Universität Ozegin in Istanbul. Alpay Er engagiert sich auch in der Türkischen Gesellschaft für Industriedesigner (ETMK), sowohl als Vorstandsmitglied als auch als Vorsitzender der Abteilung ETMK Istanbul. In Kooperation mit der ITU und der Industriekammer Istanbuls (ISO) initiierte Alpay Er das Projekt „Industriedesign für KMUs", das erste, an kleine und mittelständische Unternehmen (KMUs) gerichtete Lehr- und Förderprogramm für Design in der Türkei. Zudem ist er als Berater für zahlreiche Unternehmen und Institutionen tätig und ist Mitglied des Gastgebervorstands für den 2013 in Istanbul stattfindenden Kongress der International Design Alliance (IDA) sowie Mitglied der Design Research Society (DRS). Darüber hinaus ist Alpay Er Vorstandsmitglied des Icsid (2011–2013).

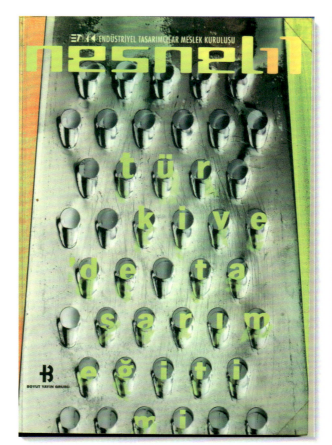

03

»Design is an indispensable part of innovation, making it more human and meaningful.«

»Design ist für Innovation unverzichtbar und macht sie menschlicher und bedeutsamer.«

What current trends do you see in the category "Life science and medicine"?
Miniaturisation continues where the function of the product allows it. Digital technologies are overtaking analogue ones. The ever increasing diversification of user needs and expectations both at global and local levels were also well represented.

What challenges do you see for the future in design?
The main challenge for design is to be able to function as one of the main drivers of innovation in competitive economic settings, while at the same time providing a meaningful contribution to significantly improving human life in a sustainable way.

Welche Trends konnten Sie im Bereich „Life Science und Medizin" in den letzten Jahren ausmachen?
Miniaturisierung setzt sich dort fort, wo es die Funktion der Produkte erlaubt, und digitale Technologien überholen die analogen. Die sich immer weiter auffächernden Bedürfnisse und Erwartungen der Anwender auf sowohl globaler wie auch lokaler Ebene wurden ebenfalls gut reflektiert.

Welche Herausforderungen sehen Sie für die Zukunft im Design?
Die größte Herausforderung für das Design besteht darin, in wirtschaftlichen Wettbewerbssituationen als ein Hauptantrieb für Innovation dienen zu können und das Leben der Menschen gleichzeitig durch bedeutende Beiträge entscheidend zu verbessern, und zwar auf nachhaltige Weise.

01

Joachim H. Faust
Germany

Jury member since 2005
Appointed five times
Jurymitglied seit 2005
Berufen fünf Mal

Joachim H. Faust, born in 1954, studied architecture at the Technical University of Berlin, the Technical University of Aachen, as well as at Texas A&M University (with Prof. E. J. Romieniec), where he received his Master of Architecture in 1981. He worked as a concept designer in the design department of Skidmore, Owings & Merrill in Houston, Texas and as a project manager in the architectural firm Faust Consult GmbH in Mainz. From 1984 to 1986, he worked for KPF Kohn, Pedersen, Fox/Eggers Group in New York and as a project manager at the New York office of Skidmore, Owings & Merrill.

In 1987, Joachim H. Faust took over the management of the HPP office in Frankfurt am Main. Since 1997, he has been managing partner of the HPP Hentrich-Petschnigg & Partner GmbH + Co. KG in Düsseldorf. He also writes articles and gives lectures on architecture and interior design.

Joachim H. Faust, 1954 geboren, studierte Architektur an der TU Berlin und der RWTH Aachen sowie – bei Prof. E. J. Romieniec – an der Texas A&M University, wo er sein Studium 1981 mit dem Master of Architecture abschloss. Faust war Entwurfsarchitekt im Design Department des Büros Skidmore, Owings & Merrill, Houston, Texas, sowie Projektleiter im Architekturbüro der Faust Consult GmbH in Mainz. Anschließend arbeitete er im Büro KPF Kohn, Pedersen, Fox/Eggers Group in New York und war Projektleiter im Büro Skidmore, Owings & Merrill in New York.

1987 übernahm Joachim H. Faust die Leitung des HPP-Büros in Frankfurt am Main und ist seit 1997 geschäftsführender Gesellschafter der HPP Hentrich-Petschnigg & Partner GmbH + Co. KG in Düsseldorf. Er ist zudem als Autor tätig und hält Vorträge zu Fachthemen der Architektur und Innenarchitektur.

02

»All those award submissions were successful which expressed a clear design idea relating to the design task.«

»All jene Einreichungen waren erfolgreich, die eine klare Designidee in Bezug zur Aufgabenstellung ausdrücken konnten.«

What trends have you noticed in the field of "Architecture and urban design" in recent years?
In some architectural solutions, the consistency in the use of materials was highly convincing. They stood out in particular for their uniformity, the absence of high contrasts and a tendency towards the use of natural materials.

What are the important criteria for you as a juror in the assessment of a product?
The idea has to be clear and convincing. It then is up to the skill of the designer to translate this idea of space and material into architecture.

How do you assess the significance of design quality for the success of a company?
Today, every company acknowledges that a product is marketable only if it is well designed. Good design translates the inherent value of a product or building into an intuitively perceivable spatial quality or use.

Welche Trends konnten Sie im Bereich „Architektur und Urban Design" in den letzten Jahren ausmachen?
Bei einigen Architekturlösungen konnte die Stringenz der Verwendung von Materialien überzeugen. Einheitlichkeit, ohne große Materia wechsel, und die Tendenz zu natürlichen Materialien sind besonders aufgefallen.

Worauf achten Sie als Juror, wenn Sie ein Produkt bewerten?
Die Klarheit der Idee muss eindeutig erkennbar sein. Es ist dann die Kunst des Designers, diese Idee von Raum und Material inhaltlich in Architektur zu übersetzen.

Wie schätzen Sie den Stellenwert des Designs für den Erfolg eines Unternehmens ein?
Heute hat jedes Unternehmen erkannt, dass nur durch gutes Design ein Produkt auch marktfähig ist. Gutes Design übersetzt den inhaltlichen Wert eines Produktes oder eines Gebäudes in intuitiv erkennbare Raumqualität bzw. Bedienbarkeit eines Produktes.

Joachim H. Faust

01

Andrea Finke-Anlauff
Germany

Jury member since 2004
Appointed four times
Jurymitglied seit 2004
Berufen vier Mal

Andrea Finke-Anlauff was born in Braunschweig, Germany. She studied industrial design at the Braunschweig University of Arts, during which time she spent a year in Barcelona (Facultat de Belles Arts) and also gained an MA in Design Leadership from an international degree course at the University of Industrial Arts in Helsinki, Finland. During her studies, she worked for various departments of Nokia in Great Britain, Japan and Finland. She graduated with a diploma in design and signed a consultancy agreement with Nokia.

Andrea Finke-Anlauff founded "mangodesign" in 1994. The agency's focus, aside from product design, is on interface design e. g. for agricultural vehicles and products from the mobile, entertainment and car industry. She went on to co-found the design manufactory "mangoobjects" in 2003. Her expertise has enabled Andrea Finke-Anlauff to teach interface design at art schools. She provides in-house trainings for customers and holds lectures at design events.

Andrea Finke-Anlauff wurde in Braunschweig geboren. Sie studierte Industrial Design an der Hochschule für Bildende Künste in Braunschweig, war währenddessen ein Jahr in Barcelona (Facultat de Belles Arts) und nahm an einem internationalen Studiengang MA in Design Leadership an der University of Industrial Arts in Helsinki teil. Bereits während des Studiums, das sie als Diplom-Industriedesignerin abschloss, arbeitete sie in verschiedenen Abteilungen von Nokia in Großbritannien, Japan und Finnland und erhielt im selben Jahr einen Beratervertrag bei Nokia.

1994 gründete Andrea Finke-Anlauff die Firma „mangodesign", deren Schwerpunkte – neben Produktdesign – auf Interfacedesign für landwirtschaftliche Nutzfahrzeuge sowie Produkte aus der Mobilfunk-, Unterhaltungs- und Autoindustrie liegen. Zudem war sie 2003 Mitbegründerin der Designmanufaktur „mangoobjects". Als Expertin unterrichtete Andrea Finke-Anlauff an Kunsthochschulen Interfacedesign. Sie führt betriebsinterne Schulungen für Kunden durch und hält Vorträge auf Designveranstaltungen.

01 **Emporia Connect**
   User interface, graphic
   and product design for
   a telephone for the elderly
   with Internet access:
   high-contrast displays and
   large lettering ensure
   good readability – the call
   is taken by opening the
   phone, closing it automatically ends the call
   User Interface, Grafik-
   und Produktdesign für ein
   Seniorentelefon mit Internetzugang: Hohe Kontraste
   und große Schriften sorgen
   für gute Lesbarkeit – Öffnen
   nimmt eingehende Anrufe
   an, Zusammenklappen
   beendet Telefonate automatisch

02 **TouchME**
   User interface and graphics
   for an ISOBUS terminal
   by Müller-Elektronik GmbH:
   the 12,1' touch display for
   monitoring and operating
   agricultural machinery
   can be installed vertically
   or horizontally
   User Interface und Grafik
   für ein ISOBUS-Terminal
   der Müller-Elektronik GmbH:
   12,1"-Touchdisplay zur
   Steuerung und Kontrolle
   von Landmaschinen,
   montierbar im Hoch- und
   im Querformat

02

»Design does not only improve the visual appearance of a product, but also how it is used.«

»Design verbessert nicht nur die visuelle Anmutung, sondern auch den Umgang mit einem Produkt.«

What trends have you noticed in the field of "Computers and information technology" in recent years?
New product categories and technologies such as tablets and mobile telephones with touch screens have made computer applications much easier to use and more mobile. Many areas are marked by a merging of product types as manufacturers try out various product combinations such as cameras with a telephone function and mobile phones that could easily pass for tablets.

What aspects of the jury process for the Red Dot Award: Product Design 2013 have stayed in your mind in particular?
I was overwhelmed by the sheer number of submissions. The design quality of the products from Asian countries has significantly improved and the high number of submissions from the Far East is proof of a heightened awareness of design.

Welche Trends konnten Sie im Bereich „Computer und Informationstechnik" in den letzten Jahren ausmachen?
Neue Produktkategorien und Technologien wie Tablets und Mobiltelefone mit Touchbedienung haben den Umgang mit Computern erheblich erleichtert und noch mobiler gemacht. In vielen Bereichen findet eine Ver-schmelzung statt: Die Hersteller testen verschiedenste Produktkombinationen, z. B. Kamera mit Telefonfunktion oder Mobiltelefone, die als Tablets gelten können.

Was ist Ihnen vor der Jurierung des Red Dot Award: Product Design 2013 besonders im Gedächtnis geblieben?
Ich war überwältigt von der Menge der eingereichten Arbeiten. Die Designqualität der Produkte aus asiati-schen Ländern ist deutlich gestiegen und die hohe Zahl der Anmeldungen aus Fernost beweist ein gesteigertes Bewusstsein für gute Gestaltung.

Andrea Finke-Anlauff

01

Prof. Lutz Fügener
Germany

Jury member since 2008
Appointed six times
Jurymitglied seit 2008
Berufen sechs Mal

Professor Lutz Fügener began his studies at the Technical University Dresden, where he completed a foundation course in mechanical engineering. He then transferred to the Burg Giebichenstein University of Art and Design in Halle/Saale, Germany, where he obtained a degree in industrial design in 1995. In the same year, he became junior partner of Fisch & Vogel Design in Berlin. Since then, the firm (today called "studioFT") has increasingly specialised in transportation design. Two years after joining the firm, Lutz Fügener became senior partner and co-owner. Since 2000, he has been chair of the BA degree course in transportation design and is a member of the board of governors of Pforzheim University.

Professor Lutz Fügener absolvierte ein Grundstudium in Maschinenbau an der Technischen Universität Dresden und nahm daraufhin ein Studium für Industrial Design an der Hochschule für Kunst und Design, Burg Giebichenstein, in Halle an der Saale auf. Sein Diplom machte er im Jahr 1995. Im selben Jahr wurde er Juniorpartner des Büros Fisch & Vogel Design in Berlin. Seit dieser Zeit spezialisierte sich das Büro (heute „studioFT") mehr und mehr auf den Bereich „Transportation Design". Zwei Jahre nach seinem Einstieg wurde Lutz Fügener Seniorpartner und gleichberechtigter Mitinhaber des Büros. Seit 2000 ist er Verantwortlicher des BA-Studiengangs „Transportation Design" und Mitglied des Hochschulrates der Hochschule Pforzheim.

»The automobile is one of those products whose design performance is explicitly perceived, judged and discussed even by customers.«

»Das Automobil ist eines der Produkte, bei dem die Designleistung auch vom Kunden explizit wahrgenommen, bewertet und diskutiert wird.«

What challenges do you see for the future in design?
Today's challenge consists of the profound integration of design into the development process of the vehicle, a process that ideally occurs not only simultaneously with the development, but which from early on represents a truly institutionalised cooperation between design and engineering.

What are the important criteria for you as a juror in the assessment of a product?
When assessing a product, for me it is important to what extent the creative work of the designers has influenced and ideally shaped the concept, development and realisation of a vehicle; that is, whether it has contributed to giving it a character easily recognisable by its users. Complementing this effect with high quality design craftsmanship in realisation yields the highest quality products.

Welche Herausforderungen sehen Sie für die Zukunft im Design?
Die heutige Herausforderung besteht in der tiefen Einbindung des Designs in den Entwicklungsprozess des Fahrzeugs, der im Idealfall nicht nur in einer Gleichzeitigkeit der Entwicklung, sondern einer wirklichen, frühzeitig institutionalisierten Kooperation von Design und Engineering besteht.

Worauf achten Sie als Juror, wenn Sie ein Produkt bewerten?
Für mich ist es bei der Bewertung wichtig, inwieweit die gestalterische Arbeit der Designer bei der Konzeption und Umsetzung eines Fahrzeugs Einfluss auf das Konzept erlangt hat und es im Idealfall prägt also zu seinem für den Nutzer wahrnehmbaren Charakter verholfen hat. Wenn dieser Effekt mit einer hohen designhandwerklichen Qualität in der Umsetzung einhergeht, entstehen die hochwertigsten Produkte.

Prof. Lutz Fügener

01

Hideshi Hamaguchi
Japan

Jury member since 2012
Appointed twice
Jurymitglied seit 2012
Berufen zwei Mal

Hideshi Hamaguchi graduated with a Bachelor
of Science in chemical engineering from Kyoto
University. Starting his business career with Panasonic
in Japan, Hamaguchi later became director of the New
Business Planning Group at Panasonic Electric Works,
Ltd. and then executive vice president of Panasonic
Electric Works Laboratory of America, Inc. In 1994, he
developed Japan's first corporate Intranet and also led
the concept development for the first USB flash drive.
Hideshi Hamaguchi has over 15 years of experience
in defining strategies and decision-making, as well
as in concept development for various industries and
businesses. As director of strategy at Ziba Design,
he is today considered a leading mind in creative
concept and strategy development on both sides of
the Pacific and is involved in almost every project this
renowned business consultancy takes on. For clients
such as FedEx, Polycom and M-System he has led the
development of several award-winning products.

Hideshi Hamaguchi graduierte als Bachelor of Science
in Chemical Engineering an der Kyoto University. Seine
Karriere begann er bei Panasonic in Japan, wo er später
zum Direktor der New Business Planning Group von
Panasonic Electric Works, Ltd. und zum Executive Vice
President von Panasonic Electric Works Laboratory of
America, Inc. aufstieg. 1994 entwickelte er Japans erstes Firmen-Intranet und übernahm zudem die Leitung
der Konzeptentwicklung des ersten USB-Laufwerks.
Hideshi Hamaguchi besitzt mehr als 15 Jahre Erfahrung
in der Konzeptentwicklung sowie Strategie- und Ent-
scheidungsfindung in unterschiedlichen Industrien und
Unternehmen. Als Director of Strategy bei Ziba Design
wird er heute als führender Kopf in der kreativen Kon-
zept- und Strategieentwicklung auf beiden Seiten
des Pazifiks angesehen und ist in nahezu jedes Projekt
der renommierten Unternehmensberatung involviert.
Für Kunden wie FedEx, Polycom und M-System leitete
er einige ausgezeichnete Projekte.

02 / 03

»My philosophy of life is: All I need is less.«

»Meine Lebensphilosophie lautet: Was ich brauche, ist weniger.«

What is, in your opinion, the significance of design quality in the product categories you evaluated?
Design is significant in these categories because it forms emotional and cognitive connections. Technology is so ingrained in our lives today that it is no longer enough to be beautiful and functional – the product must also have a story and personal resonance. Design is the means to create that.

Do you see a correlation between the economic success of a company and the design quality of its products?
Definitely. If a company has the right design for all the phases of consumer interaction – attraction, engagement and ensuing actions – it should directly affect its success.

What challenges do you see for the future in design?
The challenge is finding the sweet spot between what resonates with the consumer and what is true to the brand.

Wie schätzen Sie den Stellenwert der Designqualität in den von Ihnen beurteilten Produktkategorien ein?
In diesen Kategorien ist Design wesentlich, da es emotionale und kognitive Verbindungen bildet. Technologie ist in unserem Leben heute derart tief verwurzelt, dass es nicht mehr genügt, schön oder funktional zu sein – das Produkt muss eine Geschichte und persönliche Resonanz besitzen. Design ist das Mittel, um das zu erreichen.

Sehen Sie einen Zusammenhang zwischen dem wirtschaftlichen Erfolg eines Unternehmens und der Designqualität seiner Produkte?
Auf jeden Fall. Wenn ein Unternehmen in allen Stadien der Interaktion mit dem Verbraucher – Anziehungskraft, Einbindung, Folgehandlung – über das richtige Design verfügt, so hat das seinen Erfolg direkt beeinflussen.

Welche Herausforderungen sehen Sie für die Zukunft im Design?
Die Herausforderung ist, den Punkt zwischen dem, was beim Konsumenten auf Resonanz stößt, und dem, was die Marke ausmacht, zu treffen.

01

Prof. Renke He
China

Jury member since 2008
Appointed six times
Jurymitglied seit 2008
Berufen sechs Mal

Professor Renke He, born in 1958, studied civil engineering and architecture at Hunan University. From 1987 to 1988, he was a visiting scholar at the Industrial Design Department of the Royal Danish Academy of Fine Arts in Copenhagen and, from 1998 to 1999, at North Carolina State University's School of Design. Renke He is dean and professor of the School of Design at Hunan University in China and is also director of the Chinese Industrial Design Education Committee. Currently, he holds the position of vice-chair of the China Industrial Design Association.

Professor Renke He wurde 1958 geboren und studierte an der Hunan University Bauingenieurwesen und Architektur. Von 1987 bis 1988 war er als Gastprofessor für Industrial Design an der Royal Danish Academy of Fine Arts in Kopenhagen tätig, und von 1998 bis 1999 hatte er eine Gastprofessur an der School of Design der North Carolina State University inne. Renke He ist Dekan und Professor an der Hunan University China, School of Design, sowie Direktor des Chinese Industrial Design Education Committee. Er ist zudem stellvertretender Vorsitzender der China Industrial Design Association.

01 SANY 1000
    Ton truck crane,
    design by School of Design,
    Hunan University, China
    Kran-Lkw, Gestaltung der
    School of Design, Hunan
    University, China

02 Tourism product_1
    Design for the Xinjiang
    Uygur Autonomous Region
    by School of Design,
    Hunan University, China
    Gestaltung der School of
    Design, Hunan University,
    China, für die autonome
    Region Xinjiang Uygur

02

»Good design means good business. Nowadays, you cannot talk about companies without talking about their design.«

»Gute Gestaltung impliziert gute Geschäfte. Heutzutage kann man nicht über Unternehmen sprechen, ohne über ihre Gestaltung zu sprechen.«

What trends have you noticed in the field of "Households" in recent years?
In many countries of the world, both air and water pollution have today become serious problems that people have to deal with. This year, I have seen many air and water purifier designs that have introduced new and unique design languages to express functions and identities.

What trends have you noticed in the field of "Kitchens" in recent years?
As more and more high-tech electrical household appliances are used, kitchens have become more complicated. In order to turn kitchens back into more pleasant places, built-in design has become a trend. "Less is more" is still an important design principle.

Which product would you like to realise one day?
A mobile phone-based remote medical system for people who live far away from big cities.

Welche Trends konnten Sie im Bereich „Haushalt" in den letzten Jahren ausmachen?
In vielen Ländern der Welt sind heute Luft- und Wasserverschmutzung zu ernsten Problemen geworden, die die Menschen angehen müssen. In diesem Jahr gab es viele Gestaltungen zur Luft- und Wasserreinigung, die sich im Ausdruck ihrer Funktion und Identität neuer und einzigartiger Gestaltungssprachen bedienen.

Welche Trends konnten Sie im Bereich „Küche" in den letzten Jahren ausmachen?
Da immer mehr elektronische Hightech-Haushaltsgeräte zum Einsatz kommen, ist die Küche viel komplizierter geworden. Um sie als Ort wieder freundlicher zu machen, entwickelte sich der Trend zu Einbauentwürfen. „Weniger ist mehr" ist nach wie vor ein wichtiges Gestaltungsprinzip.

Welches Produkt würden Sie gerne einmal realisieren?
Ein per Mobiltelefon fernbedienbares medizinisches System für Menschen, die weit ab von großen Städten leben.

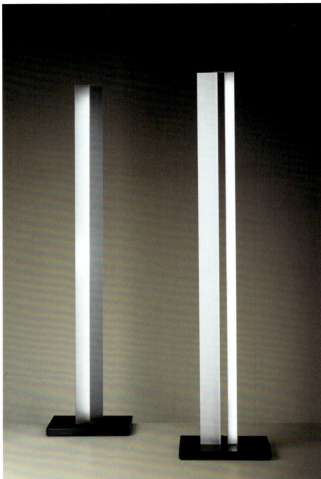

01

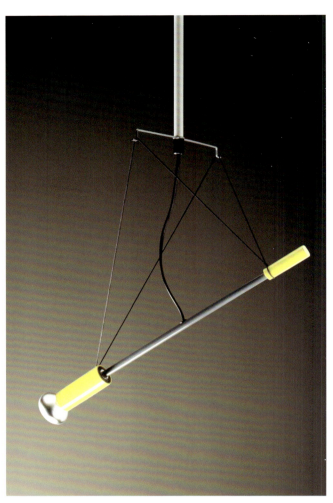

02

Prof. Herman Hermsen
Netherlands

Jury member since 1999
Appointed ten times
Jurymitglied seit 1999
Berufen zehn Mal

Professor Herman Hermsen, born in 1953 in Nijmegen, Netherlands, studied at the ArtEZ Institute of the Arts in Arnhem from 1974 to 1979. Following an assistant professorship, he began his career in teaching in 1985. Until 1990, he taught product design at the Utrecht School of the Arts (HKU), after which time he returned to Arnhem as lecturer at the Academy. Hermsen has been professor of product and jewellery design at the University of Applied Sciences in Düsseldorf since 1992. He gives guest lectures at universities and colleges throughout Europe, the United States, and Japan, and began regularly organising specialist symposia in 1998. He has also served as juror for various competitions. Herman Hermsen has received numerous international awards for his work in product and jewellery design, which is shown worldwide in solo and group exhibitions and held in the collections of renowned museums, such as the Cooper-Hewitt Museum, New York; the Pinakothek der Moderne, Munich; and the Museum of Arts and Crafts, Kyoto.

Professor Herman Hermsen, 1953 in Nijmegen in den Niederlanden geboren, studierte von 1974 bis 1979 am ArtEZ Institute of the Arts in Arnheim und ging nach einer Assistenzzeit ab 1985 in die Lehre. Bis 1990 unterrichtete er Produktdesign an der Utrecht School of the Arts (HKU) und kehrte anschließend nach Arnheim zurück, um an der dortigen Hochschule als Dozent zu arbeiten. Seit 1992 ist Hermsen Professor für Produkt- und Schmuckdesign an der Fachhochschule Düsseldorf; er hält Gastvorlesungen an Hochschulen in ganz Europa, den USA und Japan, organisiert seit 1988 regelmäßig Fachsymposien und ist Juror in verschiedenen Wettbewerbsgremien. Für seine Arbeiten im Produkt- und Schmuckdesign, die weltweit in Einzel- und Gruppenausstellungen präsentiert werden und sich in den Sammlungen großer renommierter Museen befinden, z. B. Cooper-Hewitt Museum, New York, Pinakothek der Moderne, München, und Museum of Arts and Crafts, Kyoto, erhielt Herman Hermsen zahlreiche internationale Auszeichnungen.

03

»As a juror, I essentially search for innovative aspects in form and technology.«

»Als Juror suche ich in erster Linie nach innovativen Aspekten in Form und Technik.«

What is, in your opinion, the significance of design quality in the product categories you evaluated?
The significance of design quality is very high, because these products are worn on the body. Wearers identify with them and use them to present themselves.

What current trends do you see in the category "Watches and jewellery"?
Some products were very interesting because they integrate technical innovations, for example a digital wristwatch with GPS function and a wristwatch made from recycled tin cans. Different stylistic features, which are shaped by local markets around the world, come together in this category.

What challenges do you see for the future in design?
Many issues that have been relevant for some time: environmental protection, sustainability, quality over quantity and innovation without neglecting traditional values.

Wie schätzen Sie den Stellenwert der Designqualität in den von Ihnen beurteilten Produktkategorien ein?
Der Stellenwert der Designqualität ist sehr hoch, denn diese Produkte werden am Körper getragen – und damit identifiziert und präsentiert sich der Träger.

Welche Trends konnten Sie im Bereich „Uhren und Schmuck" ausmachen?
Einige Produkte waren sehr interessant, weil sie technische Innovationen integrierten, z. B. die GPS-Funktion in einer digitalen Armbanduhr, und eine Armbanduhr etwa aus recycelten Blechdosen gefertigt war. In dieser Kategorie kommen unterschiedliche Stilmerkmale aus aller Welt zusammen, die durch die Märkte vor Ort geprägt werden.

Welche Herausforderungen sehen Sie für die Zukunft im Design?
Einige schon längst offene Türen: Umweltschutz, Nachhaltigkeit, Qualität über Quantität und Innovation, ohne die tradierten Werte zu vernachlässigen.

Prof. Herman Hermsen

01

Prof. Carlos Hinrichsen
Chile

Jury member since 2006
Appointed seven times
Jurymitglied seit 2006
Berufen sieben Mal

Professor Carlos Hinrichsen, born in 1957, graduated as an industrial designer in Chile in 1982 and went to Japan to add a Master's degree in engineering in 1991. From 2007 to 2009, he was president of the Icsid and has since served as its senator. Since then he has been heading research projects focused on innovation, design and education and, in 2010, was honoured with the distinction of "Commander of the Order of the Lion of Finland" in recognition of his valuable contribution to the development of design education, design innovation and their promotion in Chile and Finland. From 1992 to 2010, he was director of the School of Design Duoc UC, Chile. He has been a design process consultant for over two decades, and is currently design director for the Latin American Region of Design Innovation, a European design company with clients and activities across the world. Since 2002, Carlos Hinrichsen has been an honorary member of the Chilean Association of Design. He has been the keynote speaker at different conferences and seminars worldwide. He is also director of Duoc UC International Affairs.

Professor Carlos Hinrichsen, geboren 1957, schloss 1982 sein Studium als Industriedesigner in Chile ab und erwarb 1991 zusätzlich einen Ingenieurs-Masterabschluss in Japan. Von 2007 bis 2009 war er Präsident des Icsid und fungiert dort seither als Senator. Seitdem leitet er Forschungsprojekte mit Schwerpunkt auf Innovation, Gestaltung und Ausbildung und wurde 2010 in Anerkennung seiner Verdienste in Sachen Förderung der Designausbildung und -innovation sowie deren Vermittlung in Chile und Finnland mit dem „Commander of the Order of the Lion of Finland" ausgezeichnet. Von 1992 bis 2010 war er Direktor der Designschule „Instituto Profesional Duoc UC". Seit mehr als zwei Jahrzehnten ist er als Berater im Bereich Design Process tätig und derzeit der Design Director der Latin American Region of Design Innovation, einer europäischen Designfirma mit Kunden und Beschäftigungsfeldern in aller Welt. Seit 2002 ist Carlos Hinrichsen Ehrenmitglied der chilenischen Vereinigung der Designfirmen (QVID). Zudem war er Hauptredner verschiedener Konferenzen und Seminare weltweit. Darüber hinaus ist Carlos Hinrichsen Direktor der Duoc UC International Affairs.

02

»Every year, this award highlights the latest developments
in design and altogether opens
up an unprecedented field of
knowledge and expectations.«

»Jedes Jahr zeigt dieser
Wettbewerb die neuesten
Errungenschaften im Design
und eröffnet insgesamt
einen beispiellosen Wissens-
und Erwartungshorizont.«

What is, in your opinion, the significance of design quality in the product categories you evaluated?
This year, products presented themselves as realistic images of the desires and dreams of users, designers and producers brought to life. In the categories I evaluated, design quality and innovation play a key role in turning technological innovations into business success. Product quality and performance have also been improved from a consumer standpoint in a wide variety of applications.

What are the important criteria for you as a juror in the assessment of a product?
My criteria of assessment are as follows. I look at the unique ways of integrating or embedding innovation into products or services, and how designers and companies use design appropriately to implement both innovations coming from the R&D sphere as well as innovations associated with social and market changes in order to respond successfully to people's new needs and requirements.

Wie schätzen Sie den Stellenwert der Designqualität in den von Ihnen beurteilten Produktkategorien ein?
Dieses Jahr habe ich beobachtet, wie sich Produkte als realistische, lebendig gewordene Träume und Wünsche sowohl der Nutzer als auch der Gestalter und Produzenten darbieten. Designqualität und Innovation spielen eine zentrale Rolle beim Ummünzen technologischer Neuerungen in den Geschäftserfolg. Es zeigt sich eine große Bandbreite an Anwendungen, bei denen sich die Produktqualität und Leistung auch aus Sicht der Konsumenten verbessert haben.

Worauf achten Sie als Juror, wenn Sie ein Produkt bewerten?
Meine Beurteilungskriterien sind wie folgt: Ich achte auf die besondere Einbindung von Innovation in die Produkte oder Leistungen und darauf, wie Designer und Firmen Gestaltung angemessen nutzen, um sowohl die Innovation aus dem F&E-Bereich umzusetzen als auch jene Innovationen, die auf gesellschaftlichen und marktwirtschaftlichen Veränderungen basieren – mit dem Ziel, erfolgreich auf die neuen Bedürfnisse und Anforderungen der Menschen zu antworten.

01

Stephan
Hürlemann
Switzerland

Jury member since 2013
Appointed for the first time
Jurymitglied seit 2013
Berufen zum ersten Mal

Stephan Hürlemann, born in 1972, studied architecture at the ETH Zurich (Swiss Federal Institute of Technology). He then worked for some years as a musician and music manager, but was also active in the areas of Internet design and architectural visualisation.

In 2002, the designer joined the Hannes Wettstein's agency as CEO. He was made a partner in 2006. After Wettstein's death, Stephan Hürlemann converted the company to its new format. Today, he is one of Studio Hannes Wettstein's two creative directors, as well as partner and member of the three-person executive board.

Stephan Hürlemann, 1972 geboren, studierte Architektur an der ETH Zürich. Danach war er einige Jahre als Musiker und Musikmanager sowie in den Bereichen Internet-Konzeption und Architektur-Visualisierungen tätig.

2002 kam der Gestalter als CEO in das Büro von Hannes Wettstein, der ihn 2006 zu seinem Partner machte. Nach dessen Tod führte Stephan Hürlemann das Büro in seine neue Form über und ist heute einer der beiden Creative Directors des Studio Hannes Wettstein sowie Teilhaber und Mitglied der dreiköpfigen Geschäftsleitung.

02

03

»Thanks to the standardisation of light sources and their dematerialisation design comes to the fore.«

»Durch die Vereinheitlichung der Leuchtmittel und ihre Entmaterialisierung rückt das Design in den Vordergrund.«

What aspects of the jury process for the Red Dot Award: Product Design 2013 have stayed in your mind in particular?
I was very impressed with the composition of the jury and the discussions that were held in the adjudication process. I can say with a clear conscience that the jury genuinely tried to make a contribution to the promotion of good design.

What challenges do you see for the future in design?
When design is abused in order to diversify in a saturated market, that often leads to empty products which are crowned only with short term success. I'm convinced that good design requires constant challenging and optimisation using constructive dialogue. And this can only be achieved when designers and business people work together well.

Was ist Ihnen von der Jurierung des Red Dot Award: Product Design 2013 besonders im Gedächtnis geblieben?
Ich war sehr angetan von der Zusammensetzung der Jury und von den Diskussionen, die im Jurierungsprozess geführt wurden. Ich kann mit gutem Gewisser behaupten, dass die Jury aufrichtig versucht hat, einen Beitrag zur Förderung von guter Gestaltung zu leisten.

Welche Herausforderungen sehen Sie für die Zukunft im Design?
Wenn Design missbraucht wird, um sich in einem gesättigten Markt zu diversifizieren, dann entstehen oft leere Produkte, die nur von kurzem Erfolg gekrönt sind. Ich bin überzeugt, dass gutes Design das konstante Hinterfragen und Optimieren in einem konstruktiven Dialog bedingt. Und genau dies kann nur dann entstehen, wenn das Zusammenspiel zwischen Gestalter und Unternehmer funktioniert.

01

Prof. Dr. Florian Hufnagl
Germany

Jury member since 2003
Appointed ten times
Jurymitglied seit 2003
Berufen zehn Mal

Professor Dr. Florian Hufnagl, born 1948, has been director of Die Neue Sammlung – The International Design Museum, Munich, since 1990. He studied art history, classical archaeology and modern history in Munich, earning his doctorate in 1976. Thereafter, he worked for the Bavarian State Department of Monuments and Sites and, in 1980, became chief curator of Die Neue Sammlung. During this time, Hufnagl was also associate lecturer for 19th- and 20th-century art at the Institute of Art History at Ludwig-Maximilians-Universität in Munich. In 1997, he became honorary professor at the Academy of Fine Arts in Munich and, in 1998, chairman of the directors' conference of Bavaria's state museums and collections. Florian Hufnagl has written on 20th- and 21st-century architecture, painting and design in numerous publications, catalogues and essays.

Professor Dr. Florian Hufnagl, geboren 1948, ist seit 1990 leitender Sammlungsdirektor der Neuen Sammlung – The International Design Museum Munich. Er promovierte 1976 nach einem Studium der Kunstwissenschaft Klassischen Archäologie und Neueren Geschichte in München. Anschließend arbeitete er beim Bayerischen Landesamt für Denkmalpflege und seit 1980 als Museumskurator in der Neuen Sammlung. Zugleich unterrichtete Hufnagl als Lehrbeauftragter für die Kunst des 19. und 20. Jahrhunderts am Institut für Kunstgeschichte der Ludwig-Maximilians-Universität München. Seit 1997 ist er Honorarprofessor an der Akademie der Bildenden Künste in München. 1998 wurde er Vorsitzender der Direktorenkonferenz der Staatlichen Museen und Sammlungen in Bayern. In zahlreichen Publikationen Katalogen und Aufsätzen setzt sich Florian Hufnagl mit der Architektur, der Malerei und dem Design des 20. und 21. Jahrhunderts auseinander.

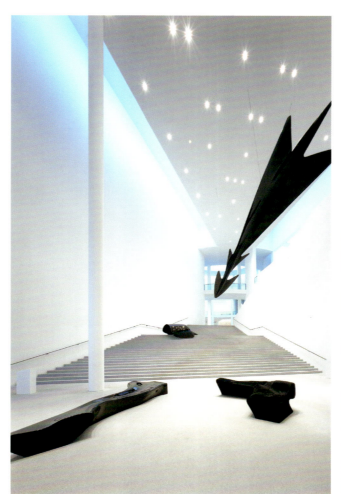

02

»The market has become global, which also means that design must more than ever account for and factor in different cultural conditions.«

»Der Markt ist global geworden, das bedeutet auch, dass Design stärker als bisher die unterschiedlichen kulturellen Voraussetzungen berücksichtigen muss.«

What trends have you noticed in the field of "Watches and jewellery" in recent years? Are there developments that have especially attracted your attention?
Watches have followed three trends for years. Role models are the high-priced, hand-made products created by Swiss manufacturers, classic sports chronometers and the kind of minimalism that one today also finds in German and Danish products.

What trends have you noticed in the field of "Tableware and decoration" in recent years?
There have been many changes in tableware in recent years. Classical sets defined only through their form have been left behind to make way for elements that can be combined freely in terms of form, colour and pattern or used individually, and which create a highly colourful and emotional impression. New and different materials are being selected. Today's consumers prefer espresso machines to coffeepots.

Do you have a philosophy toward life?
Do it better!

Welche Trends konnten Sie im Bereich „Uhren und Schmuck" in den letzten Jahren ausmachen? Gibt es Entwicklungen, die Ihnen besonders aufgefallen sind?
Bei den Uhren gibt es seit Jahren drei Richtungen. Vorbilder sind die hochpreisigen handgefertigten Produkte aus den Schweizer Manufakturen, die klassischen Sport-Chronometer und der Minimalismus, den man inzwischen auch bei deutschen und dänischer Produkten wiederfindet.

Welche Trends konnten Sie im Bereich „Tableware und Dekoration" in den letzten Jahren ausmachen?
Bei Tableware hat sich in den letzten Jahren viel verändert, weg vom klassischen, nur durch seine Form sprechenden Service hin zu in Form, Farbe und Dekor beliebig kombinierbaren und einzeln benutzbaren Elementen, die ein sehr farbenfrohes, emotionales Bild abgeben. Es gibt neue, andere Materialien, und der Konsument bevorzugt statt der Kaffeekanne die Espressomaschine.

Haben Sie ein Lebensmotto?
Do it better!

01

Gerald Kiska
Austria

Jury member since 2011
Appointed twice
Jurymitglied seit 2011
Berufen zwei Mal

Gerald Kiska studied at the University of Arts and Industrial Design in Linz, Austria. He subsequently worked for several design studios both in Austria and abroad. They included Interform Design in Wolfsburg, Germany from 1984 to 1985; Form Orange in Hard from 1985 to 1986; Agentur Idea in Stuttgart in 1986; and later Porsche Design in Zell am See. In 1991, Gerald Kiska founded his own design studio in Anif, Salzburg, which in terms of floor space (5,000 sqm) and staff (more than 110 employees from 20 nations) is one of the biggest owner-operated studios in Europe today. From 1994 to 1995, he was a visiting professor at the Offenbach Academy of Art and Design (HfG), Germany. From 1995 to 2002, he was a founding member of and lecturer at the University of Applied Sciences in Graz, Austria. Gerald Kiska became known through his work for the motorcycle manufacturer KTM. Today, he works in a wide variety of sectors, including the automotive industry, consumer goods, food and beverages, investment goods and professional tools, with a focus on the challenges posed by the development and strengthening of brands.

Gerald Kiska absolvierte die Hochschule für Gestaltung in Linz und arbeitete anschließend in verschiedenen Designbüros im In- und Ausland, darunter 1984/85 bei Interform Design in Wolfsburg, 1985/86 bei Form Orange in Hard, 1986 in der Agentur Idea, Stuttgart, und anschließend bei Porsche Design in Zell am See. 1991 gründete er sein eigenes Designunternehmen in Anif/Salzburg, das heute bezüglich seiner Fläche (5.000 qm) und Mitarbeiterzahl (über 110 Mitarbeiter aus 20 Nationen) eines der größten eigentümergeführten Studios Europas ist. Von 1994 bis 1995 lehrte er im Rahmen einer Gastprofessur an der Hochschule für Gestaltung Offenbach am Main, von 1995 bis 2002 engagierte er sich als Gründer und Dozent an der Fachhochschule für Industrial Design Graz. Bekannt wurde Gerald Kiska durch seine Arbeiten für den Motorradhersteller KTM. Heute arbeitet er für ein breites Spektrum an Branchen, darunter Fahrzeugindustrie, Konsumgüter, Nahrungsmittel & Getränke, Investitionsgüter und professionelle Werkzeuge. Der Schwerpunkt liegt auf Herausforderungen rund um die Entwicklung oder Stärkung von Marken.

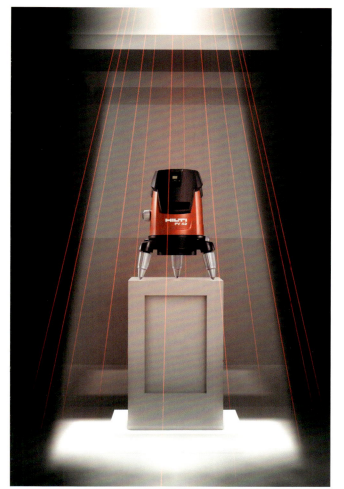

02

»Design is an important
component of branding
and thus indispensable
for a company's success.«

»Design ist ein wichtiger
Bestandteil der Markenbildung
und daher für den Erfolg von
Unternehmen unabdingbar.«

**What trends have you noticed in the field
of "Gardens" in recent years?**
I have noticed that the established brands in this
field, which have been using good design for a
long time, have managed to greatly expand their
market share, to the point where it is very difficult
for almost any competitor to launch an attack.

**What are the important criteria for you
as a juror in the assessment of a product?**
I pay attention to the usual design criteria such
as aesthetics, ergonomics and surface feel, but
particularly take into account branding effects:
Are the manufacturer's values embodied in the
product? Does the product stand out from the
competition?

**How do you assess the significance of design
quality for the success of a company?**
The strength of a brand is one of the most
important parameters for a company's economic
success. Products are a brand's most important
messengers. Therefore, design is the most
important component of branding.

**Welche Trends konnten Sie im Bereich „Garten"
in den letzten Jahren ausmachen?**
In diesem Bereich ist mir aufgefallen, dass jene etablierten Marken, die schon seit Langem auf Design gesetzt haben, einen großen Vorsprung zum Durchschnitt
aufgebaut haben, sodass es für jeden Konkurrenten
sehr schwierig ist, s e anzugreifen.

**Worauf achten Sie als Juror, wenn Sie ein Produkt
bewerten?**
Ich achte auf die üblichen Designkriterien wie Ästhetik,
Ergonomie und Haptik, aber auch stark auf Branding-
Effekte: Sind die Markenwerte des Herstellers ablesbar?
Findet eine Differenzierung vom Wettbewerb statt?

**Wie schätzen Sie den Stellenwert des Designs für
den Erfolg eines Unternehmens ein?**
Die Stärke einer Marke ist einer der wichtigsten Para-
meter für den wirtschaftlichen Erfolg des Unternehmens. Produkte sind die wichtigsten Botschafter einer
Marke. Daher ist Design der wichtigste Bestandteil
der Markenbildung.

01

Wolfgang K. Meyer-Hayoz
Switzerland

Jury member since 2003
Appointed seven times
Jurymitglied seit 2003
Berufen sieben Mal

Wolfgang K. Meyer-Hayoz, born in 1947, studied mechanical engineering, visual communication and industrial design, graduating from the Stuttgart State Academy of Art and Design. In 1985, he founded the Meyer-Hayoz Design Engineering Group with offices in Switzerland (Winterthur) and Germany (Konstanz). The company offers consultancy services for business start-ups, small- and medium-sized enterprises, as well as world market leaders in design strategy, industrial design, user interface design, temporary architecture and communication design. It has received numerous international awards.

From 1987 to 1993, Meyer-Hayoz was president of the Swiss Design Association (SDA). He is a member of various other associations as well, including the Association of German Industrial Designers (VDID) and the Schweizerische Management Gesellschaft (SMG). Aside from his work as a designer and a consultant, Wolfgang K. Meyer-Hayoz is also a guest lecturer at the University of St. Gallen and serves as juror on international design panels.

Wolfgang K. Meyer-Hayoz, geboren 1947, absolvierte Studien in den Fachbereichen Maschinenbau, Visuelle Kommunikation sowie Industrial Design mit Abschluss an der Staatlichen Akademie der Bildenden Künste in Stuttgart. 1985 gründete er die Meyer-Hayoz Design Engineering Group mit Büros in der Schweiz (Winterthur) und Deutschland (Konstanz). Das Unternehmen berät Start-ups, kleine und mittelständische Unternehmen sowie Weltmarktführer in Design Strategy, Industrial Design, User Interface Design, Temporary Architecture und Communication Design und wurde bereits vielfach international ausgezeichnet.

Von 1987 bis 1993 war Meyer-Hayoz Präsident der Swiss Design Association (SDA); er ist u. a. Mitglied im Verband Deutscher Industrie Designer (VDID) und der Schweizerischen Management Gesellschaft (SMG). Neben seiner Tätigkeit als Designer und Consultant ist Wolfgang K. Meyer-Hayoz u. a. Gastdozent an der Universität St. Gallen sowie Juror internationaler Designgremien.

01 excellence med
Vibration therapy device
by Wellengang GmbH
Schwingungs-Therapiegerät
der Wellengang GmbH

02 the modula wave
– 05 Control panel and graphical
user interface "Visko" (02),
trade fair booth concept
(03), card game to configure
the machine system (04)
and product brochure
(05) for the modular wave
soldering machine "the
modula wave" of Kirsten
Soldering AG
Bedienpanel und grafische
Benutzeroberfläche „Visko"
(02), Messestand-Konzept
(03), Kartenspiel zur
Maschinenkonfiguration (04)
und Produktbroschüre (05)
für die modulare Wellenlöt-
maschine „the modula wave"
der Kirsten Soldering AG

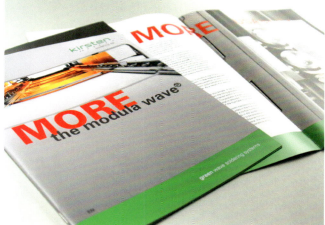

02 – 05

»My philosophy of life is to stimulate fascination for good and sustainable design.«

»Mein Lebensmotto ist, Menschen für gutes und nachhaltiges Design zu begeistern.«

What trends have you noticed in the field of "Life science and medicine" in recent years?
The obviously fast absorption of the latest research findings into new product applications (e.g., for laboratory analysis systems), but also new life-saving systems to prevent smoke poisoning. In general, a reduced language of form is coming back to the fore. This enables system solutions that provide a wide range of options in terms of application and save resources at the same time.

Do you see a correlation between the design quality of a company's products and the economic success of this company?
With the experience gathered over many industrial projects, I can say that this causality definitely exists. However, it requires a comprehensive understanding of design that includes innovation capacity, technology leadership as well as a conscious corporate management culture.

Welche Trends konnten Sie im Bereich „Life Science und Medizin" in den letzten Jahren ausmachen?
Die offensichtlich rasche Übernahme von Erkenntnissen der Forschung in neue Produktanwendungen, z. B. für Labor-Analysesysteme, aber auch neue lebenssichernde Systeme als Schutz gegen Rauchvergiftungen. Allgemein erkennt man wieder stärker eine reduzierte Formensprache. Sie erlaubt Systemdesignlösungen, welche eine große Variantenvielfalt für die Anwendung zur Verfügung stellt und gleichzeitig ressourcenschonend ist.

Sehen Sie einen Zusammenhang zwischen der Designqualität, die sich in den Produkten eines Unternehmens äußert, und dessen wirtschaftlichem Erfolg?
Aufgrund meiner in vielen Industrieprojekten gewonnenen Erfahrungen kann ich sagen, dass diese Kausalität absolut besteht. Es bedingt jedoch ein umfassendes Designverständnis, welches Innovationskraft, Technologieführerschaft sowie eine bewusste Führungskultur im Unternehmen beinhaltet.

01

Jure Miklavc
Slovenia

Jury member since 2013
Appointed for the first time
Jurymitglied seit 2013
Berufen zum ersten Mal

Jure Miklavc was born in Kranj, Slovenia, in 1970. He is a graduate in industrial design from the Academy of Fine Arts in Ljubljana, Slovenia, and has nearly 20 years of experience in the field of design. Miklavc started his career working as a freelance designer with graphic designer Barbara Šušterič, before founding his own design consultancy Studio Miklavc. Studio Miklavc works in the field of product design, visual communications, brand development and consultancy for a variety of clients mainly from the areas of business and economy, but also from government and culture.

Designs by Studio Miklavc have received many prestigious awards and have been displayed in numerous exhibitions. Miklavc is also active in promoting design and its role in Slovenian society. Since 2005, he has been involved in design education as a permanent lecturer at the Academy of Fine Arts and Design.

Jure Miklavc wurde 1970 in Kranj, Slowenien, geboren und machte seinen Abschluss in Industrial Design an der Academy of Fine Arts in Ljubljana, Slowenien. Er verfügt über beinahe 20 Jahre Erfahrung im Designbereich. Zunächst arbeitete er als freiberuflicher Designer zusammen mit der Grafikerin Barbara Šušterič und gründete anschließend sein eigenes Design-Beratungsunternehmen Studio Miklavc. Studio Miklavc ist im Bereich Produktdesign, visuelle Kommunikation, Markenentwicklung und Beratung für eine Vielzahl von Kunden vor allem aus dem Unternehmens- und Wirtschaftssektor, aber auch aus Regierungskreisen und Kultur tätig.

Viele Gestaltungen aus dem Hause Studio Miklavc erhielten angesehene Auszeichnungen und wurden in zahlreichen Ausstellungen gezeigt. Jure Miklavc ist im Bereich der Designförderung und gesellschaftlichen Relevanz von Gestaltung in Slowenien aktiv und seit 2005 als Dozent an der Academy of Fine Arts and Design tätig.

02

»Regardless of rational criteria, I'm open to the emotional charm of products.«

»Unabhängig von rationalen Kriterien bin ich offen für den emotionalen Charme von Produkten.«

What trends have you noticed in the field of "Industry and crafts" in recent years?
Tools are one of the most used products in our daily lives. They define who we are and the quality of our working environment. They are becoming more and more sophisticated and therefore more intuitive and ergonomic. As a result, we can carry out more precise operations with less effort and risk to our health. Thanks to well-designed products, working has become more focused and enjoyable.

What trends have you noticed in the field of "Automotive and transportation" in recent years?
I have noticed good design attempts in products and sectors that are not known for sophisticated solutions such as professional equipment for special vehicles. A greater level of intelligence is also noticeable in the development of the infrastructure for rechargeable electric cars.

Welche Trends konnten Sie im Bereich „Industrie und Handwerk" in den letzten Jahren ausmachen?
Werkzeuge gehören zu den meistgebrauchten Produkten unseres täglichen Lebens; sie bestimmen, wer wir sind und welchen Qualitätsgrad unsere Arbeitsumgebung hat. Sie werden immer durchdachter und dadurch intuitiver und ergonomischer. Wir sind daher in der Lage, immer präzisere Vorgänge mit weniger Anstrengung und Gefahr für unsere Gesundheit auszuführen. Dank gut gestalteter Produkte ist das Arbeiten konzentrierter und angenehmer geworden.

Welche Trends konnten Sie im Bereich „Automotive und Transport" in den letzten Jahren ausmachen?
Mir sind gute Gestaltungsbestrebungen bei Produkten und in Bereichen aufgefallen, die nicht für ausgeklügelte Lösungen bekannt sind, wie etwa in der professionellen Ausstattung für Spezialfahrzeuge. Mehr Intelligenz zeigt sich auch in der Entwicklung der Infrastruktur für aufladbare Elektrofahrzeuge.

Jure Miklavc

01

Prof. Ron A. Nabarro
Israel

Jury member since 2005
Appointed five times
Jurymitglied seit 2005
Berufen fünf Mal

Professor Ron A. Nabarro is an industrial designer, entrepreneur, researcher and educator. Since 1970, he has designed more than 700 products, mainly in the field of advanced technologies, and in 2009, he received the World Technology Network Award in the field of design. From 1992 to 2009, he was a professor of industrial design at the Technion Israel Institute of Technology where he founded and led the Graduate Program in Advanced Design Studies and Design Management. Currently, Ron A. Nabarro teaches at Beijing DeTao Masters Academy, China. After having been an executive board member of the Icsid from 1999 to 2003, he now acts as an Icsid regional advisor. He has lectured in over 20 countries, has acted as a consultant for a wide variety of organisations and is a frequent keynote speaker at conferences. Ron A. Nabarro is co-founder and partner of the four start-up companies Scentcom Ltd., Cellomate Ltd., MedImprove and Balance, and is also co-founder and CEO of Senior-Touch Ltd., an age-friendly R&D company and design consultancy. Its areas of research and interest are age-friendly design, design management and design education.

Professor Ron A. Nabarro ist Industriedesigner, Unternehmer, Forscher und Lehrender. Seit 1970 gestaltete er mehr als 700 Produkte, hauptsächlich im Bereich der modernen Technologien, und erhielt 2009 den World Technology Network Award im Bereich Design. Von 1992 bis 2009 war er Professor für Industriedesign am Technion Israel Institute of Technology, wo er das Graduate Program in Advanced Design Studies and Design Management begründete und leitete. Derzeit lehrt Ron A. Nabarro an der Beijing DeTao Masters Academy in China. Nachdem er von 1999 bis 2003 als Vorstandsmitglied des Icsid diente, ist er aktuell als regionaler Berater von Icsid tätig. Er hat in mehr als 20 Ländern Vorträge gehalten, ist als Berater für eine Vielzahl an Organisationen tätig und ist oft Hauptredner auf Konferenzen. Ron A. Nabarro ist Mitbegründer und Partner der vier Start-up-Unternehmen Scentcom Ltd., Cellomate Ltd., MedImprove und Balance und ebenso Mitbegründer und CEO von Senior-Touch Ltd., einem Beratungsunternehmen für altersfreundliche F&E sowie Gestaltung. Dessen Forschungs- und Interessensbereiche liegen in altersfreundlichem Design, Designmanagement und Designerziehung.

02

»My abiding impression from this jury session as well as others over the years is the respectful and conscientious attitude of all jury members throughout the adjudication process.«

»Mein bleibender Eindruck dieser Jurysitzung und anderer über die Jahre hinweg ist die respektvolle und gewissenhafte Haltung aller Juroren im Jurierungsprozess.«

What trends have you noticed in the field of "Bathrooms, spas and air conditioning" in recent years?
A dramatic change is taking place in this field. The old approach of trying to find a "practical" solution is giving way to an understanding that time spent in the bathroom and spa is supposed to be quality time for body and soul. The new designs reflect this trend, allowing the user to experience comfort and relaxation in this environment.

What current trends have you noticed in the category "Gardens"?
In the "Gardens" category, a determined effort to align design with the needs of the gardening community is becoming apparent. There are no compromises; the solutions encompass all aspects.

What challenges do you see for the future in design?
The ageing population is increasingly becoming one of the most significant social, economic and demographic phenomena of our times. I see this as one of the most important challenges for design.

Welche Trends konnten Sie im Bereich „Bad, Wellness und Klima" in den letzten Jahren ausmachen?
In diesem Bereich findet eine dramatische Veränderung statt. Der alte Ansatz, eine „praktische" Lösung zu finden, weicht dem Verständnis, dass die Zeit im Badezimmer oder Wellnessbereich eine wertvolle Zeit für Körper und Seele ist. Die neuen Gestaltungen spiegeln diesen Trend wider und ermöglichen es dem Nutzer, in dieser Umgebung Entspannung und Komfort zu erfahren.

Welche Trends konnten Sie im Bereich „Garten" ausmachen?
In der Kategorie „Garten" fällt das starke Bestreben auf, Design auf die Bedürfnisse der Gärtner abzustimmen. Es gibt keine Kompromisse, die Lösungen umfassen alle Aspekte.

Welche Herausforderungen sehen Sie für die Zukunft im Design?
Altern wird zunehmend zu einem der wichtigsten sozialen, wirtschaftlichen und demografischen Phänomene unserer Zeit. Ich halte es für eine der wichtigsten Herausforderungen im Design.

Prof. Ron A. Nabarro

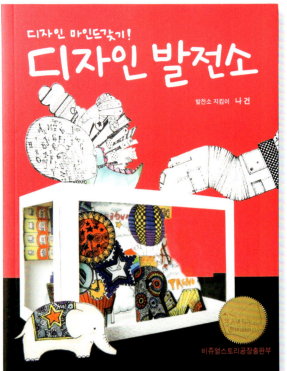

01

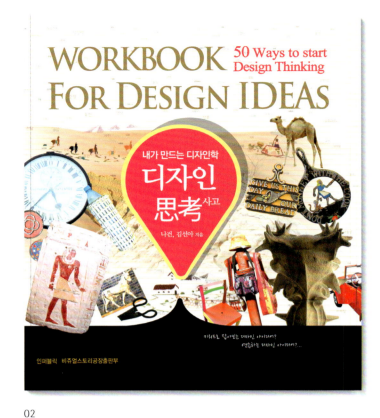

02

Prof. Dr. Ken Nah
Korea

Jury member since 2012
Appointed twice
Jurymitglied seit 2012
Berufen zwei Mal

In 1983, Professor Dr. Ken Nah received his B.A. from Hanyang University, South Korea, majoring in Industrial Engineering. His interest in human factors/ergonomics led him to pursue a Master's Degree from Korea Advanced Institute for Science and Technology (KAIST) in 1985. He gained his Ph.D. in 1996 in Engineering Design from Tufts University. Ken Nah is also a USA certified professional ergonomist (CPE). From 2002 to 2006, he was the dean of the International Design School for Advanced Studies (IDAS).

Currently, he is professor of Design Management and also the director of the Human and Experience Research (HER) Lab at IDAS. He has also held the post of director at the International Design Trend Center (IDTC) since 2002 and was the director general of "World Design Capital Seoul 2010". Alongside his teaching career, Ken Nah is also the vice president of the Korea Association of Industrial Designers (KAID), the Ergonomics Society of Korea (ESK), the Korea Institute of Design Management (MIDM), as well as the chairman of the Design and Brand Committee of the Korea Consulting Association (KCA).

Professor Dr. Ken Nah, der im Hauptfach Industrial Engineering studierte, graduierte 1983 an der Hanyang University in Südkorea als Bachelor of the Arts. Sein Interesse an Human Factors/Ergonomie vertiefte er 1985 mit einem Master-Abschluss am Korea Advanced Institute for Science and Technology (KAIST) und promovierte 1996 an der Tufts University. Darüber hinaus ist Ken Nah ein in den USA zertifizierter Ergonom (CPE). Von 2003 bis 2006 war er Dekan der International Design School for Advanced Studies (IDAS).

Aktuell ist er als Professor für Design Management tätig und zudem Direktor des „Human and Experience Research (HER)"-Labors an der IDAS. Von 2002 an war er Leiter des International Design Trend Center (IDTC). Ken Nah war Generaldirektor der „World Design Capital Seoul 2010". Neben seiner Lehrtätigkeit ist er Vice President der Korea Association of Industrial Designers (KAID), der Ergonomics Society of Korea (ESK), des Korea Institute of Design Management (MIDM) sowie Vorsitzender des „Design and Brand"-Komitees der Korea Consulting Association (KCA).

»I was very impressed by the diversity of products and excellent design quality of the entries this year.«

»Die Produktvielfalt und herausragende Gestaltungsqualität der diesjährigen Einre chungen hat mich sehr beeindruckt.«

What current trends do you see in the category "Computers and information technology"?
The form factors "simple" and "attractive". It was clear to me how difficult it must have been for designers to compete in this award category with innovative design products.

Do you see a correlation between the design quality of a company's products and the economic success of this company?
Definitely. Nowadays, design, especially at the sensual level, is the single most important criterion for the consumers' choice of products. It easily captivates the consumers' attention and makes them love the company's products.

What challenges do you see for the future in design?
I am a strong believer in design, since design is the most effective tool for innovation and a significant value creator. Therefore, design will continue to be increasingly important strategically and will become more inclusive in the future.

Welche Trends konnten Sie im Bereich „Computer und Informationstechnik" in den letzten Jahren ausmachen?
Die Formfaktoren „schlicht" und „schön". Daher kann ich nachempfinden, wie schwierig es für die Gestalter gewesen sein muss, in dieser Wettbewerbskategorie mit innovativen Designresultaten anzutreten.

Sehen Sie einen Zusammenhang zwischen der Designqualität, die sich in den Produkten eines Unternehmens äußert, und dem wirtschaftlichen Erfolg dieses Unternehmens?
Definitiv. Heutzutage ist Gestaltung das wichtigste Kriterium für die Produktwahl der Konsumenten, insbesondere auf der sinnlicher Ebene. Sie zieht die Aufmerksamkeit unmitte bar auf sich und bewirkt, dass die Konsumenten die Produkte des Unternehmens lieben.

Welche Herausforderungen sehen Sie für die Zukunft im Design?
Ich bin ein großer Designverfechter, da es das effektivste Instrument zur Innovation und ein wesentlicher Wertschöpfer ist. Daher wird Design weiterhin strategisch immer wichtiger und in Zukunft integrativer werden.

Prof. Dr. Ken Nah

01

Ken Okuyama
Japan

Jury member since 2010
Appointed four times
Jurymitglied seit 2010
Berufen vier Mal

Ken Kiyoyuki Okuyama, industrial designer and CEO of KEN OKUYAMA DESIGN, was born in Yamagata, Japan, 1959. He has worked as Chief Designer for General Motors, as Senior Designer for Porsche AG, and as Design Director for Pininfarina S.p.A., being responsible for Ferrari Enzo, Maserati Quattroporte and many other cars. He is also known for many different product designs such as motorcycles, furniture, robots and architecture.

KEN OKUYAMA DESIGN was founded in 2007 and provides business consultancy services to numerous corporations. Ken Okuyama also produces cars, eyewear and interior products under his original brand. He is currently a visiting professor at several universities and also frequently publishes books.

Ken Kiyoyuki Okuyama, Industriedesigner und CEO von KEN OKUYAMA DESIGN, wurde 1959 in Yamagata, Japan, geboren. Er war als Chief Designer bei General Motors, als Senior Designer bei der Porsche AG und als Design Director bei Pininfarina S.p.A. tätig und zeichnete verantwortlich für den Ferrari Enzo, den Maserati Quattroporte und viele weitere Automobile. Zudem ist er für viele unterschiedliche Produktgestaltungen wie Motorräder, Möbel, Roboter und Architektur bekannt.

KEN OKUYAMA DESIGN wurde 2007 als Beratungsunternehmen gegründet und arbeitet für zahlreiche Unternehmen. Ken Okuyama produziert unter seiner originären Marke auch Autos, Brillen und Inneneinrichtungsgegenstände. Derzeit lehrt er als Gastprofessor an verschiedenen Universitäten und publiziert zudem Bücher.

02

»My abiding impression of this year's jury session is the increased demand for more prominent corporate identity.«

»Mein bleibender Eindruck aus der diesjährigen Jurysitzung ist das Bedürfnis nach mehr hervorstechender Unternehmensidentität als zuvor.«

What trends have you noticed in the field of "Automotive and transportation" in recent years?
Driving performance is no longer a sales point. Transport design should propose not only mobility but also a new lifestyle. A car's design reflects its owner's character and lifestyle more than ever.

Do you see a correlation between the design quality of a company's products and the economic success of this company?
The correlation is the result of a clear vision and the teamwork that made it happen, plus the personalities of individual team members.

What are the important criteria for you as a juror in the assessment of a product?
A juror has to determine a product's value to society and the market. Therefore, an objective view and wide ranging knowledge of technology, materials, manufacturing, etc. are necessary.

Welche Trends konnten Sie im Bereich „Automotive und Transport" in den letzten Jahren ausmachen?
Fahr-Performance ist kein Verkaufsargument mehr. Im Segment „Transport" sollte Gestaltung nicht nur auf Mobilität abzielen, sondern auch auf einen neuen Lebensstil. Das Design eines Autos spiegelt mehr denn je den Charakter und Lebensstil seines Besitzers wider.

Sehen Sie einen Zusammenhang zwischen der Designqualität, die sich in den Produkten eines Unternehmens äußert, und dem wirtschaftlichen Erfolg dieses Unternehmens?
Diese Wechselwirkung ist das Ergebnis einer klaren Vision und der ihr zugrunde liegenden Teamarbeit – plus der Persönlichkeiten der einzelnen Teammitglieder.

Worauf achten Sie als Juror, wenn Sie ein Produkt bewerten?
Ein Juror muss der Wert bestimmen, den ein Produkt für die Gesellschaft und den Markt hat. Daher sind eine objektive Sichtweise und eine große Bandbreite an Wissen über Technik, Werkstoffe, Herstellung etc. notwendig.

Ken Okuyama

01

Simon Ong
Singapore

Jury member since 2006
Appointed six times
Jurymitglied seit 2006
Berufen sechs Mal

Simon Ong, born in Singapore in 1953, holds a Master's degree in design (MDes) from the University of New South Wales and an MBA from the University of South Australia. He is the group managing director and co-founder of Kingsmen Creatives Ltd., a leading communications design and production group in the Asia-Pacific region and in the Middle East. His work has been distinguished by several awards, such as the Eddie Award, the VM&SD/ISP Design Award and the A.R.E. Design Award, the Singapore Promising Brand Award, the SRA Best Retail Concept Award, and the Annual Outdoor Advertising Award.

From 1995 to 1997, Simon Ong held the position of president of the Interior Designers Association of Singapore and from 1998 to 2007 he was a member of the advisory committee of the School of Design of the Temasek Polytechnic, Singapore. Currently he is an IDP member of the Design Singapore Council and, among others, chairman of the Design Cluster of the Singapore Workforce Development Agency.

Simon Ong, geboren 1953 in Singapur, absolvierte sein Designstudium an der University of New South Wales, Australien, mit der Promotion und erwarb einen MBA an der University of South Australia. Er ist Geschäftsführer und Mitbegründer von Kingsmen Creatives Ltd., einer führenden Gruppe von Unternehmen für Kommunikationsdesign und Fertigung im Asien-Pazifik-Raum und im Nahen Osten. Für seine Arbeiten wurde er mehrfach ausgezeichnet, darunter mit dem Eddie Award, VM&SD/ISP Design Award und A.R.E. Design Award, Singapore Promising Brand Award, SRA Best Retail Concept Award und Annual Outdoor Advertising Award.

Von 1995 bis 1997 war Simon Ong Präsident der Interior Designers Association of Singapore und gehörte von 1998 bis 2007 dem Beraterausschuss der Temasek Polytechnic, School of Design, Singapur, an. Derzeit ist er IDP-Mitglied des Design Singapore Council und u. a. Vorsitzender des Design-Clusters der Singapore Workforce Development Agency.

02

»Design quality is one of the most important tools to differentiate products from their competitors in a global marketplace.«

»Designqualität wird in einem globalisierten Markt als eines der wichtigsten Instrumente zur Unterscheidung der eigenen Produkte von denen der Konkurrenz eingesetzt.«

What is, in your opinion, the significance of design quality in the product categories you evaluated?
New technologies have not only transformed the way we lead our lives, but have opened up new opportunities for creativity. This is evident in the entries submitted by designers this year.

What are the important criteria for you as a juror in the assessment of a product?
The first impression of a product is very important. It must impress the jury enough to warrant a second look. It has to be appealing and, naturally it must possess ergonomic, innovative and functional qualities. Most of all, it must be pleasant to use.

Do you have a philosophy toward life?
Be sensitive to people and the world around us. But most of all, be happy!

Wie schätzen Sie den Stellenwert der Designqualität in den von Ihnen beurteilten Produktkategorien ein?
Die neuen Technologien haben nicht nur unsere Lebensweise verändert, sondern auch die Möglichkeit zu neuen Entwürfen eröffnet. Dies zeigt sich in den Antworten der Designer bei den diesjährigen Einreichungen.

Worauf achten Sie als Juror, wenn Sie ein Produkt bewerten?
Sehr wichtig ist der erste Eindruck, den das Produkt hinterlässt. Es muss die Jury so weit beeindrucken, dass es einen zweiten Blick rechtfertigt. Es muss seine weitere Betrachtung empfehlen und selbstverständlich ergonomische, innovative und funktionale Qualitäten besitzen. Vor allem aber muss es eine positive Nutzerfahrung bieten.

Haben Sie ein Lebensmotto?
Sei den Menschen und unserer Umwelt gegenüber aufgeschlossen. Und allem voran: Sei glücklich!

Simon Ong

01

Max Ottenwälder
Germany

Jury member since 1999
Appointed five times
Jurymitglied seit 1999
Berufen fünf Mal

Max Ottenwälder, born in 1954, graduated with a diploma in industrial design from the Schwäbisch Gmünd University of Applied Sciences (Design) in 1979. Since 1980 he has been working as a self-employed designer and, in 1990, founded the design agency Ottenwälder und Ottenwälder together with Petra Kurz-Ottenwälder. His successful work as a designer and design consultant for renowned companies in different industries and product fields has won him several international design and innovation prizes. Max Ottenwälder writes articles on design philosophy and product semantics for books and specialist magazines and gives lectures and seminars at universities, trade fairs and private institutions. As a guest lecturer for product semantics and design language he has developed a theory for the assessment of the emotional value proposition of objects. He has been a member of the jury of the Red Dot Design Award several times. In his artistic work Max Ottenwälder has initiated performances and exhibitions with "linomorph" iron wire sculptures since 1990.

Max Ottenwälder, geboren 1954, erhielt sein Diplom als Industrial Designer 1979 an der Hochschule für Gestaltung in Schwäbisch Gmünd. Seit 1980 ist er selbstständig als Designer tätig und gründete 1990 gemeinsam mit Petra Kurz-Ottenwälder das Designbüro Ottenwälder und Ottenwälder. Die erfolgreiche Tätigkeit als Gestalter und Designberater für renommierte Unternehmen in unterschiedlichsten Branchen und Produktbereichen wurde vielfach mit internationalen Design- und Innovationspreisen ausgezeichnet. Max Ottenwälder verfasst Aufsätze zu Designphilosophie und Produktsemantik in Büchern und Fachzeitschriften und hält Vorträge und Seminare an Hochschulen, auf Messen sowie an privaten Institutionen. Als Gastdozent für Produktsemantik und Formensprache entwickelte er eine Lehre zur Beurteilung des emotionalen Werteversprechens von Objekten. Er war mehrfach Jurymitglied im Red Dot Design Award. In seiner künstlerischen Tätigkeit initiiert Max Ottenwälder seit 1990 Performances und Ausstellungen mit linomorphen Eisendrahtskulpturen.

02

»The responsibility for future products includes an awareness of high-quality design and the aspects ecology and humane production.«

»Die Verantwortung für die Produkte der Zukunft umfasst eine Bewusstseinsbildung für hochwertiges Design sowie die Aspekte Ökologie und menschenwürdige Produktion.«

What trends have you noticed in the field of "Kitchens" in recent years?
What stands out in this field is that the leading manufacturers are increasingly adjusting their product design to their target markets. They achieve this through a targeted use of materials and surfaces as well as the growing use of decorative components.

What are the important criteria for you as a juror in the assessment of a product?
I approach a product from the consumer's perspective first and then from that of the manufacturer or the brand respectively. Then I assess to what extent the design fulfils its value proposition. Good function, a significant use of form and high quality materials in an adequate value-for-money ratio are the basic prerequisites for a good product.

How do you assess the significance of design quality for the success of a company?
High quality product design directly benefits a company's economic success. The significance of design is becoming increasingly important for global competition.

Welche Trends konnten Sie im Bereich „Küche" in den letzten Jahren ausmachen?
In diesem Bereich ist auffällig, dass die führenden Hersteller das Design ihrer Produkte immer mehr ihren Zielmärkten anpassen. Dies geschieht mithilfe gezielter Material- und Oberflächenauswahl sowie mit zunehmend dekorativen Komponenten.

Worauf achten Sie als Juror, wenn Sie ein Produkt bewerten?
Ich nähere mich dem Produkt zuerst aus der Perspektive des Verbrauchers und dann aus der des Herstellers bzw. der Marke. Dann beurteile ich, inwiefern das Design sein Gebrauchswertversprechen erfüllt. Gute Funktion, signifikante Formensprache und hohe Materialqualität in einem angemessenen Preis-Leistungs-Verhältnis sind die Grundvoraussetzungen für ein gutes Produkt.

Wie schätzen Sie den Stellenwert des Designs für den Erfolg eines Unternehmens ein?
Hohe Designqualität der Produkte fördert direkt den wirtschaftlichen Erfolg von Unternehmen. Der Stellenwert des Designs wird zunehmend wichtiger im globalen Wettbewerb.

Max Ottenwälder

01

Oana Radeş
Netherlands

Jury member since 2013
Appointed for the first time
Jurymitglied seit 2013
Berufen zum ersten Mal

Oana Radeş studied architecture in Bucharest, Romania, and graduated with honours in 2005 from the Technical University in Eindhoven, Netherlands. Between 2006 and 2010, she worked at MVRDV as a project leader for several large-scale public buildings and urban planning in Switzerland, Japan, India, Singapore, Germany and the Netherlands.

Since January 2011, she has been a partner of Shift architecture urbanism, together with Thijs van Bijsterveldt and Harm Timmermans. Shift operates both beyond the traditional boundaries of architecture – through self-initiated studies on current societal issues with spatial implications – and within the very core of architecture, through the continuous pursuit of craftsmanship and performance in each building project. Since 2009, Oana Radeş has taught at various architecture and design academies in the Netherlands.

Oana Radeş studierte Architektur in Bukarest, Rumänien, und schloss ihre Ausbildung 2005 an der Technischen Universität in Eindhoven, Niederlande, ab. Von 2006 bis 2010 arbeitete sie bei MVRDV als Projektleiterin für verschiedene, groß angelegte öffentliche Gebäude und Städteplanungen in der Schweiz, Japan, Indien, Singapur, Deutschland und den Niederlanden.

Seit Januar 2011 ist sie zusammen mit Thijs van Bijsterveldt und Harm Timmermans Partner von Shift architecture urbanism. Shift operiert sowohl über die traditionellen Grenzen von Architektur hinaus – mit selbst initiierten Studien zu derzeitigen gesellschaftlichen Problemen mit räumlichen Implikationen – als auch mit dem zentralen Kern von Architektur, indem sie bei jedem Bauprojekt kontinuierlich Handwerkskunst und Performance weiterverfolgen. Seit 2009 lehrt Oana Radeş an verschiedenen Architektur- und Design-Akademien in den Niederlanden.

02

»I was impressed by the earnestness with which design was discussed and evaluated.«

»Ich war von der Ernsthaftigkeit beeindruckt, mit der Gestaltungen diskutiert und bewertet wurden.«

What is, in your opinion, the significance of design quality in the product categories you evaluated?
Well-designed spaces are crucial for our well-being, not only in terms of physical comfort, but especially in regard to mental and social aspects. At the community level, high-quality urban spaces are meaningful spaces, which can facilitate encounters, stimulate interaction and trigger a sense of belonging.

What trends have you noticed in the field of "Architecture and urban design" in recent years?
There is an increasing engagement in the architectural discourse with the social, cultural and economic needs of society. There seems to be less of a focus on iconic, spectacular forms, and more of a preoccupation with formulating creative and relevant responses to various urgent local and global issues.

Wie schätzen Sie den Stellenwert der Designqualität in den von Ihnen beurteilten Produktkategorien ein?
Gut gestaltete Plätze sind entscheidend für unser Wohlbefinden, nicht nur im Sinne körperlicher Annehmlichkeit, sondern insbesondere im Hinblick auf soziale und seelische Aspekte. Auf der Ebene von Gemeinschaft sind qualitativ hochwertige urbane Plätze bedeutsame Orte für die Begegnung, die Stimulation von Interaktion und das Ermöglichen eines Zugehörigkeitsgefühls.

Welche Trends konnten Sie im Bereich „Architektur und Urban Design" in den letzten Jahren ausmachen?
Im Architekturdiskurs herrscht ein verstärkter Dialog mit den sozialen, kulturellen und ökonomischen Bedürfnissen der Gesellschaft. Der Fokus scheint dabei weniger auf ikonischen, spektakulären Formen zu liegen als vielmehr auf der Verpflichtung, relevante und kreative Antworten auf dringliche globale und lokale Probleme zu entwerfen.

01

Dirk Schumann
Germany

Jury member since 2006
Appointed five times
Jurymitglied seit 2006
Berufen fünf Mal

Dirk Schumann, born in 1960 in Soest, studied product design at Münster University of Applied Sciences. After graduating in 1987, he joined oco-design as an industrial designer, moved to siegerdesign in 1989, and was a lecturer in product design at Münster University of Applied Sciences until 1991. In 1992, he founded his own design studio Schumanndesign in Münster, developing design concepts for companies in Germany, Italy, India, Thailand and China. For several years now, he has focused on conceptual architecture, created visionary living spaces and held lectures at international conferences. Dirk Schumann has taken part in exhibitions both in Germany and abroad with works that have garnered several awards, including the Gold Prize (Minister of Economy, Trade and Industry Prize) in the International Design Competition, Osaka; the Comfort & Design Award, Milan; the iF product design award, Hanover; the Red Dot, Essen; the Focus in Gold, Stuttgart; as well as the Good Design Award, Chicago and Tokyo.

Dirk Schumann, 1960 in Soest geboren, studierte Produktdesign an der Fachhochschule Münster. Nach seinem Abschluss 1987 arbeitete er als Industriedesigner für oco-design, wechselte 1989 zu siegerdesign und war bis 1991 an der Fachhochschule Münster als Lehrbeauftragter für Produktdesign tätig. 1992 eröffnete er sein eigenes Designstudio Schumanndesign in Münster und entwickelt Designkonzepte für Unternehmen in Deutschland, Italien, Indien, Thailand und China. Seit einigen Jahren beschäftigt er sich mit konzeptioneller Architektur, entwirft visionäre Lebensräume und hält Vorträge auf internationalen Kongressen. Dirk Schumann nimmt an Ausstellungen im In- und Ausland teil und wurde für seine Arbeiten mehrfach ausgezeichnet, u. a. mit dem Gold Prize (Minister of Economy, Trade and Industry Prize) des International Design Competition, Osaka, dem Comfort & Design Award, Mailand, dem iF product design award, Hannover, dem Red Dot, Essen, dem Focus in Gold, Stuttgart, sowie dem Good Design Award, Chicago und Tokio.

02

»Design is a versatile marketing tool that has a significant influence on the economic success and perception of a company on the market.«

»Design ist ein vielschichtiges Marketinginstrument, das den wirtschaftlichen Erfolg und die Wahrnehmung eines Unternehmens am Markt signifikant beeinflusst.«

What trends have you noticed in the field of "Industry and crafts" in recent years?
There are developments that improve on the function, usability and safety of products. Especially in the professional realm, this creates high added-value for products and a positive perception of the manufacturing company, which thus documents a user-oriented focus.

What are the important criteria for you as a juror in the assessment of a product?
Individual impressions of course take over at first contact with a new product. They are followed by more in-depth considerations of aspects such as functional and formal innovation and originality, utility value and user orientation, functional intelligibility, ecological criteria and how the product relates to the manufacturer.

Which project would you like to realise one day?
Projects that are located at the interface of design, technology and architecture.

Welche Trends konnten Sie im Bereich „Industrie und Handwerk" in den letzten Jahren ausmachen?
Hier gibt es Entwicklungen, die die Funktionen, die Handhabung und die Sicherheit der Produkte verbessern. Speziell im professionellen Bereich verleiht dies den Produkten einen hohen Mehrwert und eine positive Wahrnehmung der Herstellerunternehmen, die dadurch eine auf den Nutzer bezogene Fokussierung dokumentieren.

Worauf achten Sie als Juror, wenn Sie ein Produkt bewerten?
Beim ersten Kontakt mit dem Produkt steht natürlich das eigene Empfinden im Vordergrund. Dann folgen vertiefende Betrachtungen von Aspekten wie funktionale und formale Innovation und Eigenständigkeit, Gebrauchswert und Nutzerbezogenheit, die Funktionsverständlichkeit, ökologische Kriterien und der Bezug des Produktes zum Herstellerunternehmen.

Welches Projekt würden Sie gerne einmal realisieren?
Projekte an der Schnittstelle von Design, Technologie und Architektur.

Dirk Schumann

01

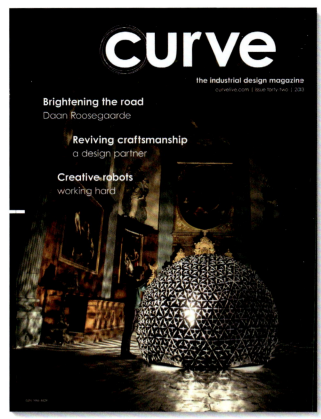

02

Belinda Stening
Australia

Jury member since 2013
Appointed for the first time
Jurymitglied seit 2013
Berufen zum ersten Mal

Belinda Stening is the founder and publisher of the industrial design magazine Curve, as well as its online counterpart "curvelive.com", which features the world's best in industrial design and product development. Since 2001 Curve aims to promote the industrial design profession, as well as manufacturing businesses developing and producing innovative products. As an internationally recognised publication, Curve is a member of the International Design Media Network, an initiative of the International Design Alliance (IDA). Belinda Stening holds a Bachelor Degree in Industrial Design and a Master of Fine Art – Sculpture from the Royal Melbourne Institute of Technology, Australia. As a professional industrial designer, she has many years of experience in design consulting, including consumer appliance design, biomedical product design, as well as packaging and furniture design. She has been a juror for design awards programs in the USA and Australia, and has worked extensively in the design education sector.

Belinda Stening ist die Gründerin und Herausgeberin des Industriedesign-Magazins „Curve" sowie seines Online-Pendants „curvelive.com", das die weltweit Besten im Bereich Industriedesign und Produktentwicklung vorstellt. Seit 2001 ist es das Ziel von Curve, den Beruf des Industriedesigners sowie die Industrieunternehmen zu fördern, die innovative Produkte entwickeln und produzieren. Als international anerkannte Publikation ist Curve Mitglied des International Design Media Network, einer Initiative der International Design Alliance (IDA). Belinda Stening hat einen Bachelor-Abschluss in Industrial Design und einen „Master of Fine Art – Sculpture" vom Royal Melbourne Institute of Technology, Australien. Als professionelle Industriedesignerin verfügt sie über langjährige Erfahrung im Design Consulting, einschließlich Consumer Appliance Design, im biomedizinischen Produktdesign sowie im Verpackungs- und Möbeldesign. Sie war als Jurorin verschiedener Designwettbewerbe in den USA und Australien sowie in der Designausbildung tätig.

03

»As a designer I'd like to think that I could develop a product that innovates the way we communicate design to the world.«

»Als Designerin gefällt mir die Vorstellung, mal ein Produkt zu entwickeln, das für die Kommunikation von Design in der Welt eine Neuerung darstellt.«

What trends have you noticed in the field of "Households" in recent years?
There are many products for compact spaces. In small homes, kitchen and laundry spaces are often integrated into living and entertainment areas. That is why household products for the laundry and kitchen are becoming more flexible and modular.

What trends have you noticed in the field of "Kitchens" in recent years?
There is a marked reduction in the use of superfluous embellishments. More "humble", simpler materials such as wood, glass and stainless steel are giving kitchen products a warmer and nostalgic style. The most appealing products are pared back and forms are plain but friendly.

What challenges do you see for the future in design?
I think it is very important for designers to remember who they are designing for. It is the designer's responsibility to be the advocate for the ultimate end-user of a product.

Welche Trends konnten Sie im Bereich „Haushalt" in den letzten Jahren ausmachen?
Es gibt viele Produkte für kompakte Räume. In kleinen Wohnungen sind Küche und Waschbereich oft in den Wohn- und Aufenthaltsraum integriert, weshalb Haushaltsgeräte für diese Anwendungen immer flexibler und modularer werden.

Welche Trends konnten Sie im Bereich „Küche" in den letzten Jahren ausmachen?
Es zeigt sich eine deutliche Reduktion im Einsatz überflüssiger Verzierungen. Einfachere und „zurückhaltendere" Materialien wie Holz, Glas und Edelstahl verleihen den Küchenprodukten ein wärmeres und nostalgisches Aussehen. Die ansprechendsten Produkte sind auf freundliche und einfache Formen zurechtgestutzt.

Welche Herausforderungen sehen Sie für die Zukunft im Design?
Ich denke, als Designer ist es sehr wichtig, sich stets zu vergegenwärtigen, für wen man eigentlich gestaltet. Die Verantwortung der Designer besteht in ihrer Rolle als Anwälte für die letztlichen Endverbraucher der Produkte.

01

02

Nils Toft
Denmark

Jury member since 2006
Appointed seven times
Jurymitglied seit 2006
Berufen sieben Mal

Nils Toft, born in Copenhagen in 1957, graduated as an architect and designer from the Royal Danish Academy of Fine Arts in Copenhagen in 1985. He also holds a Master's degree in Industrial Design and Business Development. Starting his career as an industrial designer, Nils Toft joined the former Christian Bjørn Design in 1987, an internationally active design studio in Copenhagen with branches in Beijing and Ho Chi Minh City. Within a few years, he became a partner of CBD and, as managing director, ran the business. Today, Nils Toft is the founder and managing director of Designidea. With offices in Copenhagen and Beijing, Designidea works in the following key fields: communication, consumer electronics, computing, agriculture, medicine, and graphic arts, as well as projects in design-strategy, graphic and exhibition design.

Nils Toft, geboren 1957 in Kopenhagen, machte seinen Abschluss als Architekt und Designer 1985 an der Royal Danish Academy of Fine Arts in Kopenhagen. Er verfügt zudem über einen Master im Bereich Industrial Design und Business Development. Zu Beginn seiner Karriere als Industriedesigner trat Nils Toft 1987 bei dem damaligen Christian Bjørn Design ein, einem international operierenden Designstudio in Kopenhagen, das mit Niederlassungen in Beijing und Ho-Chi-Minh-Stadt vertreten ist. Innerhalb weniger Jahre wurde er Partner bei CBD und leitete das Unternehmen als Managing Director. Heute ist Nils Toft Gründer und Managing Director von Designidea. Mit Büros in Kopenhagen und Beijing operiert Designidea in verschiedenen Hauptbereichen: Kommunikation, Unterhaltungselektronik, Computer, Landwirtschaft, Medizin und Grafikdesign sowie Projekte im Bereich Designstrategie, Grafik- und Ausstellungsdesign.

01 **Jabra HALO2**
The Jabra HALO2 headset
with advanced Bluetooth®
technology ensures improved connectivity; thanks to
Multiuse™ it connects more
than one device at a time
Das mit erweiterter
Bluetooth®-Technologie
ausgestattete Headset Jabra
HALO2 ermöglicht eine op-
timierte Connectivität; dank
Multiuse™ ist der Anschluss
von mehr als einem Gerät
gleichzeitig möglich

02 **GD1000**
Professional vacuum cleaner
for institutional use as part
of a large series of machines
for professional cleaning,
designed for Nilfisk-Advance
Profistaubsauger für den
institutionellen Einsatz als Teil
einer großen Serie professioneller Reinigungsmaschinen,
gestaltet für Nilfisk-Advance

03 **Wittenborg 7100**
Espresso and fresh brew
coffee machine for
professional office use
Espresso- und Brühkaffee-
maschine für den professionellen Einsatz in Büros

03

»The great satisfaction of working as an industrial designer lies in the constant challenge presented by new industries and new products.«

»Die große Genugtuung in der Arbeit als Industriedesigner liegt in der ständigen Herausforderung durch neue Branchen und neue Produkte.«

What trends have you noticed in the field of "Bathrooms, spas and air conditioning" in recent years?
This category is focused on high-level design and consumers characterised by growing consumer expectations for design innovation. It has left its design infancy behind and entered a stage where design innovation and differentiation is important in order to maintain a competitive edge. Design just for the sake of being different is not enough; consumers are too smart and have much higher expectations.

What aspects of the jury process for the Red Dot Award: Product Design 2013 have stayed in your mind in particular?
A jury session becomes memorable when you overcome the prejudices and preconceptions that are based on a first quick glance of a product. In one jury session, I was blessed with a jury team that kept an open mind and discovered that scepticism can turn into excitement if one takes the time to understand a design.

Welche Trends konnten Sie im Bereich „Bad, Wellness und Klima" in den letzten Jahren ausmachen?
Diese Kategorie stellt eine auf hochwertige Gestaltung und auf Verbraucher fokussierte Kategorie mit wachsenden Erwartungen der Konsumenten an Designinnovation dar. Sie hat ihre gestalterische Kindheit hinter sich gelassen und eine Ebene erreicht, auf der Designinnovation und Differenzierung für die Wettbewerbsfähigkeit zentral sind. Gestaltung nur um des Andersseins willen reicht nicht aus; die Konsumenten sind dafür zu erfahren und haben höhere Erwartungen.

Was ist Ihnen von der Jurierung des Red Dot Award: Product Design 2013 besonders im Gedächtnis geblieben?
Eine Jurysitzung wird dann zum unvergesslichen Moment, wenn man seine Voreingenommenheit überwindet, die nach einem ersten kurzen Blick auf ein Produkt entstehen kann. Ich hatte das Glück, mit einem Jurorenteam zusammenzuarbeiten, das aufgeschlossen war und ebenfalls entdeckte, dass sich Skeptizismus in Begeisterung wandelt, wenn man sich für sein Verständnis eines Designs Zeit gibt.

01

Prof. Danny Venlet
Belgium

Jury member since 2005
Appointed eight times
Jurymitglied seit 2005
Berufen acht Mal

Professor Danny Venlet was born in 1958 in Victoria, Australia and studied interior design at Sint-Lukas, the Institute for Architecture and Arts in Brussels. Back in Australia in 1991, Venlet joined up with Marc Newson and Tina Engelen to form "Daffodil design". Venlet then started to attract international attention with large-scale interior design projects such as the Burdekin Hotel in Sydney and Q Bar, an Australian chain of nightclubs. His design projects range from private mansions, lofts, bars and restaurants all the way to showrooms and offices of large companies. Danny Venlet has taught at several schools and universities in Australia, as well as in Belgium. Today, he is professor at the Royal College of the Arts in Ghent and at the independent College of Advertising and Design in Brussels where he also is the Artistic Director from MAD Brussels (Mode And Design Center).

Professor Danny Venlet wurde 1958 in Victoria, Australien, geboren und studierte Interior Design am Sint-Lukas, dem Institut für Architektur und Kunst in Brüssel. 1991 kehrte er nach Australien zurück und gründete zusammen mit Marc Newson und Tina Engelen „Daffodil design". Erste internationale Aufmerksamkeit erlangte er durch die Innenausstattung großer Projekte wie dem Burdekin Hotel in Sydney sowie der Q Bar, einer australischen Nachtclub-Kette. Seine Design-Projekte reichen von privaten Wohnhäusern über Lofts, Bars und Restaurants bis zu Ausstellungsräumen und Büros großer Unternehmen. Danny Venlet lehrte an zahlreichen Schulen und Universitäten sowohl in Australien als auch in Belgien. Heute ist er Professor am Royal College of the Arts in Gent und am privaten College of Advertising and Design in Brüssel, wo er zudem Artistic Director des MAD Brussels (Mode And Design Center) ist.

02

»In order to be a success for the company a product needs to have many qualities such as uniqueness, financial viability, emotional value, and transportability.«

»Um einem Unternehmen Erfolg zu bringen, muss ein Produkt viele Qualitäten haben, z. B. Einzigartigkeit, finanzielle Realisierbarkeit, emotionalen Wert und Transportierbarkeit.«

What are the important criteria for you as a juror in the assessment of a product?
If I had to narrow it down to three criteria they would be: the product's level of innovation or uniqueness, its emotional value, and the quality of its execution.

What aspects of the jury process for the Red Dot Award: Product Design 2013 have stayed in your mind in particular?
As always, it has been a very professional event. Despite being in a recession, companies see the value of winning a Red Dot more than ever.

What challenges do you see for the future in design?
The challenge we face as designers is to make sure that design will still make sense in the future. Everything is design today and we need to make sure that we don't lose the real significance of design.

Worauf achten Sie als Juror, wenn Sie ein Produkt bewerten?
Wenn ich es auf drei Kriterien beschränken müsste, wären es folgende: der Grad der Innovation oder Einzigartigkeit des Produkts, sein emotionaler Wert und die Qualität der Ausführung.

Was ist Ihnen von der Jurierung des Red Dot Award: Product Design 2013 besonders im Gedächtnis geblieben?
Wie immer war es eine sehr professionelle Veranstaltung. Und obwohl wir uns in einer Rezession befinden, erkennen Unternehmen den Wert, den der Gewinn eines Red Dot mit sich bringt, mehr als je zuvor.

Welche Herausforderungen sehen Sie für die Zukunft im Design?
Die Herausforderungen, vor denen wir als Designer stehen, bestehen darin zu gewährleisten, dass Gestaltung auch in Zukunft noch sinnvoll ist. Heutzutage ist alles Design, und wir müssen sichergehen, dass wir seine wahre Bedeutung nicht verlieren.

Prof. Danny Venlet

493

01

Cheng Chung Yao
Taiwan

Jury member since 2012
Appointed twice
Jurymitglied seit 2012
Berufen zwei Mal

Cheng Chung Yao studied at the Pratt Institute New York and graduated with a Master's degree in architecture. In 1991, he founded the Department of Interior Space Design at Shih Chien University and has worked as a lecturer at the Graduate School of Architecture at Tam Kang University as well as at the Graduate School of Architecture at Chiao Tung University. In 1999, he founded "t1 design" where he heads a team of architects and interior designers as well as exhibition and graphic designers.

The company's best-known products currently include the City Plaza of Taiwan Pavilion of the 2010 Shanghai Expo, the Taiwan Design Museum and the Taiwan Design Centre. Furthermore, Cheng Chung Yao curated and designed the International Interior Design Exhibition for the Expo, was president of the Chinese Society of Interior Designers, chief executive of the Asia Pacific Space Designers Association, board member of the International Federation of Interior Architects/Designers and founder of the Taiwan Interior Design Award.

Cheng Chung Yao studierte Architektur am Pratt Institute New York und schloss sein Studium mit dem Master ab. 1991 gründete er die Fakultät für Interior Space Design an der Shih Chien University und war als Dozent an der Graduate School of Architecture der Tam Kang University sowie an der Graduate School of Architecture der Chiao Tung University tätig. 1999 gründete er „t1 design" und leitet dort ein Team aus Architekten, Innenarchitekten sowie Ausstellungs- und Grafikdesignern.

Zu den aktuell bekanntesten Projekten des Büros zählen der City Plaza of Taiwan Pavilion der Expo 2010 in Shanghai, das Taiwan Design Museum und das Taiwan Design Center. Zudem kuratierte und gestaltete Cheng Chung Yao die International Interior Design Exhibition für die Expo, war u. a. Präsident der Chinese Society of Interior Designers, Hauptgeschäftsführer der Asia Pacific Space Designers Association und Vorstandsmitglied der International Federation of Interior Architects/Designers und gründete den Taiwan Interior Design Award.

01 MG-5879
   The exhibition area of
   "concept design" in the
   Taiwan Design Museum
   Der Ausstellungsbereich
   „concept design" im
   Taiwan Design Museum

02 MG-6266
   The exhibition area of
   "product design" in the
   Taiwan Design Museum
   Der Ausstellungsbereich
   „product design" im
   Taiwan Design Museum

02

»The selection process of the Red Dot Design Award is a great testimony of contemporary design history, aesthetics and philosophy.«

»Der Auswahlprozess im Red Dot Design Award ist eine großartige Bekundung zeitgenössischer Designgeschichte, Ästhetik und Philosophie.«

What is, in your opinion, the significance of design quality in the product categories you evaluated?
The substance of design is not just to address the specific requirements of the time; it also promotes the vision that everyone has the ability to shape their own life in a distinctive way.

What trends have you noticed in the field of "Interior design" in recent years?
Today's interior design reflects the development of lifestyle, the pursuit of innovation in our way of life, as well as the penetration of society and the quest for consolidation with contemporary technology.

What challenges do you see for the future in design?
The divergence and the convergence of the design profession, as the future of design is crossing interdisciplinary boundaries. In addition, the influence and confluence of different cultures, as the future of design is crossing intercultural boundaries.

Wie schätzen Sie den Stellenwert der Designqualität in den von Ihnen beurteilten Produktkategorien ein?
Das Wesen des Designs ist nicht nur, Antworten auf die spezifischen Anforderungen seiner Zeit zu geben; es fördert zudem die Vision, dass jeder damit sein Leben auf unverwechselbare Weise formen kann.

Welche Trends konnten Sie im Bereich „Interior Design" in den letzten Jahren ausmachen?
Das Interior Design von heute reflektiert die Entwicklung des Lifestyles, das Bestreben, das Leben innovativer zu machen, ebenso wie die Durchdringung der Gesellschaft und die Suche nach einer Verschmelzung mit zeitgenössischer Technologie.

Welche Herausforderungen sehen Sie für die Zukunft im Design?
Die Divergenz und Konvergenz der Designprofession, da die Zukunft des Designs interdisziplinäre Grenzen überschreitet; außerdem die Beeinflussung und das Zusammenfließen verschiedener Kulturen, da die Zukunft des Designs interkulturelle Grenzen überschreitet.

Cheng Chung Yao

Imprint
Impressum

Editor | Herausgeber:
Peter Zec

Project management | Projektleitung:
Sabine Wöll

Project assistance | Projektassistenz:
Jennifer Bürling
Theresa Falkenberg
Sora Lina Loesch
Anna Kraatz
Stefanie Riechert
Anne Kämmerling
Anamaria Sumic
Lars Hofmann

Editorial work | Redaktion:
Bettina Derksen, Simmern, Germany
Kirsten Müller, Essen, Germany
Mareike Ahlborn, Essen, Germany
Klaus Dimmler, Essen, Germany
Burkhard Jacob
(Red Dot: Design Team of the Year),
Krefeld, Germany
Karin Kirch, Essen, Germany
Karoline Laarmann, Dortmund, Germany
Bettina Laustroer, Rosenheim, Germany
Astrid Ruta, Essen, Germany
Martina Stein, Otterberg, Germany

Proofreading | Lektorat:
Klaus Dimmler, Essen, Germany
Mareike Ahlborn, Essen, Germany
Jörg Arnke, Essen, Germany
Sabine Beeres, Leverkusen, Germany
Dawn Michelle d'Atri, Kirchhundem, Germany
Die Schreibweisen, Castrop-Rauxel, Germany
Annette Gillich-Beltz, Essen, Germany
Karin Kirch, Essen, Germany
Regina Schier, Essen, Germany

Translations | Übersetzung:
Heike Bors, Tokyo, Japan
Patrick Conroy, Larnaka, Cyprus
Stanislav Eberlein, Tokyo, Japan
Bill Kings, Wuppertal, Germany
Cathleen Poehler, Montreal, Canada
Ian Stachel-Williamson,
Christchurch, New Zealand
Bruce Stout, Grafenau, Germany
Philippa Watts, Exeter, Great Britain
Andreas Zantop, Berlin, Germany
Christiane Zschunke,
Frankfurt/Main, Germany

Layout | Gestaltung:
Gruschka Kramer
Visuelle Kommunikation,
Wuppertal, Germany
Lena Gruschka, Johannes Kramer

Photographs | Fotos:
Markus Benz, Walter Knoll
(portrait Foster + Partners)
Gandia Blasco/Odosdesign
(product photo juror Stefan Diez)
Gordon Bruce (product photos juror Gordon Bruce),
Gordon Bruce Design LLC, USA
Thomas De Boever (portrait Alain Gilles), Gent, Belgium
Ingmar Kurth (product photo juror Stefan Diez),
Frankfurt/Main, Germany
Flo Maeght (portrait Philippe Starck), Paris, France
Melkus Sportwagen GmbH
(product photo juror Lutz Fügener), Dresden, Germany
Ralph Richter (product photos juror Joachim H. Faust),
Düsseldorf, Germany
Rainer Viertlböck (product photos juror Florian Hufnagl),
Gauting, Germany

In-company photos | Werksfotos der Firmen

Jury photographs | Jurorenfotos:
Simon Bierwald, Dortmund, Germany

Production and lithography | Produktion und Lithografie:
tarcom GmbH, Gelsenkirchen, Germany
Bernd Reinkens, Gregor Baals,
Jonas Mühlenweg, Gundula Seraphin

Printing | Druck:
Dr. Cantz'sche Druckerei Medien GmbH,
Ostfildern, Germany

Bookbindery | Buchbindung:
BELTZ Bad Langensalza GmbH,
Bad Langensalza, Germany

Publisher + worldwide distribution | Verlag + Vertrieb weltweit:
Red Dot Edition | Fachverlag für Design
Contact | Kontakt:
Sabine Wöll
Gelsenkirchener Str. 181, 45309 Essen
Germany
Phone +49 201 8141-822
Fax +49 201 8141-810
E-mail edition@red-dot.de
www.red-dot.de
www.red-dot-store.de
Verkehrsnummer: 13674

Red Dot Design Yearbook 2013/2014

Softcover
Living (978-3-89939-145-9)
Doing (978-3-89939-146-6)
Working (978-3-89939-151-0)
Set = Living & Doing & Working (978-3-89939-144-2)

Hardcover
Living (978-3-89939-148-0)
Doing (978-3-89939-149-7)
Working (978-3-89939-152-7)
Set = Living & Doing & Working (978-3-89939-147-3)

Bibliographic information published by the Deutsche Nationalbibliothek:
The Deutsche Nationalbibliothek lists this publication in the Deutsche Nationalbibliografie; detailed bibliographic data are available on the Internet at http://dnb.ddb.de
Bibliografische Information der Deutschen Nationalbibliothek:
Die Deutsche Nationalbibliothek verzeichnet diese Publikation in der Deutschen Nationalbibliografie; detaillierte bibliografische Daten sind im Internet über http://dnb.ddb.de abrufbar.

The Red Dot Award: Product Design competition is the continuation of the Design Innovations competition.
Der Wettbewerb „Red Dot Award: Product Design" gilt als Fortsetzung des Wettbewerbs „Design Innovationen".